The Development & Meaning of Eddington's 'Fundamental Theory'

INCLUDING A COMPILATION FROM EDDINGTON'S UNPUBLISHED MANUSCRIPTS

BY

NOEL B. SLATER

M.A., Ph.D., F.R.S.E.

Senior Lecturer in Mathematics in the University of Leeds

CAMBRIDGE
AT THE UNIVERSITY PRESS
1957

PUBLISHED BY
THE SYNDICS OF THE CAMBRIDGE UNIVERSITY PRESS
Bentley House, 200 Euston Road, London, N.W. 1
American Branch: 32 East 57th Street, New York 22, N.Y.

Printed in Great Britain at the University Press, Cambridge
(Brooke Crutchley, University Printer)

PREFACE

This book has two closely related aims. One aim is to make available those parts of Sir Arthur Eddington's unpublished writings in 1940–4 which have survived and are of scientific importance. The larger aim is to elucidate the argument of his posthumous book *Fundamental Theory*. His later work was so much concentrated on this theme that his manuscripts are here arranged as variations on it. They show the development of his theory towards coherence; but in and around the main theme there are many episodes to provoke thought about the bases of physics.

I regret that the time taken over the collation, arrangement and selection has kept students of *Fundamental Theory* waiting so long for this supplementary material; but I have linked the new material with an exposition of the whole argument, in order to clarify the essentials, and also to make this work self-contained for those who have not read *Fundamental Theory*. The exposition adheres to Eddington's own standpoint and terminology; the paucity of references to his earlier *Relativity Theory of Protons and Electrons* is in accordance with his later attitude to the physical arguments (as apart from the E-number theory) of that work.

The bulk of the manuscripts was entrusted to me by the late Miss Winifred Eddington at the suggestion of Sir Edmund Whittaker (who was eventually to choose the title of this book, as earlier he had chosen that of *Fundamental Theory*). A few other manuscripts of value were found in 1954 amongst Miss Eddington's papers by Dr A. Vibert Douglas and Professor F. J. M. Stratton, who was holding them on behalf of the Royal Astronomical Society. I owe a deep debt of gratitude to these persons, especially to the late Sir Edmund Whittaker, the prime instigator of this work. For encouragement at difficult times I owe thanks also to Professors G. Lemaître, H. S. Ruse, W. M. Smart and the late E. A. Milne. Professors P. Kusch, A. T. Nordsieck and N. F. Ramsey have been helpful concerning some recent developments. For the plan of the book, the method of exposition and the selection of quotations I have sole responsibility.

My thanks are due to the Royal Astronomical Society for permission to use the manuscripts bequeathed by Miss Eddington; to the University Press for permission to quote many formulae from *Fundamental Theory* and also for their great care in the printing; and to Dr E. M. Patterson for vigilant assistance in reading proofs.

N.B.S.

November, 1956

NOTES ON ABBREVIATIONS AND CONVENTIONS

Abbreviations. The following abbreviations, which occur occasionally in Eddington's MSS., are used systematically throughout this book:

a.m.	angular momentum
d.f.	degree of freedom
E.B.	Einstein-Bose
e.m.	electromagnetic
F.D.	Fermi-Dirac
l.h.	left-hand(ed)
p.d.	probability distribution
R.C.	Riemann-Christoffel (in 'R.C. tensor')
r.h.	right-hand(ed)
s.d.	standard deviation (root mean square)

Summation conventions: These are little used before Chapter **9**. The standard tensor summation over repeated suffixes is defined there on p. 148, and on pp. 147–8 also a convention (used in Chapter **9** onwards) concerning the permutation of 1, 2, 3, when occurring in round brackets.

Symbols, descriptive and mathematical, are listed on p. 291.

Chapter and section headings. Chapters of this book are numbered **1**, **2**, **3**, ... and sections of chapters are numbered **1·1**, **1·2**, ... and referred to (without the paragraph sign) similarly. Chapters of Eddington's book *Fundamental Theory* in its final or earlier forms are headed I, II, ..., and sections 1, 2, ... (or 1·1, 1·2, ... in some cases) with the paragraph sign § in references.

The published version of *Fundamental Theory* is denoted (from **2·3** onwards) by the letter **F**, and earlier versions by **A**, **B**, **C**, **G**, **H** as explained in **2·3** (p. 11). Individual draft chapters are also identified by *italic* numerals *1–35* as explained in **2** (pp. 10, 13). Thus, for example, sections of the present book headed **6·2** Draft **B** *26* III and **6·3** Draft **B** *24* III describe two versions, *26* and *24*, of a Chapter III of the second extant draft. Such duplications are, however, rare, and the italic numeration is of minor importance.

Other relevant publications by Eddington are listed on pp. 9 and 10, and identified by bold-face letters. A Bibliography of recent work is given at the end (p. 287).

CONTENTS

Chapter **10**. SUMMARY OF **F** CHAPTERS IX–XI

Chapter **11**. DRAFTS OF **F** VI–XI: (i) DRAFT **A**

Chapter **12**. DRAFTS OF **F** VI–XI: (ii) DRAFT **G**

Chapter **13**. DRAFTS OF **F** VI–XI: (iii) DRAFT **H**

CONTENTS

CHAPTER 1

INTRODUCTION

1·1. Plan and purpose

Sir Arthur Eddington, shortly before his death in November 1944, separated from his other papers the nearly completed manuscript of a book on what may be termed structural physics. This was published by the Cambridge University Press in 1946 under the supervision of Sir Edmund Whittaker, who gave it the title *Fundamental Theory*. The book is a bold synthesis of the diverse accepted principles of physics, leading to calculated values of many of the fundamental pure ratios of physical constants in agreement with observation. Many have tried to read the book; but all have found it difficult.

Eddington found the book difficult to write; for it appears that he wrote most of it at least five or six times. He preserved a substantial part of these earlier versions in the form of neat manuscript chapters; a general account of these is given in Chapter **2**. I collated these manuscripts in detail, with each other and with the final version; and this book is based on the collation.

My purpose is to throw light on the meaning of *Fundamental Theory*, by giving detailed quotations from the manuscripts and tracing the development of the theory into its final form. I give in Chapters **3**, **9** and **10** a concise summary of the final version, with comments on obscurities and cross-references to the manuscripts. I give in Chapters **4–8**, **11–14** an account of the previous manuscripts, quoting verbatim all passages which (*a*) contribute materially to the understanding of the final version or (*b*) contain developments which were later discarded but are of intrinsic interest. Examples of (*b*) are the planoidal calculation on Saturn's rings (**7·1** § 37*a*) and the elementary theory of radiant energy (**6·2** § 3·9). Passages of type (*a*) are mainly rudimentary forms of arguments which were more highly developed later; these often attract by the vigour of the style. Other passages of type (*a*) are discordant in detail with the final version; but the discordances can be equally illuminating.

In choosing this form of presentation, I have been guided by the intention of clarifying Eddington's own conceptions, and also of making the main ideas accessible to readers with a broad interest in the structure of physics and with some acquaintance with the notions of quantum principles. Some specialists (and I would mention in particular Professor G. Lemaître) would have preferred the manuscripts to be reproduced *in toto*. The extent of the manuscripts would make this uneconomic; and in view of the duplication of many details of the argument in different manuscript versions, all readers would then have in effect to repeat my laborious collation. I should assure the specialists, however, that the quotations I give have been carefully selected to ensure that few important novelties of ideas have been omitted.

An alternative form of presentation would consist (like my original collation) of a summary of the final version section by section, appending to each section all relevant material to be found in all the manuscripts. This, however, would destroy all sense of development of the general conceptions. Each manuscript version of the theory is an organic whole—each was at the time it was written a 'final version'—and so carries on from one view of one topic to the related view of the next. The form of presentation chosen will in fact enable the reader to reconstruct successive versions of the theory.

As a third procedure, it has been suggested that the final version should be accepted, and that one should not delve into the earlier forms. This view would be reasonable if the final version were clearly intelligible. Moreover, the fact that Eddington preserved certain drafts (but by no means all of his drafted material) suggests that he attached value to some conceptions which were omitted in the final compression of the work. The late E. A. Milne described *Fundamental Theory* as 'a notable work of art'; here the artist is seen shaping his work, and illuminating by strokes of insight many fundamental concepts of physics.

Finally, it should be observed that the aim of this exposition is to clarify the theory from Eddington's viewpoint, and not to reshape its logical or algebraical foundations. Some valuable work on the foundations now in progress is briefly described in the Bibliography.

1·2. The spirit of *Fundamental Theory*

An attempt is made here to interpret and discuss the spirit and intention of *Fundamental Theory*.

Physical science, the study by man of the structure and behaviour of inanimate matter, is over-simply divided into experiment and theory; experiment, the numerical and geometrical measurement of matter; and theory, the mental construction of a synthesis and an 'explanation' of the regularities found in experiment. But the interplay of experiment and theory is close; a theory will suggest new experiments, and, moreover, theory must include the rational analysis of the methods of experiment. At one time, physical science could be divided into parallel branches named according to the types of matter and behaviour studied. Such branches were mechanics, astronomy, electricity, heat; each had its observational side and its own theory. In the present century, however, many branches have been tied back into composite stems, so that we have three main branches of experiment: (i) nuclear physics and the behaviour of 'fundamental' particles; (ii) 'molar' physics (and chemistry) of objects seen by eye and microscope; and (iii) astronomy and cosmical physics. The branches are not clearly separated; but in addition to some mingling of experimental foliage, they are linked by theoretical strands; for quantum theory appears largely in (i) and (ii), Newtonian mechanics in (ii) and (iii), and relativity in (iii) and (i).

The human mind seeks an all-embracing theory of all physics—a 'fundamental theory'. Eddington has not attempted to provide this in the form of new general equations to be verified in the behaviour of all matter. Eddington's theory is rather a synthesis of theories; he accepts concepts of quantum, Newtonian and relativity theory, concepts designed for *particular* branches of physics, and applies these concepts to the whole range (for example, the treatment of gravitation as due to exclusion, or to interchange). At the same time he examines the theory of experiment, and considers its selective effect on the results found. The practical result of his synthesis is the calculation of many physical constants which are pure numbers. His principal tool in calculation is the enumeration of the appropriate dimensionality (number of 'observable' degrees of freedom) of the elementary system concerned; and his E-number calculus serves largely to assist this enumeration.

To what extent is Eddington's theory an epistemology—a theory of knowledge? This it clearly is in intention; but the word must be more broadly interpreted than he himself would have liked.[a] He infers the numerical constants from the broad principles of physics; these principles he would like to ascribe to 'the system of thought by which the human mind interprets to itself the content of its sensory experience' (*Relativity Theory of Protons and Electrons*, p. 327). But the sensory experience is a vital part of the knowledge; and even if the physical principles assume a qualitative appearance when formulated, for example, as 'principles of impotence' (Whittaker, *From Euclid to Eddington*, p. 58, Cambridge, 1949), their background is in quantitative measurement. The child, in Shaw's *Back to Methuselah*, born with a 'simple direct sense of space-time and quantity' (or quanta?), would hail Eddington as her spiritual ancestor and accept his epistemological claim; we are not so advanced.

Eddington's theory loses much of its awesome nature if it is regarded as *a* theory (not as *the* theory!), one which has developed much in the

[a] There is an important discussion by Dingle, *The Sources of Eddington's Philosophy*, Cambridge, 1954.

same way as many theories of narrower scope and intention. Thus, like other theorists, he makes new hypotheses; but his hypotheses may not appear as additions to the customary principles, because they consist in applying old principles, or 'systems of thought', to situations for which the principles were not intended—as, for example, in his use of indeterminacy, exclusion and interchange. There is no logical compulsion behind this extension; but Eddington proceeds much more by analogy than by logic. This 'creation' of new hypotheses by misapplying old principles is, however, a familiar and fruitful scientific procedure. Again, especially in the early days of his theory, Eddington amended his calculations, where they did not fit closely to observation, by refining and complicating his (mainly dimensional) arguments until they did fit. This refinement of theory also is a procedure basic to many sciences; it is usually regarded as fair, but has been a ground of reproach to Eddington because of his epistemological claim. If we feel it necessary (as Eddington surmised we would) to weigh the theory mainly by its numerical accordances, we are denying that claim.

The value of Eddington's work, however, is not thereby impaired for the scientist. That the constants of physics can be calculated by argument from accepted physical principles (even if the principles have to be applied in novel ways) is a stimulus to a broadening of viewpoints. The mere notion that these constants may be calculable has in fact considerably affected physics in the last twenty years. Moreover, Eddington's examination both of the detailed meaning of the principles of modern physics, and of the interpretation of experimental determinations of the constants, has been followed by an increasing interest in these matters. Whether he lit the spark for this train of thought, or whether he was a largely unheeded prophet of inevitable developments in the trend of physics, is a matter for conjecture.

1·3. Outline of *Fundamental Theory*

A very brief outline of Eddington's book will now be given. The book has two main parts:

(*a*) Chapters I–V, the 'statistical theory', resulting in the calculation of some natural constants from considerations of dimensionality and extensions of accepted concepts of quantum mechanics;

(*b*) Chapters VI–VIII, an exposition of 'E-number calculus', followed by Chapters IX–XII on applications of the calculus combined with results of (*a*).

Eddington did not carry his final version of the book far enough to reach the Evaluation of the Cosmical Number (N, of particles in the universe); a paper with this title is printed as an Appendix to

Fundamental Theory. This is not summarized here, but is described briefly in Chapter **14** below.

In this short summary it is impossible to include an adequate account of the tools of dimensionality, and of physical reality of the mathematical vectors, on which the work so largely relies; for those aspects the reader is referred to Chapters **3**, **9** and **10** below.

I. *The Uncertainty of the Reference Frame* (preliminary concepts, particularly the measurement of position and the idealization of the background of observation). (i) Physical measurements of objects relate them to a concrete physical frame, which may be represented by the centroid of the matter in the universe. The uncertainty of position of the N particles leads to a small uncertainty, with standard deviation σ in any direction, in the position of the centroid. This 'uncertainty constant', σ, is regarded as fixing the scale of all physical structures by the way it is related to microscopic (quantum) magnitudes. (ii) For theoretical purposes we represent the universe, as the background of observation, as a uniform static (Einstein) distribution of N particles, calling this a (zero-temperature) *uranoid*. (iii) We may measure either *all* the physical characteristics of an object, or only *some*, regarding the others as *stabilized*, i.e. fixed and known (e.g. mass); those which are measurable are subject to uncertainty. The number of observable characteristics, or the dimensionality of the probability distribution of characteristics, is called the relevant *multiplicity* of the object.

II. *Multiplicity Factors* (gravitational-inertial theory of the distribution of energy, according to 'multiplicity', between particles and field. *Main result*: mass ratio of proton and electron). The disturbance created in the uranoid by an object-particle is the 'field' of the object; this field, added to the particle energy, gives the total energy. Following wave mechanics in selecting possible states as those which give a rigid (self-consistent) field, we find that the part of the total energy, etc., assigned to the particle varies inversely as the multiplicity; added 'test' or 'transition' energy is, however, assigned wholly to the object-particle. An object-particle may be complex—for example, the hydrogen atom is the simplest with enough mechanical degrees of freedom—136—to represent a general energy-tensor element. By applying the multiplicity rule to hydrogen, and viewing it also as a pair of subparticles, a quadratic equation is set up determining the mass ratio of these, i.e. of proton and electron.

III. *Interchange* (electrical theory, viewed as the theory of two-particle systems. *Main results*: corrections to the mass ratio and other microscopic results). The electrical energy of the two particles in hydrogen is identified with energy of 'interchange of identity'. This energy adds 1 to the 136 mechanical freedoms; thus a 'Bond' correction factor $\beta = \frac{137}{136}$ appears because of the effect of this on the multiplicity factor in comparing results which do or do not involve electrical effects.

IV. *Gravitation and Exclusion* (the effect of uncertainty of the origin on momentum distributions, and the representation of gravitation by exclusion. *Main result*: the absolute determination of particle mass). The uncertainty of the physical origin introduces a 'weighting factor' reducing the spread of *physical* (observable) momentum distributions. The proper mass of an atom is calculated (i) by reducing an infinite temperature uranoid (particles with zero rest energy but unlimited *geometrical* momenta) to the standard zero-temperature state, and alternatively (ii) by assigning a distribution of particles to the lowest (momentum-) energy levels allowed by a generalized principle of exclusion.

V. *The Planoid* (the use of flat instead of spherical space. *Main results*: the non-Coulombian energy of two protons, the constant of gravitation). In a flat-space universe or *planoid* (an ordinary sphere suitably related to the curved-space *uranoid*) we can represent particle energy as arising from angular momentum of interchange of identity, thus adding a third description of rest energy to (i) and (ii) of IV. In the planoid we also calculate the induction effect of unbalanced electrical charges, and so the 'non-Coulombian' energy of two like particles. The numerical results obtained so far are summarized (cf. **8·4** §40 below).

VI. *The Complete Momentum Vector* (the symbolic calculus of E-numbers). The mechanical and electrical properties of a particle correspond to a symbolic 'complete momentum vector' or E-number $P = \sum_{1}^{16} E_\mu p_\mu$, where the E_μ are symbolic coefficients (akin to the unit vectors **i**, **j**, **k** of quaternion or vector theory) which obey

$$E_\mu E_\nu = \pm E_\nu E_\mu, \quad E_\mu{}^2 = -1, \quad E_{16} = \sqrt{-1}.$$

(A double-suffix notation $E_{\mu\nu}$ is also used, where $0 \leqslant \mu < \nu \leqslant 5$.) Transformations qPq^{-1} (q an E-number) correspond to relativistic rotations, including Lorentz transformations when certain suffixes are linked with space-time axes. The symbol E_{45} is particularly associated with time (or energy); and the E-number $S = -PE_{45}$ (called a *strain vector*) associates P with an observable system of simultaneity.

VII. *Wave Vectors* (individuality, matrix forms of E-numbers, quantum-classical analogies). A sufficient analysis of matter is into *idempotent* particles, i.e. those with $P^2 = P$. If an E-number is represented by a 4 by 4 matrix, $P_{\alpha\beta}$, idempotency is equivalent to factorizability of the matrix, i.e. $P_{\alpha\beta} = \psi_\alpha \phi_\beta$. The four-component entities ψ, ϕ

are *wave vectors*; this starts a new tensor analysis in which, for example, second-rank space-time tensors yield fourth-rank wave tensors.

Problems of quantal theory involve *correlations* of particles as simultaneous systems. Thus in passing from classical to quantal theory we drop 'time' as an individual characteristic, replacing it by a 'phase' variable conjugate to the scale which is a basic quantum entity in microscopic systems (cf. I above). We naturally make this scale correspond to classical *energy* (conjugate to time as scale is to phase), thereby setting up an analogy between quantal and molar physics. The nomenclature of quantal physics is largely based on this analogy; for convenience we associate with it a re-labelling of E-number suffixes.

VIII. *Double Frames* (calculus of EF-numbers, interchange operators, the energy tensor). The relativity energy tensor, as an outer product of momenta or E-numbers, is represented by an EF-number, where the F-frame is a duplicate E-frame, so that there are 256 EF_μ symbols. An EF-number or symbol is representable by a double matrix $T_{\alpha\beta\gamma\delta}$ of 4^4 elements, $\alpha\beta$ referring to an E-matrix and $\gamma\delta$ to an F-matrix. The suffix permutation $(\alpha\beta)\leftrightarrow(\gamma\delta)$ is an interchange of the E, F-frames (effected by the operation ITI, where $I=\frac{1}{4}\sum_1^{16} E_\mu F_\mu$ is the *interchange operator*). The permutations $\beta\leftrightarrow\delta$ and $\beta\leftrightarrow\gamma$ yield the *dual* $\tilde{T}$ and *cross-dual* $T^\times$ of $T_{\alpha\beta\gamma\delta}$; the former corresponds to a change from space to strain tensors in an empty (De Sitter) universe, whereas the strain tensor $TE_{45}F_{45}$ (analogous to S in VI above) refers to a uniform Einstein universe or uranoid.

In general relativity the symmetrical tensor $T_{\mu\nu}$ $(\mu, \nu=1,2,3,4)$ of mechanical energy in space-time is related to the geometrical Ricci tensor $G_{\mu\nu}$, which is contracted from the Riemann-Christoffel tensor $B_{\mu\nu\lambda\rho}$. By using the additional suffixes $\mu, \nu=0,5$ of E-numbers, and a certain identity applying to EF-numbers symmetrical in two pairs of *matrix* suffixes, the *extended R.C. tensor* (suffixes $0, \ldots, 5$) is identified with the extended energy tensor. As the R.C. tensor is a measure of recoil of the physical frame for displacement of the object-system, this identification links the energy of object and frame or comparison system.

The final emphasis is on the strain tensor Z identified with the *cross-dual* $T^\times$ of the energy tensor. An energy tensor may be regarded as describing either related occupants of two states, or the flow of probability carried by interstate 'oscillators'; in this essentially quantum view the strain tensor Z is fundamental. The symmetry restrictions on T then forbid oscillators connecting symmetrical and antisymmetrical states, as in quantum theory.

IX. *Simple Applications.* First, the 'metastable' states of hydrogen are found to result from rotations in one of the additional dimensions provided by a single E-frame. Then the multiplicity concepts of II are combined with the allowed dimensionalities of VIII for multiple systems to calculate the masses of the neutron (a 'co-spin' bound hydrogen atom), deuterium, helium and 'mesotron'. Approximate methods based on the exclusion theory in IV are applied to the nucleus to estimate mass defects.

X. *The Wave Equation* (the hydrogen atom; e.m. potential). In VII a quasi-algebraic wave identity was found to hold relating the momentum vector of an idempotent particle to a wave-vector factor ψ of the momentum vector. Momentum as a differential operator ($p = i\hbar\partial/\partial x$) corresponds to field momentum (in the sense of II above); the (linear) differential wave equation arises when this operates on a wave function to express conservation of the field. An additional term is required in the wave equation of hydrogen, to allow for the extra-spatial momentum of interchange of the internal particles; this term is identified with the Coulomb energy e^2/r, and the wave equation is solved for the hydrogen states. This interchange momentum is then generalized for a system of charged particles, to yield the effective potential of an electron in an electromagnetic field.

XI. *The Molar Electromagnetic Field* (gauge transformations; magnetic moments). The interpretation of gauge transformations of relativistic metrics in terms of e.m. field potentials is discussed in the light of the last results of X. The additional energy appearing when a natural magnetic field is superposed on an artificial (gauge-induced) field is attributed to the magnetic moment of the object-particle. By considering the effective 'multiplicity' of the various particles, magnetic moments are calculated for the hydrogen atom and its 'internal particle' (usually regarded as the electron moment) and for the neutron.

XII. *Radiation* (fragment). The concept of the interstate oscillator (VIII, end) is applied to determine dipole transition frequencies and Compton (electron) scattering cross-sections. The infinity of 'transverse' self-energy of the electron (i.e. of energy of interaction with its own radiation) is attributed to the neglect of the finiteness of the available states, or equivalently to neglect of the momentum weighting factor of IV.

CHAPTER 2

PRELIMINARY AND MANUSCRIPT VERSIONS OF 'FUNDAMENTAL THEORY'

The account in **2·2–2·4** below, of manuscripts of *Fundamental Theory*, is preceded in **2·1** by a list of Eddington's relevant publications. A complete survey of his work is given in *The Life of Arthur Stanley Eddington*, by A. Vibert Douglas (Nelson, 1956).

2·1. Earlier publications of Eddington on his theory

Eddington's previous mathematical treatises on relativity and quantum theory were[a] **MTR**, *The Mathematical Theory of Relativity* (Cambridge, 1923, 1924), **RTPE**, *Relativity Theory of Protons and Electrons* (Cambridge, 1936). The former contains some germs of Eddington's later ideas, embedded in an account of Einstein's theory. The latter book is purely Eddingtonian—a first version of his 'fundamental theory'. It begins with a 'leisurely' account of E-number theory; but Eddington regarded the later part, on physical interpretation, as almost wholly superseded by his subsequent work. There is a list on p. 8 of **RTPE** of the papers, beginning in 1928, in which the theory was developed.

Eddington's physical and philosophical approach is described in his four books (Cambridge): *Space, Time and Gravitation* (1920), *The Nature of the Physical World* (1928), *The Expanding Universe* (1933) and *New Pathways in Science* (1935); and his final philosophical position in **PPS**, *Philosophy of Physical Science* (Cambridge, 1939).

Of the later development of his theory after **RTPE**, he gave two accounts at conferences, in Warsaw (1938) and Dublin (1942); these are published as:

'Applications cosmologiques de la théorie des quanta', in *Les nouvelles théories de la physique* (Institut International de Coopération Intellectuelle, Paris, 1939), and

D, 'The combination of relativity theory and quantum theory' (*Communications of the Dublin Institute for Advanced Studies*, Series A, no. 2, Dublin, 1943). The former is noteworthy also for comments by Bohr, Fowler, Gamow, Kramers, von Neumann, Rosenfeld and Wigner. The latter is the basis of the 'statistical part' of *Fundamental Theory*.

[a] *Literal* abbreviations in bold type are used to identify these books and papers throughout this book: *numbers* in bold type refer to chapter and section numbers of the present book.

The following *research papers*, written by Eddington after **RTPE**, are of importance:

α, 'Theory of scattering of protons by protons', *Proc. Roy. Soc.* A, **162**, 155 (1937).

β, 'The problem of n bodies in general relativity theory' (with G. L. Clark), *Proc. Roy. Soc.* A, **166**, 465 (1938).

γ, 'Lorentz invariance in quantum theory', *Proc. Camb. Phil. Soc.* **35**, 186 (1939) and **38**, 201 (1942).

δ, 'A new derivation of the quadratic equation for the masses of the proton and electron', and 'The masses of the neutron and mesotron', *Proc. Roy. Soc.* A, **174**, 16, 41 (1940).

ε, 'The physics of white dwarf matter', *M.N.R.A.S.* **100**, 582 (1940).

ζ, 'On the interaction potential in the scattering of protons by protons' (with H. M. Thaxton), *Physica*, **7**, 122 (1940).

η, 'The theoretical values of the physical constants', *Proc. Phys. Soc.* **54**, 491 (1942).

θ, 'The evaluation of the cosmical number', *Proc. Camb. Phil. Soc.* **40**, 37 (1944). This is reproduced as an Appendix in *Fundamental Theory*, and will usually be referred to as **FA**.

ι 'The recession-constant of the galaxies', *M.N.R.A.S.* **104**, 200 (1944).

2·2. The manuscripts of *Fundamental Theory*

In two conversations, in October 1942 and March 1944, Eddington told me of the progress of his new book; on the former occasion he stressed that the mathematics—the E-number calculus—of **RTPE** stood, but the interpretation had altered. On the latter occasion he remarked that he was shaping the book to appeal to physicists, by reserving the E-number theory for the later part. On my next visit to Cambridge, for an evening in June 1945, Miss Winifred Eddington showed me some papers which she had found, some of them that day. First, there was a large pile of *loose sheets*, each partly written, ending abruptly anywhere from near the top to near the bottom of the page. Dr G. L. Clark went through these in some detail, after the publication of *Fundamental Theory*, and found nothing he considered noteworthy. These sheets are not extant.

Secondly, there was an equal pile of manuscripts, headed and clipped in *chapters*. I had time during my visit merely to list the titles and to number the chapters *1, 2, 3*, ... downwards from the top. These were obviously versions of the manuscript which Sir Edmund Whittaker was then preparing for publication as *Fundamental Theory*. At the end of 1947 Miss Eddington sent most of these early manuscripts to me. My original numbering had generally survived; its interest lies in the probability that the order in the pile indicates the degree to which Eddington had referred to these manuscripts in his final writing.

Nearly all these chapters are complete, written in Eddington's precise hand,[a] with numeration and underlinings ready for the printer. Each chapter had been intended to be a final version. There are few alterations of the nature of words or phrases crossed out (occasionally symbols have been erased with a knife and replaced). Evidently, as the 'loose sheets' mentioned earlier suggest, Eddington rewrote the page rather than make an alteration. On the reverse sides of the pages, one often finds complete pages of earlier versions; this suggests that the extant pile of chapters (which include a very early version) were kept purposely by Eddington for further reference.

After a preliminary examination of these 'draft' chapters, I classified them in order of composition (see **2·3**), and then collated them word by word with the published version and with each other.

In September 1954, Dr A. V. Douglas and Professor F. J. M. Stratton sought out amongst the papers left by Miss Eddington some further MSS. of the same nature, and sent them to me. These are mainly (MS. **a** in the list below is an exception) fragmentary early chapters of *Fundamental Theory* or complete versions of other publications, and will not be described in detail. The most interesting of these fragments is quoted in **4·3**.

2·3. Chronological arrangement of the drafts

The thirty-two manuscript chapters are listed in **2·4** below, arranged as five partial '*drafts*' of the book; one or two assignments of two or three similar chapters to one 'draft' are slightly dubious, but the sequence on the whole is clear. I use the following notation throughout: *italic numerals* for the original order of the draft chapters in the 1945 pile; and **A**, **B**, **C**, **G**, **H** for the drafts in order of composition, **D** for the Dublin lectures as published, **F** for the final (printed) version of *Fundamental Theory*. Some draft chapters bear dates; but the general order of composition was inferred from the contents.

The general scope of the drafts will now be outlined.

A: three chapters, II–IV, on E-number theory. The first of these had been originally a Chapter I; thus Eddington had begun (as in **RTPE**) with E-number theory, and had then decided to preface it by a chapter on particles with linear momentum but no spin. This prefatory chapter may well have been substantially the same as the first chapter of **B** below.

B: three chapters, I–III, having roughly the content of **F** I, II, IV

[a] Reproductions of Eddington's handwriting will be found in A. V. Douglas's book, *The Life of Arthur Stanley Eddington*.

§§ 37–40, and a Chapter V on electrical theory. This version marks the inception of the general arrangement of **F**, placing the full 'statistical' theory (and not merely part of it as in **A**) before *E*-number theory.

C: four chapters, I–IV, covering **F** I–IV. This was written before Eddington's Dublin visit. Like **B**, it differs from the later versions of the 'statistical' theory by placing the connexion of coordinate and momentum distributions (**F** §§ 37, 38) at the very beginning, and introducing the 'Bernoulli fluctuation' (**F** § 3) much later, as confirming a relation, of the uncertainty constant and the particle number *N*, already inferred from a curved universe. **C** ends near the beginning of **F** V, with a calculation of the Newtonian potential of Saturn's rings as an example of 'planoidal' treatment.

D: here in order come the Dublin lectures. From notes supplied by Dr N. Symonds of the Dublin Institute and confirmed by Eddington's rough notes (received in 1954), it appears that the oral lectures (July 1942) followed much the order of **C**. The published version of **D**, written August–September 1942, follows largely the order of **F** I–V, however; in particular, the 'Bernoulli fluctuation' is shifted to the head of the statistical argument. A correlation of the sections of **D** and **F** is given in **F**, Editorial Note, p. 284. In accordance with my policy of concentrating on hitherto unpublished work, I give no account of **D**, although it will be referred to as bearing on other versions.

G: this begins (March–May 1943) with Chapters V, VI and VII (two versions) covering **F** VI—VIII on *E*-number theory. It continued with a Chapter VIII, 'Occupation Symbols' (July 1943), on the number *N*, a topic not met in any other manuscript or even in the final version. Eddington in fact filed this chapter with his final version which was sent to Sir Edmund Whittaker, but Sir Edmund subsequently returned it to Miss Eddington, preferring to represent the topic by the slightly later paper **θ** of **2·1**. I was unaware of this MS. until I received it amongst the papers recovered by Dr Douglas and Professor Stratton. Meanwhile I had assigned to **G** two later untitled incomplete Chapters VIII (dating from about October 1943); these treat topics of **F** VII, X and IX § 93. Much internal evidence supports this assignment; it indicates also that Eddington had decided to postpone the Evaluation of the Cosmical Number, and also that the plan for the later part of the book was under revision. The untitled Chapters VIII also indicate that the Wave Equation (**F** X) then came before the main 'Simple Applications' of **F** IX, if indeed all these applications were then envisaged. An interesting fragmentary Chapter IX on Molar Electromagnetic Fields (**F** XI) may also belong to **G**.

H: ten chapters covering all of **F** except the latter part of **F** XI and **F** XII. Chapters I–IV (August 1943) are close to **F** I–V; Eddington decided to split **H** IV into IV and V, when he had nearly finished **H**. Chapters V–VII (changed in pencil to VI–VIII), VIII and IX were written December 1943–April 1944. A fragment, XI, on gauge theory was written in June 1944.

F: this final (published) version was written in 1944 June–November, the month of Eddington's death. He left a plan (see **F**, p. 264) for the conclusion of Chapter XII on Radiation Theory, and for a Chapter XIII on Epistemological Theory (represented in the drafts by MS. **Ga** and in the published book by the Appendix) and a Summary, Chapter XIV.

Unclassified Manuscripts. I have omitted from the table below a few fragmentary manuscripts, which do not form recognizable parts of drafts **A–G** and (except for *29* I) were not available when the main collation was performed. These fragments are openings of early chapters; several resemble MS. *11* (**4·3**). Another, MS. *29* I, 'The Principles of Measurement', contains ideas expounded in **PPS** and elsewhere; one aphorism is

'Our principle is to measure room by what there is room for.'

The remaining MS., *12* II, 'Einstein-Bose and Fermi-Dirac Particles', contains ideas developed in **Ga** (**14·3**).

2·4. List of the Draft Versions

The table below gives the titles of the draft chapters, identified (in italic numerals) by their original order in Eddington's pile, and by my assignment, **A**, **B**, ..., of the order of composition. The corresponding chapters of **F** and the sections of this book where each draft is described or quoted, are also listed. Section headings of the drafts and their correlation with **F** sections will be found in the relevant later chapters of this book.

Chapters **3**, **9** and **10** below are summaries of the final version of **F**, with elucidations and references to the drafts, which are described and quoted in **4–8** and **11–14**. The general method of handling the drafts is described in **4·0** (p. 50). Two points of arrangement should be noted to avoid confusion. (i) When two or three drafts of one chapter occur in one overall draft, the later or latest version is placed first in the text. (ii) In the statistical theory drafts **C** and **B** are described together, chapter by chapter, in **4–7**, and draft **H** in **8**. In the E-number theory part, drafts **A**, **G** and **H** are described in turn in **11**, **12** and **13**, but the large and recently found Chapter VIII, **Ga**, is reserved for **14**.

Chronological list of the drafts

Draft no.		Chapter	Title	Date	Chapters of **F**	Sections here
A	*16*	II	The Complete Momentum Vector	Early	VI, VII	**11·1**
	35	III	Elementary Particles	—	VII, X	**11·2**
	21	IV	The Energy Tensor	—	VIII	**11·3**
B	*17*	I	The Uncertainty of the Origin	Early	I, IV	**4·2**
	32	II	The Uncertainty of Scale	—	I	**5·2**
	31	III	Degeneracy Factors (fragment)	—	II	**6·4**
	24	III	Multiplicity Factors	—	II	**6·3**
	26	III	Multiplicity Factors	—	II	**6·2**
B?	*19*	V	Electric Charge	—	(III, IV)	**7·2**
C	*33*	I	The Uncertainty of the Origin	Before July 1942 (covers drafts 33, 34, 27 and 25)	I, IV	**4·1**
	34	II	The Uncertainty of Scale		I	**5·1**
	27	III	Multiplicity Factors		II	**6·1**
	25	IV	Exclusion and Interchange		IV, V	**7·1**
G	*28*	V	The Complete Momentum Vector	March 1943?	VI, VII	**12·1**
	30	VI	Wave Vectors, VII The Hydrogen Atom and the Neutron (fragment)	March 1943?	VII, X	**12·2**
	22	VII	Double Frames	April 1943	VIII	**12·4**
	15	VII	Double Frames	May 1943	VIII	**12·3**
	a	VIII	Occupation Symbols	July 1943	Appx.	**14·3**
	18	VIII	No title	?	VII, X	**12·6**
	3	VIII	No title	Oct. 1943	VII, X	**12·5**
G?	*14*	IX	The Molar Electromagnetic Field (fragment)	—	XI	**13·7**
H	*13*	I	The Uncertainty of the Reference Frame	Aug. 1943	I	**8·1**
	10	II	Multiplicity Factors	Aug. 1943	II	**8·2**
	9	III	Electrical Theory	Aug. 1943	II, III	**8·3**
	8	IV	Gravitation, Exclusion and Interchange	Aug. 1943	IV, V	**8·4**
	7	V ('VI')	The Complete Momentum Vector	Dec. 1943	VI	**13·1**
	6	VI ('VII')	Wave Vectors	Dec. 1943	VII	**13·2**
	5	VII ('VIII')	Double Frames	Jan. 1944	VIII	**13·3**
	4	VIII	Simple Applications	Feb. 1944	IX	**13·4**
	2	IX	Wave Functions	Apr. 1944	X	**13·5**

CHAPTERS 3–8

THE STATISTICAL THEORY

CHAPTER 3

SUMMARY OF F CHAPTERS I–V

This chapter is a brief summary of the 'statistical theory' **F** I–V, section by section; the main equations are quoted with Eddington's numbering. Comments are added after each section, referring to relevant passages in the draft MSS. which are quoted or summarized in Chapters **4–8** below. There is a survey of **F** I–V in **1·3**.

3·1. F I. The Uncertainty of the Reference Frame

For relevant drafts, see Chapters **4**, **5** and **8·1**.

1. **The uncertainty of the origin**. By the principles of relativity and of quantum theory, physical observables are relations of pairs of entities, each entity having Heisenberg uncertainty in an unobservable mathematical-geometrical frame. If particles ($r=1,2,\ldots$) have geometrical coordinates x_r, y_r, z_r in this frame, and a *physical origin* has coordinates x_0, y_0, z_0, then the relative coordinates

$$\xi_r=x_r-x_0, \quad \eta_r=y_r-y_0, \quad \zeta_r=z_r-z_0 \tag{1·1}$$

are observables—'physical coordinates'. The transformation from x_r to ξ_r involves the unobservable distribution $f(x_0,y_0,z_0)$ (where $f\,dx_0dy_0dz_0$ is the probability of a value x_0 in (x_0, x_0+dx_0), etc.) of the uncertain position of the physical origin. Current wave mechanics uses 'physical coordinates'; what, then, is the physical origin and what is f?

Drafts **B**, **C**, **H** begin similarly; but in 1954 I received five earlier drafts of the opening passage. Three resemble **F**; but one other very early draft is of great interest, linking 'relatedness' with pairs of wave functions. This is quoted in **4·3**.

2. **The physical origin**—this must be the centroid of a very large number of particles distributed with spherical symmetry; for then only is the form of f definite, namely, the Gaussian

$$f(x_0,y_0,z_0)=(2\pi\sigma^2)^{-\frac{3}{2}}\exp\{-(x_0^2+y_0^2+z_0^2)/2\sigma^2\}. \tag{2·1}$$

As this is implicitly used in quantum theory, we can take over the main calculations of that theory. We shall find that σ (the '*uncertainty constant*' of the reference frame) 'puts the scale' into all structures of physics, from nucleus to cosmos.

The use of the centroid also separates 'internal' and 'external' motions, and eliminates correlation effects; see **4·2** § 1·2 (*b*).

3. **The Bernoulli fluctuation.** In a uniform p.d. of particles, the proportional fluctuation $\delta n/n$ (given by James Bernoulli's theorem) of the number found in a fixed volume is less if the total number of particles is finite than if it is infinite; the variance (mean-square deviation) of $\delta n/n$ is less by $1/N$, where N is the total number. This decrease is an 'extraordinary' (negative) fluctuation, to be removed from the 'ordinary' fluctuation due to the finiteness of n.

If the extraordinary fluctuation is applied to a fixed number n_0 of particles in a volume whose linear scale has a proportional uncertainty ϵ (so that the fluctuating volume is $V_0/(1+\epsilon)^3$), then the corresponding s.d. of the extraordinary fluctuation of the linear scale ϵ is [see notes below]

$$\sigma_\epsilon = 1/(2\sqrt{N}). \tag{3·5}$$

We use throughout a *σ-metric*, in which the unit of length is defined by the local uncertainty of distance in the direction of measurement. This is primarily a flat space, with unit of length the uncertainty σ of (2·1) due to the uncertainty of the physical origin; but there is also an uncertainty of the scale of distance corresponding to (3·5). If the scale is to be regarded as exact, we must remove the fluctuation $\sigma_\epsilon r$ from the *radial* measurement of distance r from the physical origin, so that the radial unit becomes $\sqrt{(\sigma^2-\sigma_\epsilon^2 r^2)}$. The length-element then becomes

$$ds^2 = \frac{dr^2}{1-\sigma_\epsilon^2 r^2/\sigma^2} + r^2 d\theta^2 + r^2 \sin^2\theta\, d\phi^2, \tag{3·7}$$

corresponding to a spherical space of radius $R_0 = \sigma/\sigma_\epsilon$. Thus a universe of N particles with an exact scale of length (based on the uncertainty of the physical origin) is uniformly curved; the relation of the radius to the uncertainty constant σ is, by (3·5),

$$\sigma = R_0/(2\sqrt{N}). \tag{3·8}$$

Comment. M. S. Bartlett and G. Lemaître agree with my remark[a] that formula (3·5) is erroneous, and that it should be

$$\sigma_\epsilon = 1/(3\sqrt{N}). \tag{3·5$'$}$$

Eddington's argument is, briefly: if ζ denotes the proportionate fluctuation ($\delta n/n_0$) of the number of particles in a fixed volume V_0, the extra-

[a] Compare Slater, *Phil. Mag.* (7), 38, 299 (1947).

ordinary part has s.d. $\sigma_\zeta = \sqrt{(\overline{\zeta^2})} = 1/\sqrt{N}$. If we transfer the fluctuation to the size of the volume V containing a fixed number n_0 of particles and write $V = V_0/(1+\epsilon)^3$ (cf. summary above (3·5)), then $(1+\zeta)$ is proportional to $(1+\epsilon)^3$. The argument is agreed up to this point. Eddington now says: for discrete values the relation would be $1+\zeta=(1+\epsilon)^3$. 'But in transforming a continuous distribution function, discrete values are replaced by constant ranges, and we have to insert a factor proportional to $d\epsilon/d\zeta$ to transform constant ranges of ϵ into the non-constant ranges of ϵ which correspond to constant ranges of ζ. The relation is therefore $(1+\zeta)\,d\zeta = \text{const.}\,(1+\epsilon)^3\,d\epsilon$, which gives on integration

$$(1+\zeta)^2 = (1+\epsilon)^4.\text{'} \tag{3·43}$$

As ζ is small, this gives $\sigma_\epsilon = \frac{1}{2}\sigma_\zeta = 1/(2\sqrt{N})$.

The error is that the factor $d\epsilon/d\zeta$ is not required, whether the distributions are discrete or continuous. Thus the correct relation is $1+\zeta=(1+\epsilon)^3$, $\sigma_\epsilon = \frac{1}{3}\sigma_\zeta = 1/(3\sqrt{N})$, as in (3·5)′. The central formula (3·8) becomes, since $R_0 = \sigma/\sigma_\epsilon$,

$$\sigma = R_0/(3\sqrt{N}). \tag{3·8′}$$

Effects of this change will be noted under § 5 below and discussed further at the end of §§ 40, 50, 51.

Variations of the sentences quoted above (3·43) will be found below in **5·1** § 11 and **5·2** § 2·1; these are of interest, but do not strengthen Eddington's argument.[a] For the 'σ-metric' see **5·2** § 2·2.

4. **The standard of length.** Ultimate standards of length must be constructed from a specification using only pure numbers, and so must be *quantum-specified*; such lengths are multiples of $\hbar/m_e c$ and so give equivalent metrics. We shall find, for example, that one such length $\mathfrak{R}^{-1}$ ($\mathfrak{R}$ the Rydberg constant for hydrogen[b]) is a multiple of the uncertainty constant σ:

$$\mathfrak{R}^{-1} = \frac{16\pi\sqrt{5}}{3}\,136^2 . 137\sigma. \tag{4·1}$$

Hence the σ-metric (defined after equation (3·5)) is the recognized metric of physics.

Comment. The essential point is that $\hbar/m_e c$ is (like $\mathfrak{R}^{-1}$) a multiple of σ. Such a form of $\hbar/m_e c$ can be constructed (although Eddington does not state it) from the proton-electron ratio (**F** §§ 18, 29–32) and the 'genesis of proper mass' (§§ 39, 40) arguments. The basic connexion of mass, $\hbar$, σ is $m \sim \hbar/\sigma$ (compare (39·9)).

On the standard of length cf. **5·1** §12 and **5·2** § 2·2.

[a] In the lately received MS. **Ga**, Eddington uses $\epsilon = \frac{1}{3}\zeta$ (equation (74·52), p. 267 below); this agrees with my correction, but he maintains that $\epsilon = \frac{1}{2}\zeta$, i.e. (3·43), is correct in the present connexion; see p. 267.

[b] See §§ 29, 50 below.

5. **Range of nuclear forces and the recession of the galaxies.** Two particles r, s have two sets of 'relative coordinates': (i) $\xi_{rs}=\xi_s-\xi_r$ (ξ_r coordinates measured from the physical origin (x_0, y_0, z_0) as in (1·1)) and (ii) ξ'_{rs}, the components of the distance measured directly from r to s. The former contains measurements from *two* random points in the distribution (2·1) of (x_0, y_0, z_0), so that

$$\xi_{rs}=\xi'_{rs}\pm\sigma\sqrt{2} \quad [\text{i.e. } \overline{\xi^2_{rs}}=\overline{\xi'^2_{rs}}+2\sigma^2]. \tag{5·1}$$

In quantum theory the non-Coulombian energy of two particles is associated with actual coincidence, i.e. with $r'_{12}=\sqrt{(\xi'^2_{12}+\eta'^2_{12}+\zeta'^2_{12})}=0$, and has the form (cf. **3·5** §49 below) $B\delta(r'_{12})$. But 'coincidence' is spread in terms of the ξ_{12}, so that by (5·1) the energy takes the form $A\,\mathrm{e}^{-r^2_{12}/k^2}$ $(r_{12}=\sqrt{(\xi^2_{12}+\ldots)})$, where

$$k=2\sigma=R_0/\sqrt{N} \quad (\text{by (3·8)}) \tag{5·2}$$

is the '*range constant*' of nuclear force.

If we treat the universe as an Einstein universe of radius R_0 and of $\frac{1}{2}N$ hydrogen atoms, each of mass M,

$$R_0/N=\kappa M/\pi c^2=3·95\times 10^{-53}\,\text{cm.}, \tag{5·41}$$

while from the empirical proton-proton scattering constant k

$$R_0/\sqrt{N}=k=1·9\times 10^{-13}\,\text{cm.} \tag{5·42}$$

Hence we obtain N and R_0; and the limiting speed (i.e. speed/distance) $V_0=c/(R_0\sqrt{3})$ of galactic recession is found to be 585 km./sec. per megaparsec. If we determine σ from (4·1) we obtain $V_0=572·4$. The observed value [in 1944] is 560.

Conversely, we deduce from this agreement that non-Coulombian energy *is* associated with $r'_{12}=0$. We thus reject 'meson fields', which give energy $A\,\mathrm{e}^{-\lambda r_{12}}$. Experiments on the shape of the non-Coulombian potential well will provide a crucial test.

Comments: (i) This 'crucial test' has yet (as far as I am aware) to be settled, although some theorists use Eddington's e^{-r^2} form pragmatically. (ii) If (3·8)′ is used for (3·8) in (5·42), this multiplies R_0 and also N by $\frac{9}{4}$ and V_0 (recession velocity) by $\frac{4}{9}$.[a] Recent work of Baade has brought the observational value down by a factor of 2 or $2\frac{1}{4}$. Thus the formula (3·8)′ is supported.

6. **Spherical space.** If N particles are uniformly distributed over a hypersphere $x^2+y^2+z^2+u^2=R_0^2$, the s.d. of a coordinate of their centroid (corresponding to the 'physical origin') is $R_0/(2\sqrt{N})$, which equals σ by (3·8). This is also the s.d. for the projection of the

[a] Slater, *Nature, Lond.*, **174**, 321 (1954).

distribution on a tangent flat space. Thus the results of §3 are in agreement with this simple picture of a hyperspherical distribution projected (locally) into a flat space.

We must not attempt to derive the *local* irregularities of curvature of space-time from local statistical fluctuations, and link these up with quantum theory. Quantum theory links up with the 'general' relativity of uniform curvature which is intermediate between the irregular curvature of Einstein's full general relativity on the one hand and the flat space-time of the 'special' theory on the other.

Comment. The correction (3·8)′ is discordant with this simple picture, which, however, Eddington regards as merely supporting the prime argument of §3; cf. **8·1** §6. In earlier versions, nevertheless, the transition from flat to spherical space on the grounds of *relativistic curvature* precedes the argument of §3. See **4·1** §7 for a full discussion, and **4·2** §1·7.

For the latter part on relativity and quantum theory, see **5·1** §13 and **5·2** §2·3.

7. **Uranoids.** We divide the universe into the *object-system* studied, and its *environment*—which is all the rest. An ideal environment is a '*uranoid*' (cf. the 'geoid' for the ideal 'earth')—an Einstein universe as a steady distribution of particles, usually at zero temperature, and electrically neutral. These particles determine the *metric* (for measurements of the object), or equivalently interact mechanically with the object, providing the inertial-gravitational field. In a non-uniform field the standard uranoid provides the uniform *inertial* part, and so is the background for quantum theory.

For early accounts (with developments that come later in **F**) see **4·1** §6 and **4·2** §1·6. For a later amplification see **8·1** §7.

8. **The extraneous standard.** We use 'natural units' such that (c being the velocity of light, κ the constant of gravitation and $2\pi\hbar = h =$ Planck's constant)

$$c=1, \quad 8\pi\kappa\hbar^2=1; \tag{8·1}$$

taking $c=1$ removes the obsolete distinction between mass and energy. These relations leave one unit—one *extraneous standard*, of length, time or mass for example—at our disposal; any physical quantity will then have one *dimension index*, y, in terms of this standard. If, for example, the unit is one of length and so has the standard uncertainty $1 \pm \sigma_\epsilon$ (cf. §3), the general quantity has uncertainty $1 \pm y\sigma_\epsilon$.

The relativity energy tensor $T_{\mu\nu}$ and the quantum momentum vector p_μ are given by

$$-8\pi\kappa T_{\mu\nu} = G_{\mu\nu} - \tfrac{1}{2}g_{\mu\nu}G, \quad p_\mu = -i\hbar\,\partial/\partial x_\mu, \tag{8·2}$$

where the Ricci tensor $G_{\mu\nu}$ is, like $g_{\mu\nu}$, of dimensions (length)$^{-2}$, and $G = \Sigma G^{\mu}_{\mu}$. Thus with $8\pi\kappa\hbar^2 = 1$, dimensionally and tensorially

an energy tensor is the product of two momentum vectors. (8·31)

Taking the $\mu = \nu = 4$ (time) component, density (T_{44}) is of dimension (mass)^2, so that volume is (mass)$^{-1}$ and

particle (*i.e. number*) *density is of dimensions mass or momentum*; (8·32)

it has also the same tensorial character.

In relativity, density is the best extraneous standard; then other dimension indices are

$$\left.\begin{array}{llll} \text{mass, momentum, energy,} & \tfrac{1}{2}; & \text{length and time,} & -\tfrac{1}{6}; \\ \text{angular momentum, action,} & \tfrac{1}{3}; & \text{electric charge,} & \tfrac{1}{6}. \end{array}\right\} \quad (8\cdot4)$$

These results (8·3, 8·4) are important later on.

For the genesis of (8·1) see § 3 of **2·1 δ**. For the role of the extraneous standard, **5·1** § 14 and **5·2** § 2·4. On 'scale', **8·1** § 8. For the italicized sentence (8·31), **11·3** § 4·5.

9. **Scale-free physics.** The linear characteristics σ and R_0 of the uranoid do not enter *scale-free physics*—the physics of structures which are unaffected by a change of the extraneous standard used; this includes classical molar theory and the 'non-quantized' parts of quantum theory. The other main branches are *cosmical* physics (involving R_0, not σ) and *quantal* or 'scale-fixed' physics (involving σ, not R_0)—the physics of discrete eigenstates.

For 'scale-free system' see **5·1** § 15.

10. **Pseudo-discrete states.** Wave functions in quantal (scale-fixed) theory are *discrete* and self-normalizing (e.g. harmonic oscillator), unlike the wave functions of scale-free theory, typified by 'infinite plane waves', which are normalized by choosing an arbitrary volume V_n as the 'volume per particle'; the wave functions so normalized are called *pseudo-discrete.* The occupant of such a wave function is not, of course, restricted to be in the volume V_n, but is anywhere in the distribution; it is essentially an *unidentified member of a large assemblage.* Pseudo-discrete states are created by breaking up the range of a parameter α (e.g. momentum) into small elements, say $\delta\alpha_r$ for the rth pseudo-discrete state. If the state of the whole assemblage is in one such element, it is said to be *almost exact*; we shall make much use of the state of 'almost exact rest'.

The proper mass (and momentum) of a particle are 'scale-fixed'; the proper density (and energy tensor) are 'scale-free'; these characteristics typify quantal and scale-free physics respectively. This corresponds to the distinction between correlation (atomic) and distribution wave functions, or broadly between electrical and mechanical theory.

There is a subtle distinction between splitting the ranges of parameters α as they occur in *distribution functions* (i.e. probability densities) and wave functions. See **5·1** §16 and **5·2** §2·5. The late version **8·1** §10 also discusses the relation of electrics and mechanics.

11. **Stabilization.** A quantity whose value is assumed known (e.g. the mass of an electron taken from tables) is called a *stabilized characteristic*; it is then not 'observable' in the experiment and has no probability scatter.

A particle whose p.d. (after any stabilization of characteristics) has k dimensions will be called a V_k. If the momentum 4-vector $p_1, \ldots, p_4$ and proper mass m of a particle are connected by

$$p_4^2 = p_1^2 + p_2^2 + p_3^2 + m^2,$$

we call p_4 the *hamiltonian*; the particle is a V_3 or a V_4 according as m is or is not 'stabilized'.

Stabilization may be applied to the environment (e.g. to make it a uranoid); it may be applied to a tensor (e.g. as a symmetry condition).

See comments in **5·1** §17 and **8·1** §11; also a full discussion of *stabilization of tensors* in **5·1** §18 and **5·2** §2·6.

3·2. F II. Multiplicity Factors

Drafts are in **6**, **7·1**, **8·2** and **8·3**.

12. **Complementary fields.** Particle object-systems selected for study disturb the environment, which cannot therefore be a standard uranoid. We prefer, however, to regard the environment as a standard uranoid plus a superposed disturbance—the complementary object-field due to the object-system. The object-field is described by the same particle variates—energy, momentum—as the object particles, but refers to averaged characteristics of the unidentified environment particles. This chapter treats complementary *gravitational* fields.

A contrast between this viewpoint and general relativity, **6·1** §19; and 'intra-atomic fields', **8·2** §12.

13. **The rigid-field convention.** In contrast with relativity, which has a metric varying with the matter present, wave mechanics employs a rigid metrical (or gravitational) field and constructs in

this a skeleton frame of eigenstates, leaving flexible the degree of occupation. For this to be valid,

the field must be stationary for small changes of the occupation. (13·1)

Quantum particles are defined by this convention.

The basis of this assertion is Hartree's 'self-consistent field' conception; this is described in **6·3** § 3·1, and noted in (the later) **8·2** § 13. The following development is foreshadowed in the paper **2·1 ε**.

14. **Separation of field and particle energy.** Consider first a system with *discrete* eigenstates ψ_r with occupations j_r and total energy $H^0 = H^0(j_1, j_2, j_3, \ldots)$. If the field is of the rigid type (§ 13), the energy per particle in state r is

$$E_r = \partial H^0 / \partial j_r \tag{14·1}$$

(for the total change in H^0 is $\Sigma E_r \delta j_r$ if the field is 'rigid'); the total *particle* energy

$$E^0 = \Sigma E_r j_r, \tag{14·3}$$

and the *field* energy $$W^0 = H^0 - E^0. \tag{14·4}$$

If H^0 is homogeneous of degree n in the j_r, then

$$E^0 = nH^0, \quad W^0 = (1-n)\, H^0. \tag{14·6}$$

Any other additive characteristic (e.g. momentum) must follow the same method of partition between particles and field, which allows small changes of occupation dj_r with the E_r and W^0 constant. These changes can allow complete transitions of individual systems (i.e. integral dj_r) provided (1) these are unidentified members of a large assemblage which (2) remains preponderantly in the initial state for which the field was calculated.

For a useful amplification see **6·1** § 20.

15. **Application to scale-free systems.** Let the classifying characteristics (i.e. observables plus stabilized entities) of a *scale-free* system be X_α ($\alpha = 1, 2, \ldots, n$) with dimension index 1 (i.e. of unit dimension in the extraneous standard which is varied when the scale is changed; cf. § 8). Possible states will be a k-dimensional locus—the *phase space*—embedded in the n-space of the X_α; we call k (which is the number of *observable* characteristics) the *multiplicity factor*. In the scale transformation $X_\alpha \to \lambda X_\alpha$ (due to varying the standard) the volume element $d\tau$ of phase space becomes $\lambda^k d\tau$.

The discrete j_r of § 14 are here replaced by continuous occupation factors $j(X)$, such that $j(X)\, d\tau$ is the collective occupation of states $d\tau$. Let H^0 denote the total, E^0 and W^0 ($\equiv H^0 - E^0$) the particle and field parts, of a characteristic such as energy, and let l be their dimension index. The particle energy (per unit $\int d\tau$) is the hamiltonian derivative $E = \eth H^0/\eth j$ [see notes]. By considering a small scale-variation we find

$$W^0 = -(1+k/l)\,E^0, \quad H^0 = -kE^0/l. \tag{15·51}$$

Considering the states as *pseudo-discrete* (§10) with occupations $j_r = j(X)\,d\tau_r$, we see from this and (14·6) that

the scale-free H^0 is of degree $-l/k$ in the pseudo-discrete occupation factors. (15·52)

For a scale-free particle assembly, characterized solely by a total energy tensor $T_{\mu\nu}$, $T_{\mu\nu} \equiv H^0 \equiv X_\alpha$, so that $l = 1$. The particle ($E_{\mu\nu}$) and field ($W_{\mu\nu}$) tensors are then by (15·51)

$$T_{\mu\nu} = -kE_{\mu\nu}, \quad W_{\mu\nu} = -(k+1)\,E_{\mu\nu}. \tag{15·7}$$

For rigid-field treatment we choose for an initial state

$$(T_{\mu\nu})_0 = -k(E_{\mu\nu})_0;$$

but by the rigidity (i.e. $\delta W_{\mu\nu} = 0$),

$$\delta T_{\mu\nu} = \delta E_{\mu\nu} \tag{15·8}$$

in a small *transition*. If, however, we take as classifying characteristic the '*generic energy tensor*' defined (both for initial and later states) as

$$X_{\mu\nu} = -T_{\mu\nu}/k, \tag{15·91}$$

then $(E_{\mu\nu})_0 = (X_{\mu\nu})_0$, but $\quad \delta E_{\mu\nu} = -k\delta X_{\mu\nu}$. (15·92)

Now $X_{\mu\nu}$, originally equal to the particle energy, changes in the transition as if merely the scale-free condition (15·7) [read as $X_{\mu\nu} = E_{\mu\nu}$] and not the rigid-field condition (15·8) applied; thus changes in $X_{\mu\nu}$ are calculated as if the rigid field were ignored. Hence by (15·92) the actual change of particle energy is $-k$ times the change 'expected' for a non-rigid field.

This section is one of the most important, and most obscure, in the book; the above summary changes the emphases in order to help comprehension. The main result is (15·7); the subsidiary argument on 'generic energy' should be tackled separately. For '*hamiltonian differentiation*' and the derivation of (15·51), see **6·1** §21. For the later part—initial and transition energy, and *generic energy*—see **6·1** §22 and **6·2** §3·3 and the later **8·2** §15. After these (or the full text of **F**) have been studied, it may be realized that field 'rigidity' is applied in two ways: (i) as in scale transformation, yielding (15·51) and (ii) for state transition (with now no scale variation) yielding (15·8). The 'generic energy' is the 'expected energy' if (ii) is relaxed, and so corresponds to energy 'created' by a change to a moving coordinate system (as in a Lorentz transformation).

16. **The 'top particle'.** Consider the total energy, H^0 or $T_{\mu\nu}$, of an assemblage of particles, number density s, multiplicity k, in a pseudo-discrete state. The energy of a 'top' or object-particle is $\mathfrak{H}=dH^0/ds$, and of a 'mean' particle is $\overline{H}=H^0/s$. By § 15, $s \propto j$, the occupation of the state, so that by (15·52) $T_{\mu\nu}\equiv H^0 \propto s^{-1/k}$; hence

$$\mathfrak{H}=-\overline{H}/k. \tag{16·2}$$

Identifying this relation with (15·7),

the particle ($E_{\mu\nu}$) *and total* ($T_{\mu\nu}$) *energies are those of a top* ($\mathfrak{H}$) *and a mean* ($\overline{H}$) *particle.* (16·3)

If the initial state considered is 'almost exact rest', the energy tensors reduce to densities proportional to the *proper masses* $\mathfrak{m}$ and $\overline{m}$ of top and mean particles; thus

$$\mathfrak{m}=-\overline{m}/k. \tag{16·4}$$

Molar characteristics relate to the mean particle, quantum transitions (of observed particles) to the top particle. The negative sign indicates the quantum 'inversion of energy' (§ 21).

The multiplicity k decides the partition of $T_{\mu\nu}$. If k is changed by 'stabilization' (as in § 11), the energy and mass of a top particle vary as $1/k$. Thus the masses m_1, m_2 of quantum particles V_{k_1}, V_{k_2} of multiplicities k_1 and k_2 obey

$$m_1/m_2=k_2/k_1. \tag{16·5}$$

This refers to particles in the *same environment* (i.e. of mean particles) and so represents the *observable* mass ratio.

'Multiplicity factors', particularly as appearing in (16·5), are the basic tool of **F**, and the 'top particle' is introduced to help their comprehension. The 'top particle' is not in evidence in early versions, although it appears in the previous version **8·2** § 18. The derivation of (16·5) (without appeal to the 'top particle') occurs in **8·2** § 17; the earliest account is that quoted in **6·4** § 3·3. Early views on 'multiplicity' (as 'degeneracy') are summarized in **5·2** § 2·9.

'Top particles' appear somewhat differently in the exclusion theory (**F** §§ 41–3).

17. **Standard carriers.** Including spin, we represent *mechanical* characteristics by *complete* momentum vectors and energy tensors with 10 and 136 independent components. These concepts are developed in VI, VIII (**9** below) and are there expanded to 16 and 256 components, respectively; but the further, 'chiral', components are dormant in neutral environments.

We define a 'particle' as the *conceptual carrier* (or occupant of a state) *of a set of variates*; this includes 'composite' particles. The simplest, or *standard carrier*, in scale-free physics is a V_{136}—a carrier of an element of complete energy tensor (unstabilized). An ordinary energy tensor $T^{\mu\nu}=\rho_0 v^\mu v^\nu + s^{\mu\nu}$ becomes, when the stress $s^{\mu\nu}=0$, the outer square

$$T^{\mu\nu}=v^\mu \sqrt{\rho_0}\,.\,v^\nu \sqrt{\rho_0} \tag{17·2}$$

of a 'root vector' $v^\mu \sqrt{\rho_0}$. When we pass to fixed-scale quantum physics, this vector becomes the momentum vector of a particle. The carrier of a *complete* root vector, i.e. of a complete energy tensor stabilized by (17·2) to be a square, is a *vector carrier* or V_{10}. The V_4 or V_3 particles of § 11 are fictions, since they are spinless.

For standard and vector particles see **6·1** § 25 and **6·2** § 3·6 (and **6·3** § 3·6). 'Complete' vectors are discussed in **F** § 53 (**9·1** below).

18. **Mass ratio of the proton and electron.** We may assign *initial* and *transition particles* to carry initial and transition energy (in the rigid field); the former are usually at 'almost exact rest'; the latter have initial energy zero.

We can analyse a classical two-particle system of masses m, m' into an *external particle* of mass M at the mass centre (which may be taken at rest, so that this is an *initial particle*) and an *internal particle*, associated with an internal mass μ, describing the relative orbit (but having zero rest energy, and so a *transition particle*); where

$$M=m+m', \quad \mu=mm'/(m+m'). \tag{18·1}$$

Thus a hydrogen atom may be analysed into proton + electron, *or* into external + internal particle.

Let a standard carrier of mass m_0, initially at rest, make a transition to momentum $\mathbf{p}'(\equiv p_1', p_2', p_3')$. The 'generic energy' (without regard to the rigid field) is, for moderate $\mathbf{p}'$,

$$X=m_0+p'^2/2m_0. \tag{18·21}$$

By (15·92) the particle energy is

$$E=m_0-kp'^2/2m_0=m_0-p'^2/2\mu, \tag{18·22}$$

where

$$\mu=m_0/k=m_0/136. \tag{18·23}$$

Using the *quantum* momentum $\mathbf{p}$ (defined 'to preserve a formal analogy between rigid-field and classical dynamics'), where

$$\mathbf{p}=i\mathbf{p}', \tag{18·31}$$

then

$$E=m_0+p^2/2\mu. \tag{18·32}$$

Now divide the standard carrier V_{136} into two vector V_{10} carriers of initial and transition energy, for which

$$E_e = m_0, \quad E_i = p^2/2\mu; \tag{18·33}$$

these are the 'external' and 'internal' carriers. The reduction of k from 136 to 10 multiplies the *initial* mass m_0 by $\frac{136}{10}$ (cf. (16·5)) but not the transition energy; so finally

$$E_e = M, \quad E_i = p^2/2\mu, \tag{18·4}$$

where $$M = \tfrac{136}{10} m_0, \quad \mu = \tfrac{1}{136} m_0. \tag{18·5}$$

Identifying the standard carrier with the hydrogen atom, we have as a fundamental constant

$$\eta_1 \equiv M/\mu = 136^2/10 = 1849{\cdot}6, \tag{18·6}$$

and the proton and electron masses ($m_p + m_e = M$, $m_p m_e = M\mu$ by (18·1)) are the roots of

$$m^2 - Mm + M\mu = 0, \tag{18·7}$$

i.e., by (18·5), $$10m^2 - 136 m_0 m + m_0^2 = 0, \tag{18·8}$$

with ratio $$\eta_2 = m_p/m_e = 1847{\cdot}60. \tag{18·9}$$

We call these m_p, m_e the *standard* masses, in contrast with the current masses (§ 29 below). As this investigation is scale-free, we are concerned only with *ratios* to m_0, which are determined by comparing *densities*.

The above is perhaps as clear as any earlier exposition, save for the 'quantum momentum' (18·31) which is discussed later in **F** § 21. For the previous version see **8·2** § 17. In earlier MSS. this calculation comes after more preliminary spade work, and so appears more succinct; e.g. in **6·1** § 27 and **6·2** § 3·7 (which should be read after the preceding sections of those drafts). There is a useful summary in the still earlier **6·3** § 3·7. Further light on (18·31) appears in **13·2** § 63.

19. **Rigid coordinates.** The very large field energy ((15·7) with $k = 136$) appearing in the 'rigid-field' theory implied by wave mechanics must be ascribed to the use of non-Galilean coordinates ('Galilean' implying no gravitational field). The rigidity condition may be expressed thus: rigid-field mechanics employs 'rigid' coordinates (x, y, z, t) so chosen that the field is stationary for small changes of occupation. For particles at almost exact rest, the connexion with Galilean (x', y', z', t') is verified to be

$$x' = x, \quad y' = y, \quad z' = z, \quad t' = -kt \tag{19·1}$$

by showing that this agrees with (15·7). Thus in practice the rigid-field condition is

a time t which is k^{-1} times the Galilean t' is used in wave mechanics. (19·7)

The difference between total and particle energy may be pictured (*a*) as field energy, or (*b*) as the effect of rigid coordinates, or (*c*) as the difference of mean and 'top-particle' energy.

Here is a naïve check of (19·1): in a Galilean frame (x', y', z', t') the momentum-mass (energy) vector of a particle or uniform distribution is $(0, 0, 0, E)$, where E is 'particle' energy—there is here no other energy since 'Galilean' means 'with no gravitational field'. If now we make the transformation (19·1), treating momentum-mass as a *covariant* vector as in wave mechanics (where $p_\mu = -i\hbar\,\partial/\partial x_\mu$), the vector in the new frame (x, y, z, t) is $(0, 0, 0, T)$, where

$$T = -kE. \tag{i}$$

The apparent increase in energy $W = T - E$ is thus

$$W = -(k+1)E. \tag{ii}$$

This W is ascribed to the field 'created' by the transformation. Now (i) and (ii) correspond to (15·7) for particles at rest. The relations $(x, y, z) = (x', y', z')$ are necessary to make particle and density characteristics behave similarly, as we have assumed them to do. Thus the change $t' = -kt$ of (19·1) corresponds to the change from Galilean t' to time t in a rigid field.

Although these ideas occur in **F**, the argument there also involves tensor densities. The general conception is discussed in **6·1** § 24 and **6·2** § 3·4. For the last sentence of the summary, cf. **8·2** § 18.

20. **The fine-structure constant.** For the relation of molar and microscopic theory, the basic multiplicity is that ($k = 136$) of the standard carrier, giving in (15·7)

$$T_{\mu\nu} = -136E_{\mu\nu}, \quad W_{\mu\nu} = -137E_{\mu\nu}. \tag{20·1}$$

The coefficient 137 here is the *fine-structure constant.*[a] It appears fundamentally as the ratio of 'atoms' of field and particle *action*, $\hbar$ and e^2/c. (For $137 = \hbar c/e^2$, see **3·3** § 33.) A *total* action unit $\frac{136}{137}\hbar$ is less prominent than the field unit $\hbar$ because quantum theory began with the radiation field; but we shall make much use of the *Bond factor* (first emphasized by W. N. Bond)

$$\beta = \tfrac{137}{136}. \tag{20·2}$$

The unit $\hbar$ appears first in the momentum $p_\mu = -i\hbar\,\partial/\partial x_\mu$; this operator refers to field momentum only (**10·2** § 105).

[a] Accepted usage makes the 'fine-structure constant' $\alpha \approx \frac{1}{137}$.

This section is inserted as an illustration of 'multiplicity factors'. As an introduction to the role of the Bond factor, the reader might here study **6·2** §§ 3·8, 3·9 on *radiant energy*.

21. **The inversion of energy.** Consider V_1's—particles of multiplicity $k=1$, having mass m as the only unstabilized characteristic. By (15·7), the *total* energy per particle in an (initial) rigid field is $-m$. In rigid-field dynamics we reverse this sign; this is 'the inversion of energy'. If, however, the particle is now given velocity v, we reckon the kinetic energy not as $-\frac{1}{2}mv^2$ but as $+\frac{1}{2}m(iv)^2$, attributing imaginary velocity and momentum. This is the source of $\sqrt{-1}$ in quantum theory. In problems solved by wave mechanics with real quantum momenta, the corresponding *classical* momenta are imaginary.

This is elucidated at length in **6·1** § 23 and **6·2** § 3·5. For $\sqrt{-1}$ see also **8·2** § 21 and **11·1** § 25(*b*).

22. **Mutual and self energy.** The energy of a system of two particles, such as an object-particle and a reference or '*comparison*' *particle*, is strictly a mutual property, but is usually divided into separate parts for the two particles. Suppose a simple object-particle at rest, mass m, is observed in conjunction with a comparison particle of mass m_0. The total energy tensors reduce to densities $\rho=Am^2$, $\rho_0=A_0m_0^2$ for the particles separately, and the total mutual energy tensor $M_{\mu\nu}$ [cf. p. 114] to $\rho_m=Cmm_0$. Since $\rho+\rho_0=\rho_m$,

$$Am^2+A_0m_0^2=Cmm_0. \tag{22·42}$$

Comparing this with the equation (18·8) for the mass of a simple V_{10} (proton or electron) and *identifying* the m_0's,

$$A:A_0:C=10:1:136. \tag{22·51}$$

The multiplicity factors here are 136 for the mutual energy tensor, 10 for the V_{10} and 1 for the comparison particle (as carrier merely of the density standard).

For an *internal* particle the zero rest energy makes $\rho=0$, so that $\rho_0=\rho_m$ and

$$A_0m_0^2=C\mu m_0=136A_0\mu m_0. \tag{22·61}$$

The *external* V_{10} particle has mass $M=\frac{136}{10}m_0=Cm_0/A$, so that

$$AM^2=CMm_0 \tag{22·62}$$

—as if the mutual density were attributed wholly to the object-particle; so the internal plus external particle, or hydrogen atom, requires only one comparison particle.

Collecting these formulae for the partition of a mutual (total) density into self-densities

$$\left.\begin{array}{lll} 136A_0mm_0 = 10A_0m^2 + A_0m_0^2 & \text{(proton or electron),} \\ 136A_0\mu m_0 = \qquad\qquad A_0m_0^2 & \text{(internal particle),} \\ 136A_0Mm_0 = 10A_0M^2 & \text{(external particle).} \end{array}\right\} \quad (22{\cdot}7)$$

Particles which have comparison particles (with a term $A_0m_0^2$) have electrical characteristics, whereas the external particle is neutral.

Equations (22·7) determine the masses in such a way that *specification* of an object-particle (i.e. substituting it for a stabilized comparison particle) does not affect the total energy tensor.

This section is based on the approach to the mass quadratic given in paper **δ** of **2·1**. Its function here is the comparison of the multiplicative and additive (mutual and self-energy) representations of a system. A useful introduction to §§22, 23 is provided by **7·1** §§30, 31; see also **7·1** §29 (end) and **11·3** §4·5.

23. **Comparison particles.** A *perfect* object-system is one extended to include its comparison particle; the latter then carries as sole characteristic the standard of mass, although before it is introduced into the object-system it is an (unspecified) standard carrier.

The (particle) energy tensor for *unit occupation* is

$$E_{\mu\nu} = \frac{1}{V_0}\frac{p_\mu p_\nu}{m} \qquad (23{\cdot}1)$$

(V_0 = normalization volume), since for particles at rest ($p_4 = m$) this becomes $\rho = m/V_0$.

The point of this formula will appear in §26 below. Its derivation is discussed in **8·3** §23.

3·3. F III. Interchange

Relevant drafts are in **7** and **8·3**.

24. **The phase dimension.** Using flat space for wave mechanics, we regard the uncertain scale (representing the extraneous standard of measurement) as the momentum conjugate to a variable, termed *phase*, which occupies a fifth dimension normal to space-time. Individual object-systems have in their comparison particles 'destabilized scales' with much larger fluctuations than the mean scale, s.d. σ_ϵ, of §3. If the scale as an *angular* momentum tends to exactitude, the conjugate angular phase tends to a uniform distribution over 2π (the *widening factor*)—a 'thickness' in the fifth dimension. But if we take the phase and scale as a *linear* coordinate and momentum x

and p, and suppose these separable from the other dimensions, then the x is defined primarily over a linear width $2\pi l$ and has eigenfunctions $e^{inx/l}$; the scale momenta $p = n\hbar/l$ are then proportional to the integers (24·3).

The scale and phase resemble energy and time, but are in fact space-like owing to the nature of a standard as a *divisor* in measurement.

For the relation of scale to the comparison particle, see **8·3** § 24 (with also a historical note on curvature); and for the connexion of this with the linear scale variate, see **7·1** § 29. For the widening factor, see **5·2** § 2·7 (*b*).

25. **Interchange of suffixes.** An observable (a coordinate or momentum) involves two entities; a measurement involves four, two for the object-observable and two for the comparison standard. The characteristic of four entities that is measured is a '*measurable*'.

Let the object-entities A_1, A_2 be a proton and electron, and A'_1, A'_2 the comparison particles. The transformation of the measurable $[A_1, A'_1, A_2, A'_2]$,

$$[A_1 A'_1 \, A_2 A'_2] \to [A_1 A'_2 \, A_2 A'_1], \tag{25·1}$$

in which two 'perfect' particles (§ 23) exchange comparison particles, is called *interchange*. Since comparison particles are indistinguishable, all observable results are invariant for interchange, which is a relativistic transformation like space-time rotation. Now results are also invariant for interchange of the *suffixes* introduced (mathematically) to identify particles. We amalgamate these two interchanges by making the comparison particles carry the suffix, as well as the scale-and-phase.

When we replace the proton and electron by an external and internal particle, only one comparison particle is required to carry the scale-and-phase (cf. § 22), but a permutation variate also is needed to represent the transformation (25·1). The *permutation coordinate* is an angle θ such that (25·1) corresponds to $\theta \to \theta + \pi$. The uncertainty of θ has range 2π; θ is unobservable, like an 'ignorable angle'; but the constant (permutation) angular momentum conjugate to θ gives an 'interchange energy' identifiable with Coulomb energy.

This is the starting point of the 'electrical theory'. See **8·3** § 25 for notes on interchange and **7·1** § 38. 'Measurables' are discussed further in the Appendix to **F** (cf. Chapter **14** below).

26. **The two-particle transformation.** When a system of two particles, m, m' is analysed into an external and internal particle, or an *extracule* and an *intracule*, these have classically masses

$$M = m + m', \quad \mu = mm'/(m + m'). \tag{26·11}$$

To check this for wave mechanics: if x_α, x'_α ($\alpha = 1, \ldots, 4$) are the space-time coordinates, then for extra- and intracule

$$\mathrm{x}_\alpha = \frac{m x_\alpha + m' x'_\alpha}{m + m'}, \quad \xi_\alpha = x'_\alpha - x_\alpha. \tag{26·12}$$

This gives, assuming (26·11),

$$\frac{1}{m}\frac{\partial^2}{\partial x_\alpha \partial x_\beta} + \frac{1}{m'}\frac{\partial^2}{\partial x'_\alpha \partial x'_\beta} = \frac{1}{M}\frac{\partial^2}{\partial \mathrm{x}_\alpha \partial \mathrm{x}_\beta} + \frac{1}{\mu}\frac{\partial^2}{\partial \xi_\alpha \partial \xi_\beta}, \tag{26·14}$$

or, with quantum momenta $p_\alpha = -i\hbar\partial/\partial x_\alpha$, p'_α, P_α, ϖ_α,

$$\frac{p_\alpha p_\beta}{m} + \frac{p'_\alpha p'_\beta}{m'} = \frac{P_\alpha P_\beta}{M} + \frac{\varpi_\alpha \varpi_\beta}{\mu}. \tag{26·21}$$

From (23·1), assuming equal V_0's for m, m', M and μ (now to be identified as proton, electron and hydrogen extra- and intracule), it follows that the energy tensors satisfy

$$(T_{\alpha\beta})_m + (T_{\alpha\beta})_{m'} = (T_{\alpha\beta})_M + (T_{\alpha\beta})_\mu, \tag{26·22}$$

since the equality of numbers of the four types of particle makes the energy tensors proportional to the tensors, $p_\alpha p_\beta / m V_0$, etc., for unit occupation. The result (26·22) justifies (26·11). The equality of the V_0's followed from the considerations (i) m, m', M are particles in the same space-time and have therefore equal V_0's, and (ii) the Jacobian of (26·12) is unity (with or without time coordinates), so that $V_{0(M)} V_{0(\mu)} = V_{0(m)} V_{0(m')}$ and $V_{0(\mu)}$ therefore equals the other three V_0's.

For momenta $p = \sqrt{\sum_1^3 p_\alpha^2}$, etc., (26·21) gives

$$p^2/m + p'^2/m' = P^2/M + \varpi^2/\mu. \tag{26·41}$$

For the *hamiltonians*

$$h = m + p^2/2m, \quad h' = m' + p'^2/2m', \quad H = M + P^2/2M, \quad \eta_0 = \varpi^2/2\mu, \tag{26·42}$$

we have

$$h + h' = H + \eta_0. \tag{26·43}$$

We shall treat a transformation $h + h' \to H + \eta$, where

$$\eta = \mu + \varpi^2/2\mu, \tag{26·44}$$

by changing the energy zero to add the requisite density to make the intracule hamiltonian formally similar to h, H. This transformation is called *freeing the intracule*.

Eddington's argument glosses over the role of V_0 when he first uses (23·1) here; this summary has been studied to make the argument clearer. An early version is summarized in **7·2** § 5·1, a later in **8·3** § 30.

27. **Hydrocules.** In a hamiltonian

$$E=\mu_1+p^2/2\mu_2, \tag{27·1}$$

μ_2 is the 'mass constant' (which the physical tables list) and μ_1 the 'rest mass' (differing from μ_2 by potential energy in a gravitational or electrical field, but agreeing in a pure inertial field). The particle is called '*free*' or '*bound*' according as $\mu_1=\mu_2$ or 0. Classical theory—and our discussion in § 18—uses bound intracules; quantal theory (e.g. Dirac's wave equation of hydrogen) uses free intracules. To 'free' the intracule we replace (18·33) by

$$E_e=m_0, \quad E_i=\mu+p^2/2\mu, \tag{27·2}$$

regarding E_e, E_i as initial and transition energies; so that the background particles (initial particles) have energy m_0 whereas the standard particle now has rest energy

$$m_0+\mu=\tfrac{137}{136}m_0=\beta m_0. \tag{27·3}$$

This is a change of reckoning of energy whereby the background particles are no longer standard particles but are what we call '*hydrocules*', having rest energy $1/\beta$ times that of standard particles. We distinguish

System A: standard particle uranoid; bound intracules;
System B: hydrocule uranoid; free intracules.

To compensate the energy change from A to B, we

multiply measured densities (by a change of standard) *by* β. (27·4)

This allows the gravitational constant κ to be unchanged, and $\hbar$ also ($8\pi\kappa\hbar^2=1$). By (8·4), as we are using a density standard,

in passing from A to B, lengths and times are multiplied by $\beta^{-\frac{1}{3}}$. (27·6)

As we keep the particle density s unchanged, the rest masses m_0/β, m_0 of hydrocule and standard particle in A become m_0, $m_0+\mu$ in B.

See also summary **8·3** § 26; and an early approach in **7·2** § 5·2.

28. **Separation of electrical energy.** The standard carrier, or V_{136}, of mass m_0, carries an energy tensor; if we add a permutation variate (so that the vector carriers can be *suffixed* when the V_{136} is analysed) we obtain a V_{137} of mass, by (16·5),

$$m_0'=\tfrac{136}{137}m_0=m_0/\beta. \tag{28·1}$$

By comparison with the masses in § 27, we identify this with the *hydrocule*; but by keeping the electrical (permutation) degree of freedom separate we treat the hydrocule as the V_{136} (i.e. with 136 mechanical components) of system B.

System A is the *observational system*, as observation is essentially molar, and uses the original energy zero as for 'bound' intracules. System B is the *theoretical system* of quantal physics (partly because there we have to suffix the particles in the two-particle system so that the theory essentially is of *hydrocules*). *Theoretical formulae have to be translated from B to A for comparison with observation.*

This 'translation' is the key to the β-factors in Eddington's numerical formulae. The early approach **7·2** § 5·2 should be consulted again for this section. Concerning the 'observational system', the reader should (after § 31) consult **8·3** § 28.

29. **Current masses of the proton and electron.** By the quantum theory of the *free* intracule (i.e. quantum theory in system B of § 27), the Rydberg constant for hydrogen is

$$\mathfrak{R} = \tfrac{1}{2}\left(\tfrac{1}{137}\right)^2 \frac{\mu c}{2\pi\hbar} \quad (137 \equiv \hbar c/e^2). \tag{29·1}$$

The *experimental* determination, being from molar wave-length measurements, is in system A; thus as $\mathfrak{R}^{-1}$ is a length, by (27·6) the empirical $\mathfrak{R}$ is $\mathfrak{R}_A = \beta^{-\frac{1}{6}}\mathfrak{R}_B$. To preserve the form (29·1) we write

$$\mathfrak{R}_{(A)} = \tfrac{1}{2}\left(\tfrac{1}{137}\right)^2 \frac{\mu_A c}{2\pi\hbar}, \tag{29·3}$$

where

$$\mu_A = \beta^{-\frac{1}{6}}\mu_B. \tag{29·2}$$

This μ_A, found from $\mathfrak{R}_A$ (experimental) by the *accepted* form of equation, is regarded as the observational $\mu = m_p m_e/(m_p + m_e)$, and so is the mass of the intracule in A. But extracule masses transform (§ 27) like measured densities, so that

$$M_A = \beta^{-1}M_B. \tag{29·4}$$

Thus

$$\left(\frac{M}{\mu}\right)_A = \beta^{-\frac{5}{6}}\left(\frac{M}{\mu}\right)_B = \frac{136^2}{10}\beta^{-\frac{5}{6}} \tag{29·5}$$

by (18·6); and the quadratic $m^2 - mM + M\mu = 0$ for proton and electron becomes in the observational system A

$$10m^2 - 136mm_0 + \beta^{\frac{5}{6}}m_0^2 = 0, \tag{29·6}$$

so that

$$m_p/m_e = 1836{\cdot}34. \tag{29·7}$$

We call these *current* masses, in contrast with the *standard* masses of § 18.

Note. How Eddington establishes (29·6) from (29·5) is not clear, since he does not re-define m_0. But if we couple with (29·2, 29·4) the relation

$$m_{0A} = \beta^{-1}m_{0B} \tag{29·4)'$$

(consistent with § 28) and assume (like Eddington) that the basic formulae (18·5) are now being applied in *system B*, we have by (29·2, 29·4, 29·4′)

$$M_A = \tfrac{136}{10} m_{0A}, \quad \mu_A = \frac{1}{136} \frac{m_{0B}}{\beta^{\frac{1}{6}}} = \frac{\beta^{\frac{5}{6}}}{136} m_{0A}. \qquad (29{\cdot}5)'$$

The quadratic $(m^2 - mM + M\mu)_A = 0$ merely expresses the relations $m_p + m_e = M$, $m_p m_e = M\mu$; if now we insert (29·5)′ in it, we obtain the desired (29·6), where m_0 means m_{0A}.

Mass, charge, etc., are primarily molar quantities. A characteristic of a particle is said to be *molarly controlled* if it is found as $1/n$ of the value for an aggregate of n particles. Thus M, but *not* μ, m_e, is molarly controlled.

Quantum mechanics, using ideas of mass, charge, etc., which are not directly measurable quantities, gives these names to *analogues* of molar quantities. It accepts the molarly controlled definition of extracules (and all neutral atoms); but the intracule mass is *spectroscopically controlled*, i.e. determined by (29·3).

Molar experiment (i.e. electrolysis) determines the Faraday constant $\mathfrak{F}' = e'/Mc$, where e' is the effective electronic charge molarly determined; the relation of this to the e of system B is found in § 30. In terms of the theoretical $\mathfrak{F} = e/Mc$,

$$\hbar = 137 e^2/c = 137 \mathfrak{F}^2 M^2 c. \qquad (29{\cdot}91)$$

Thus the (observational) intracule mass in (29·3) is

$$\mu_A = 4\pi \,.\, 137^3 \mathfrak{R} \mathfrak{F}^2 M^2. \qquad (29{\cdot}92)$$

The first part of **8·3** § 27 is similar.

The Rydberg constant (for hydrogen) multiplied by $n_1^{-2} - n_2^{-2}$ (n_1, n_2 integers) gives the reciprocals of spectral wave-lengths; the Faraday constant is the ratio of charge transported to mass in electrolysis. Thus $\mathfrak{R}$ and $\mathfrak{F}'$ are respectively 'spectroscopically' and 'molarly' controlled in experiment.

30. **Molarly controlled charge.** Changing from system A to B, (a) densities are multiplied by β, (b) lengths and times by $\beta^{-\frac{1}{6}}$. We define a further *system B′*, in which electrical quantities are controlled by classical electrodynamics (in which $\hbar \to 0$) and electrical and mechanical action (density × (length)^4) are additive. This additivity, requiring action to be invariant when density is multiplied by β, gives instead of rule (b) the rule (b'):

In $A \to B'$, lengths and times are multiplied by $\beta^{-\frac{1}{4}}$. (30·1)

Hence:

In $B \to B'$, lengths and times are multiplied by $\beta^{-\frac{1}{12}}$,
with densities unchanged. (30·2)

Comparing values in B' (primed) and B,

$$e'=\beta^{-\frac{1}{24}}e, \quad h'=\beta^{-\frac{1}{12}}h, \quad \mathfrak{F}'=\beta^{-\frac{1}{24}}\mathfrak{F}. \tag{30·5}$$

Using this in (29·92) with the observed $\mathfrak{R}$, $\mathfrak{F}'$ and M we obtain the observational M/μ and hence

$$m_p/m_e=1836{\cdot}56 \pm 0{\cdot}51 \text{ (observational)}$$
$$=1836{\cdot}34 \text{ (theoretical)}. \tag{30·6}$$

The latter value is that (29·7) determined from the quadratic (29·6).

A diagram in **F** shows the relations of systems B and B' via A; the quotations **8·3** § 27 should clarify these. It is seen from §§ 29, 30 here how much theory underlies the 'observational' m_p/m_e.

31. **Secondary anchors.** We have linked theory with experiment by examining the observational procedure leading to one determination of m_p/m_e; this is a 'primary anchor'. Typical 'secondary anchors' are (1) the spectroscopic determination of m_p/m_e (comparing the Rydberg constants of hydrogen and deuterium); (2) the deflexion method for e/m_ec; (3) the direct determination of h/e. The importance of these depends on the precision of observational results, and equally on the firmness with which β-factors can be assigned in the reduction of the results.

The last part of **8·3** § 28 is similar.

32. **Calculated values of the microscopic constants.** To represent the c.g.s. units we use the accurately known values of c, $\mathfrak{R}$ and $\mathfrak{F}'$. Combining these with our theoretical values for the 'fine-structure constant' ($\hbar c/e^2=137$), m_p/m_e and e'/e, as in (29·7) and (30·5), we obtain values of M_H, m_e, m_p, e', h', h/e which agree with the correctly reduced experimental values. [The results are tabulated with assessed experimental probable errors.]

Like **8·3** § 29. Earlier comparisons are in paper η of **2·1** above. The vital constant $1/\alpha=\hbar c/e^2$ has now been placed at $137{\cdot}0377 \pm 0{\cdot}0016$ (DuMond and Cohen, *Rev. Mod. Phys.* **25**, 691 (1953)); a β-factor used as in **F** would reduce this to 137·017.

33. **The Coulomb energy.** The mathematical treatment of the two-particle (hydrogen) system introduces suffixing of the particles, and so gives the intracule an interchange a.m. conjugate to the permutation coordinate (of § 25). We take this a.m. to be $\hbar$. (The usual quantum $\frac{1}{2}\hbar$ of a.m. would be unsuitable, for this a.m. is of a type which cannot be coupled with any compensating 'recoil' a.m. of the uranoid. The present $\hbar$ is made up in effect of two balancing internal a.m.'s of $\frac{1}{2}\hbar$.) As the permutation coordinate is analogous to *time*, the

transformation from the rigid coordinates of wave mechanics to the Galilean frame of observation (§ 19) reduces this a.m. to $\hbar/137$ (the multiplicity k being 137 since the electrical (interchange) coordinate is here on a par with the 136 mechanical d.f.'s). The corresponding linear momentum of the intracule, with r the separation of the 'particles', is $\hbar/137r$. In the current wave equation of the intracule, the momentum term in the direction of the 'time analogue' or phase coordinate is e^2/cr (the *Coulomb energy*); the identification of these terms gives $e^2 = \hbar c/137$.

This section, although represented as a general idea of 'how the Coulomb energy appears in the equations of the intracule', is of basic importance. The concepts employed appear in many places (recoil, time analogue, rigid frames). A late version occurs at the end of **8·3** § 30. See also **7·2** § 5·5.

3·4. F IV. Gravitation and Exclusion

Drafts are in **4**, **5·2**, **6·1**, **7**, **8·2** and **8·4**.

34. **Unsteady states.** [The relativistic-gravitational result on which this is based is suspect; see editorial footnote in **F**. The theory is not developed further; only the following wave-mechanical result of Darwin is required.] A Gaussian wave packet, with initial spread given by an s.d. σ in each coordinate, has after time t the s.d.'s $\sigma_t^2 = \sigma^2 + \hbar^2 t^2/4m^2\sigma^2$; but the momenta p_α have constant mean square.

A similar calculation was found in **H** § 20 (**8·2**).

35. **Under-observation.** A system is in its 'normal' state when it is subjected to a *conventional* (i.e. an agreed) *amount of observational probing*. The standard probing determines the 'multiplicity' k (§ 11). A system is '*fully observed*' in a coordinate x_α if the uncertainty of (x_α, p_α) is at the Heisenberg minimum. If the probing ceases, the uncertainty increases (cf. § 34) with time; thus time here serves as a 'coefficient of under-observation'. There is no point in extending spatial p.d.'s to time p.d.'s, either time in this special sense or in its normal sense. The *time analogue* or phase in quantum mechanics is conjugate to the scale (§ 24, § 33) which as a divisor has a 'negative fluctuation' (§ 3); thus phase can be assigned a p.d. Moreover, it combines with space measurements in the same way as 'real' time combines with space in Lorentz transformation; it is this *time analogue* which appears in the 'Lorentz-invariant' wave equation. 'Time' thus appears as a coordinate t, a measure of under-observation τ, or as a time analogue.

Discussions on time coordinates, time indicators, time analogues are extensive; compare **4·1** § 9 (for 'conventional probing'), **4·2** § 1·9 (*b*), **5·2** § 2·7 (*a*), (*b*) and **7·2** § 5·3. The vital role of the phase is prominent in **F** X.

36. **Structural and predictive theory.** The problems of wave mechanics are (i) structural or (ii) predictive. In the former, 'full observation' (as in § 35) is postulated. The latter are mainly 'problems of decay', i.e. of dispersion of probability, with the time coming in, as in § 35, as coefficient of under-observation. We are concerned mainly with (i). Further, ignoring dispersing wave packets, we shall in cases of predictive theory consider wave functions for steady states only, any time changes being in the occupation of the states.

This (and § 35) is represented in **8·4** § 31.

37. **Physical and geometrical distribution functions.** Let x be a geometrical coordinate of a particle precisely as in § 1, x_0 the (uncorrelated) corresponding coordinate of the physical origin and $\xi = x - x_0$ the 'physical' coordinate of the particle. Let $f(x_0)$, $g(x)$, $h(\xi)$ be their distribution functions (i.e. $f(x_0)\,dx_0$ = chance that x_0 is in $(x_0, x_0 + dx_0)$, etc.). Then

$$h(\xi) = \int_{-\infty}^{\infty} g(x_0 + \xi)\, f(x_0)\, dx_0. \tag{37·1}$$

Let F, G, H be the Fourier transforms of f, g, h, e.g.

$$F(q) = \frac{1}{2\pi} \int_{-\infty}^{\infty} \mathrm{e}^{-iqx} f(x)\, dx. \tag{37·2}$$

By (37·1),
$$H(q) = 2\pi G(q)\, F(-q). \tag{37·3}$$

As $f(x_0)$ is Gaussian with s.d. σ (§ 2), $2\pi F(-q) = \mathrm{e}^{-\frac{1}{2}\sigma^2 q^2}$; so

$$H(q) = \mathrm{e}^{-\frac{1}{2}\sigma^2 q^2}\, G(q). \tag{37·5}$$

Now the (complex) *wave function* $\mathfrak{f}(x)$ of the coordinate x implies a probability $|\mathfrak{f}|^2\, dx$ of observing x in $(x, x + dx)$, so that the modulus $|\mathfrak{f}|^2 \sim f$, the distribution function. Further, the *momentum* wave function $\mathfrak{F}(p)$ is the Fourier transform of $\mathfrak{f}$:

$$\mathfrak{F}(p) = (2\pi\hbar)^{-\frac{1}{2}} \int_{-\infty}^{\infty} \mathrm{e}^{-ipx/\hbar}\, \mathfrak{f}(x)\, dx. \tag{37·72}$$

Comparison with (37·2) implies an 'identification'

$$q = 2p/\hbar, \tag{37·8}$$

which would suggest the interpretation of (37·3) as a relation between distributions F, G of conjugate momenta of origin and of particle in the geometrical frame, and H of the 'physical' momentum of the particle (relative to the origin). We shall see that a certain convention of wave mechanics is implied in this.

The development begun here leads to the striking 'absolute determination' of mass in § 40; but the arguments of § 37 are tentative. The MSS. are listed after § 38.

38. **The weight function.** Suppose $g(x)$ is Gaussian (a 'wave packet') with s.d. s. By (37·1), $h(\xi)$ is also Gaussian with s.d. s', where

$$s'^2 = s^2 + \sigma^2. \tag{38·1}$$

For the Gaussian wave packet s, current theory finds the momentum distribution function to be

$$G(p) = (\pi\hbar^2/2s^2)^{\frac{1}{2}}\,\mathrm{e}^{-2s^2p^2/\hbar^2}. \tag{38·2}$$

This is for the 'geometrical' momentum; analogy gives similar momentum distributions F (the momentum of the origin) and H (particle physical momentum) with σ, s' for s. Hence by (38·1)

$$H(p) = \text{const.} \times G(p)\,F(-p), \tag{38·3}$$

or,

the probability of physical momentum p is the combined probability of geometrical momentum p of the particle and −p of the physical origin. (38·4)

Regard p', p as momenta of a two-particle system (origin and particle), and transform to 'external and internal' momenta P, ϖ as in § 26; then ϖ is the particle 'physical' momentum. If $K(P)$ is the distribution of P, the phase-space conservation (§ 26) implies

$$K(P)\,H(\varpi) = F(p')\,G(p). \tag{38·5}$$

Let $K(P) = \delta(P)$ (δ the Dirac function which is zero if $P \neq 0$). Then for all 'probable' states, $P=0$, and by § 26, $p = -p' = \varpi$. This reduces (38·5) to (38·3). Thus the result (38·3) of 'current wave mechanics' implies the *convention*

$$K(P) = \delta(P), \tag{38·6}$$

or, 'the probability of momentum p, without an equal recoil momentum of the physical origin, is zero'. This convention applies generally in wave mechanics and must not be restricted to Gaussian wave packets.

We prefer to regard $F(-p)$ as a *weight function* $w(p)$:

$$w(p) = (2\pi\varpi^2)^{-\frac{1}{2}}\,\mathrm{e}^{-p^2/2\varpi^2}, \tag{38·71}$$

where

$$\varpi = \hbar/2\sigma. \tag{38·72}$$

Then for (38·3) we have

$$H(p)\,dp = \text{const.}\,G(p)\,w(p)\,dp, \tag{38·8}$$

which shows physical momenta as geometrical momenta *weighted* by this factor $w(p)$. In three dimensions

$$w(p_1, p_2, p_3) = (2\pi\varpi^2)^{-\frac{3}{2}} \exp\left\{-\frac{p_1^2+p_2^2+p_3^2}{2\varpi^2}\right\}. \tag{38·9}$$

This factor, 'allowing for the recoil of the physical reference frame', reduces the frequency of large momenta, and avoids the divergence of some energy integrals.

The argument here is fairly cogent, the main difficulty lying in accepting Darwin's result (38·2) as applying to *both* physical and geometrical momenta. The topics of §§ 37, 38 come in Chapter I of early drafts; the full mathematical treatment quoted in **4·1** §§ 3–5 should be studied. The idea of 'weight' appears first in the earlier **4·2** § 1·5. Later versions are **D** § 21 and **8·4** §§ 31, 32; 'bridge passages' quoted in **8·4** are important.

39. **The genesis of proper mass.** In an *infinite-temperature uranoid* of particles with geometrical momenta ranging uniformly from $-\infty$ to $+\infty$, the *physical* momentum distribution is given by (38·9), with mean values

$$\overline{p_1^2} = \overline{p_2^2} = \overline{p_3^2} = \varpi^2 = \gamma^2/4\sigma^2. \tag{39·1}$$

(Planck's constant $h = 2\pi\hbar$ is here replaced by $2\pi\gamma$, a constant to be discussed in § 40.) Consider particles of *zero* proper mass; their energy $E = \sqrt{(p_1^2+p_2^2+p_3^2)}$, so that the average

$$\overline{E^2} = 3\varpi^2 = 3\gamma^2/4\sigma^2. \tag{39·2}$$

In uniformly curved space of radius R_0, the pressure P and density ρ satisfy

$$8\pi\kappa P = -R_0^{-2} + \lambda, \quad 8\pi\kappa\rho = 3R_0^{-2} - \lambda, \tag{39·3}$$

and the *proper* density $\rho_0 = \rho - 3P$. We *define temperature* Θ as the ratio of thermal to proper energy density; thus

$$\Theta = (\rho - \rho_0)/\rho_0 = 3(\lambda - R_0^{-2})/(6R_0^{-2} - 4\lambda). \tag{39·4}$$

Keeping the particle number N and, hence, $R_0 \equiv 2\sigma\sqrt{N}$ constant, as we pass from zero to infinite temperature the parameter λ increases. Distinguishing infinite-temperature values by primes,

$$\Theta' = \infty: \quad \lambda' = \tfrac{3}{2}R_0^{-2}, \quad 4\pi\kappa\rho' = \tfrac{3}{4}R_0^{-2}, \quad \rho_0' = 0, \tag{39·51}$$

$$\Theta = 0: \quad \lambda = R_0^{-2}, \quad 4\pi\kappa\rho = R_0^{-2}, \quad \rho_0 = \rho. \tag{39·52}$$

Hence

$$\rho_0 = \rho = \tfrac{4}{3}\rho'. \tag{39·6}$$

Thus, beginning with zero proper mass ($\rho_0'=0$) at infinite temperature, we 'create' rest mass ρ_0 by lowering the temperature to zero. We shall now find the proper particle-mass m in the zero-temperature uranoid. The molar energy tensor $T_{\mu\nu}$ is $N\overline{Ap_\mu p_\nu}$, where $\overline{Ap_\mu p_\nu}$ is the mean particle self-energy tensor (§ 22). At infinite temperature the density T_{44} is

$$\rho' = NA\overline{p_4^2} \equiv NA\overline{E^2} = 3NA\varpi^2 \tag{39·72}$$

by (39·1). At zero temperature the momentum is $(0, 0, 0, m)$ so that

$$\rho = NAm^2. \tag{39·73}$$

Comparing the ratio ρ/ρ' here with (39·6), $m^2 = 4\varpi^2$, so that

$$m = 2\varpi = \gamma/\sigma = 2\gamma\sqrt{N}/R_0. \tag{39·9}$$

(Also (39·52) gives $\qquad A = 1/(16\pi\kappa\gamma^2N^2)$.) $\qquad$ (39·82)

This 'genesis of mass' is a brilliant conception. For the meaning of 'temperature' in relation to mass, see **8·4** § 33, also extensive quotations in the earlier **4·1** §§ 8, 10 and **4·2** § 1·8; these show difficulties also concerning a factor 2 (e.g. in (39·82)) which is ascribed to 'double reckoning' of energy; this difficulty is transferred in **F** to § 40.

The result (39·82) would of course be affected by the change from (3·8) to (3·8)′ proposed in **3·1** § 3 above.

40. **Absolute determination of $\mathbf{m_0}$.** In § 39 we had V_3 particles (the only 'observables' being p_1, p_2, p_3). The actual universe is of V_{10} particles of mass $M = \frac{3}{10}m$ ((16·5)), so that (39·9) gives

$$M = \tfrac{3}{5}\gamma\sqrt{N}/R_0. \tag{40·2}$$

To relate the γ of § 39 to the Planck $\hbar$ of wave mechanics, we observe that we used

$$E = \sqrt{\{(M^2 + \Sigma p_\alpha^2)\}} \doteqdot M + \Sigma p_\alpha^2/2M = M - \frac{\gamma^2}{2M}\Sigma\partial^2/\partial x_\alpha^2, \tag{40·31}$$

the definition of γ being by the operational form $p_\alpha = -i\gamma\partial/\partial x_\alpha$. In the rigid field of wave mechanics,

$$E = M + \Sigma p_\alpha^2/2\mu = M - \frac{\hbar^2}{2\mu}\Sigma\frac{\partial^2}{\partial x_\alpha^2}; \tag{40·32}$$

the internal mass μ appears as in (18·33) from the separation into external carriers of the 'initial' energy M and internal carriers of the 'transition' energy; also the momentum p_α equals $-i\hbar\partial/\partial x_\alpha$ since we are now in orthodox mechanics. Comparing these E's

$$\gamma = \hbar\sqrt{(M/\mu)} = 136\hbar/\sqrt{10} \tag{40·4}$$

by (18·6). But (40·2) requires amendment; the N particles ($\frac{1}{2}N$ protons, $\frac{1}{2}N$ electrons) form $\frac{1}{2}N$ extracules, so that N should be halved in

(40·2). Further, halving N doubles σ^2; and γ, as an a.m. of dimensions $ML^2T^{-1}=ML=L^{-2}$ (cf. §8) in natural units, is thereby halved. Thus (40·2) should be

$$M=\frac{3}{5}\frac{\gamma}{2}\frac{\sqrt{(\frac{1}{2}N)}}{R_0}. \qquad (40·5)$$

Then by (40·4),

$$M=\frac{136}{10}\frac{3}{4}\frac{\hbar\sqrt{(\frac{4}{5}N)}}{R_0}, \qquad (40·6)$$

$$m_0=\frac{3}{4}\frac{\hbar\sqrt{(\frac{4}{5}N)}}{R_0}. \qquad (40·7)$$

(The 'standard particle' mass m_0 is $(\frac{10}{136})M$ by (18·5).) For practical measurement, we have to pass from the theoretical system 'B' (here used) to the observational system 'A' (§27), thus replacing $R_{0(B)}$ by $R_{0(A)}\beta^{-\frac{1}{6}}$. Thus we have finally for the *absolute mass of a hydrogen atom* (*extracule*)

$$M=\tfrac{136}{10}\cdot\tfrac{3}{4}\cdot\beta^{\frac{1}{6}}\frac{\hbar\sqrt{(\frac{4}{5}N)}}{cR_0} \quad (\text{and } m_0=\tfrac{10}{136}M). \qquad (40·8)$$

'This is the central formula of unified theory.' As $\hbar/Mc$ is accurately known, this gives $R_0/\sqrt{N}$ and σ.

The complex arguments of **F** have been abbreviated. It is significant that the penultimate version (**8·4** §34) is almost identical with **F**. For earlier versions, see **4·1** §10, continued in **6·1** §28; and **4·2** §1·8.

In assessing the effect of the change from (3·8) to (3·8′) on this work, it should be remembered that the combination $R_0/2\sqrt{N}$ represents σ (3·8) for Eddington (cf. (39·9) in particular). Thus, provided the final formulae *are* expressed in terms of σ, the change from (3·8) to (3·8′) does not affect them. The reader may then interpret σ in the final formulae as $R_0/3\sqrt{N}$, to be consistent with the reinterpretation of the galactic recession in §5.

41. **Exclusion.** The exclusion principle for hydrogen, transferred to extra- and intracules, becomes:

In a steady state the maximum number of particles is two extracules per cell h^3 *of xp-space* ('the mechanical principle') *and two intracules per cell* h^3 *of* $\xi\varpi$*-space* ('the electrical principle'). (41·12)

We show that the mechanical principle leads to the result (40·8) obtained by gravitation; thus *exclusion is a wave-mechanical substitute for gravitation*, although only for steady states.

Consider a unit volume of 3-space (x, y, z) so that the cell of 6-space (x, p) corresponds to a cell $dp_1dp_2dp_3=h^3$ of momentum space. Let there be n extracules per unit (x, y, z) volume, each of mass constant μ_0, so that the classical energies are $E=p^2/2\mu_0\left(p^2=\sum_1^3 p_\alpha^2\right)$. For zero temperature we make $\sum_1^n E$ minimum subject to a density (in p-space) of two particles per cell; thus we fill a sphere of radius $p = \mathfrak{p}$, where

$$\tfrac{4}{3}\pi\mathfrak{p}^3 = \tfrac{1}{2}nh^3. \tag{41·21}$$

The 'top' and 'mean' energies are

$$\mathfrak{E} = \mathfrak{p}^2/2\mu_0 = \left(\frac{3n}{8\pi}\right)^{\frac{2}{3}} \frac{h^2}{2\mu_0}, \tag{41·22}$$

$$\bar{E} = \tfrac{3}{5}\mathfrak{E} \tag{41·23}$$

(cf. 'white-dwarf matter' theory). We apply this to the $\frac{1}{2}N$ extracules in the uranoid, for which $n = \frac{1}{2}N/(2\pi^2 R_0^3)$. Defining

$$\mu_1 = \mu_0(\tfrac{3}{4}N)^{\frac{1}{3}}, \tag{41·3}$$

we have

$$\mathfrak{E} = \frac{3}{4}\frac{N}{2\mu_1}\left(\frac{\hbar}{R_0}\right)^2 \tag{41·41}$$

or

$$2\mu_1\mathfrak{E} = \tfrac{3}{4}\varpi^2, \tag{41·42}$$

where ϖ is the weight constant $\hbar/2\sigma = \hbar\sqrt{N}/R_0$ of (38·72).

A top particle (at zero temperature) has rest mass $\mathfrak{E}$, which is here shown as exclusion energy due to competition for lower states; this picture we call *sub-threshold theory*.

Compare **D** §§ 25, 26; a later comment on 'exclusion for gravitation' is in **8·4** § 35. See also **7·1** § 32.

42. **The negative energy levels.** For 'sub-threshold' theory, regarding $\mathfrak{E}$ as zero level, the uranoid particles fill the finite number (contrast Dirac's 'holes') of negative energy levels. A 'super-threshold' or object particle is formed by exciting an $\mathfrak{E}$-level particle, leaving an $\mathfrak{E}$-level hole which serves as a comparison particle (§§ 22, 23). The extracules in § 41 are unspecialized elements of energy tensor with original multiplicity 136 (although the object and comparison particles may later be stabilized as V_{10} and V_1). Thus the effective rest mass $\mathfrak{E}$ is to be identified with the mass m_0 of a background particle.

Wave mechanics, with its rigid environment, makes us use in its 'super-threshold' theory a complete background of 'top' particles $\mathfrak{E}$ of sub-threshold theory. As these have $\frac{5}{3}$ of the mean energy, we magnify the density by $\frac{5}{3}$ in passing to super-threshold theory. To keep R_0 fixed, the constant of gravitation, κ, must be $\frac{3}{5}$ that, κ_1, of sub-threshold theory. Thus from (8·1) the Planck constant $\hbar_1$ of sub-threshold theory is

$$\hbar_1^2 = \tfrac{3}{5}\hbar^2, \tag{42·2}$$

where $\hbar$ is the observational or super-threshold value. This $\hbar_1$ is the value implied in (41·41), which therefore requires this correction.

See note after §43. Considerable amplifications of §42 appear in **7·1** §§33, 34.

43. **Determination of m_0 by exclusion theory.** We compare uranoids with different values of $\frac{1}{2}N$, *keeping the top-level object-particle constant*, with fixed m_0 $(=\mathfrak{E})$, M, μ, $\hbar$, σ, ϖ. R_0 $(\propto N^{\frac{1}{2}})$ and μ_0 are variable; but μ_1 is constant by (41·42).

The eigenstates of the particles correspond to surface harmonics of the hypersphere of space; for the kth level there are k^2 harmonics and so $2k^2$ extracules. The top quantum number $\mathfrak{k}$ is given by

$$\tfrac{1}{2}N=2(1+2^2+\ldots+\mathfrak{k}^2)=\tfrac{1}{3}\mathfrak{k}(2\mathfrak{k}+1)(\mathfrak{k}+1)\approx\tfrac{2}{3}\mathfrak{k}^3,$$

so

$$\mathfrak{k}\approx(\tfrac{3}{4}N)^{\frac{1}{3}}. \tag{43·1}$$

We replace (41·3) as a *definition* of μ_1 by

$$\mu_1=\mathfrak{k}\mu_0 \tag{43·2}$$

((41·3) being now regarded as an *approximation* valid for large $\mathfrak{k}$). Since the scales of the various levels $(k=1,2,3,\ldots)$ in the uranoids are proportional to the integers (24·3), any mechanical characteristic varies as a power of k. A characteristic such as μ_0 of a *whole* uranoid varies as some power of the 'total' quantum number $\mathfrak{k}$ of that uranoid. As by (41·42) μ_0 varies as $\mathfrak{k}^{-1}$, represented by $(\frac{3}{4}N)^{-\frac{1}{3}}$ for large uranoids, this determines the power as minus one; so that for all uranoids

$$\mu_0=\mu_1/\mathfrak{k}\quad(\mu_1\text{ constant}), \tag{43·32}$$

and (41·41) takes the universal form

$$2\mu_1\mathfrak{E}=\mathfrak{k}^3(\hbar_1/R_0)^2, \tag{43·4}$$

remembering that $\hbar_1$ (§42) not $\hbar$ is here appropriate.

In the smallest uranoid, $\mathfrak{k}=1$, of $\frac{1}{2}N=2$ extracules, the 'top energy' $\mathfrak{E}$, being kept fixed as a quantal energy, must include for either of the particles the *exclusion* energy of the other and so is $2\mu_0$. Hence for $\mathfrak{k}=1$ by (43·32),

$$\mathfrak{E}=2\mu_1. \tag{43·5}$$

This holds for all uranoids, as $\mathfrak{E}$ and μ_1 are constant. Thus for the actual universe (with N and $\mathfrak{k}$ large)

$$\mathfrak{E}^2=2\mu_1\mathfrak{E}=\tfrac{3}{4}N(\hbar_1/R_0)^2 \tag{43·6}$$

by (43·4, 43·1); so by (42·2)

$$\mathfrak{E}^2=\frac{3}{4}\frac{3}{5}\frac{\hbar^2N}{R_0^2}. \tag{43·7}$$

This agrees with (40·7) as $m_0\equiv\mathfrak{E}$; so exclusion and gravitational theory agree.

For clarity we follow **D** § 25 (q.v.) in using $\mathfrak{k}$ as well as k in discussing *scales*, before (43·32). Earlier versions (**8·4** § 35 and **7·1** § 32) resemble **F**, but postpone the $\frac{3}{8}$ factor (**F** (42·2)) until after the full exclusion calculation.

44. **Super-dense matter.** In super-dense matter in a steady state, deviations from perfect uniformity can be treated as a superposition of waves of two types: (*a*) mechanical waves (sound) and (*b*) electrical waves (polarization changes). These may be described from the microscopic viewpoint as probability distributions of coordinates of (*b*) the unidentified doublet (intracule) ξ_1, ξ_2, ξ_3 and (*a*) the unidentified neutral particle ('extracule') x_1, x_2, x_3. An alternative analysis into waves of (*a'*) positively and (*b'*) negatively charged matter is inadmissible, since too much of the energy is in transition ('interstates') to permit a steady-state analysis.

This topic receives only casual mention in previous MSS., and is off the main track. See paper **2·1** ϵ for an earlier discussion.

45. **The degeneracy pressure.** Applying (41·23) to the intracules in super-dense hydrogen, the mean exclusion energy per intracule is

$$\bar{E}=\frac{3}{5}\left(\frac{3\sigma}{8\pi}\right)^{\frac{2}{3}}\frac{h^2}{2\mu} \qquad (45·1)$$

(σ = particle density of intracules), and the pressure is

$$P=K\sigma^{\frac{5}{3}}, \quad K=\frac{1}{5}\left(\frac{3}{8\pi}\right)^{\frac{2}{3}}\frac{h^2}{\mu}. \qquad (45·2)$$

This is the degeneracy (zero-temperature) pressure of white-dwarf matter (R. H. Fowler). Since $h^2/\mu=h^2/m_e+h^2/m_p$ the pressure appears as if it were almost wholly due to the electrons of mass m_e.

The wave functions of the steady states considered are standing waves; p_1^2, p_2^2, p_3^2 reduce to eigenvalues, but not the momenta p_1, p_2, p_3, which have expectation zero. Thus the investigation deals with energy tensors; it is scale-free and holds for large p^2.

For *standing waves*, such as were used implicitly in the 'exclusion' calculations above, the correct energy form is $E=m+p^2/2m$, and *not* $\sqrt{(m^2+p^2)}$ which, being Lorentz-invariant, is unsuited to standing waves.

For an incidental discussion of this last point see **7·2** § 5·4.

3·5. F V. The Planoid

Drafts are in **7·1** and **8·4**.

46. **Uranoid and planoid.** We consider as alternative environments for a small object-system:

(*a*) a zero-temperature (curved-space) uranoid, radius R_0, of N particles,

(*b*) a zero-temperature sphere in *flat* space or 'planoid', radius R_1, of N_1 particles.

At first we use the same quantum-specified units in both, and so the same uncertainty constant σ. For a coordinate of the centroid of the N_1 particles in (*b*) the s.d. is $\frac{1}{5}(R_1^2/N_1)^{\frac{1}{2}}$; hence (cf. (3·8))

$$\frac{R_1^2}{5N_1}=\sigma^2=\frac{R_0^2}{4N}. \tag{46·1}$$

Also by (40·7)
$$m_0=\tfrac{3}{4}\hbar\,\frac{\surd N_1}{R_1}. \tag{46·2}$$

This is represented in **7·1** §35 and **8·4** §37; see **D** §27. The planoid as a 'flat uranoid' appeared in early versions before the 'curved' uranoid; cf. **4·1** §6 and **4·2** §1·6 (*a*).

47. Interchange of extracules. If [...] denotes *averaging over the planoid volume*, and r distance from the centre,

$$[1/r^2]=3/R_1^2,\quad [1/r]=3/(2R_1). \tag{47·1}$$

These hold also for r measured from an object *near* the centre. By (46·2),

$$m_0=[\tfrac{1}{2}\hbar/r]\surd N_1, \tag{47·21}$$

$$m_0^2=\tfrac{3}{16}[\hbar^2/r^2]N_1=\frac{3}{2}\left[\sum_1^{\frac{1}{2}N_1}(\tfrac{1}{2}h/r_s)^2\right] \tag{47·22}$$

(the summation being over the $\frac{1}{2}N_1$ planoid extracules). Neglecting the fluctuation of any r_s, we write this

$$m_0^2=\tfrac{3}{2}\sum_s p_s^2\quad (p_s=\tfrac{1}{2}\hbar/r_s). \tag{47·3}$$

The rest mass of a V_3 extracule is

$$m_3=\tfrac{136}{3}m_0=m_0^2/(3\mu)=\sum_s p_s^2/(2\mu), \tag{47·4}$$

where $\mu=m_0/136$ is the intracule mass-constant.

Assuming the V_3 object-particle (near the centre) has a.m. $\frac{1}{2}\hbar$ about the *s*th planoidal extracule gives it linear momentum $\frac{1}{2}\hbar/r_s=p_s$; this 'interprets' (47·3) as the energy of such a momentum. The a.m. is, however, extra-spatial (the particles being at rest) and so represents an *interchange circulation* as in §33.

Taking the planoid particles as V_3's (like the object-particle), and letting the object make a transition to momentum p_x, p_y, p_z, we have now for its energy

$$E_3 = m_3 + (p_x^2 + p_y^2 + p_z^2)/2\mu \tag{47·51}$$

$$= (p_1^2 + \dots + p_{\frac{1}{2}N_1}^2 + p_x^2 + p_y^2 + p^2)/2\mu. \tag{47·52}$$

This exhibits *all* the energy as 'transition' energy; what was 'initial' energy (m_3) arises from the interchange momenta $\frac{1}{2}\hbar$. As, however, only the three *spatial* momenta p_x, p_y, p_z are measurable, the mass so found refers to a V_3.

We have now three representations of rest energy: (i) curvature, (ii) exclusion, (iii) interchange. If we try to distinguish the original uranoid particles of (i) by suffixes, continuous interchange appears (as (iii)) to give energy replacing the gravitational mass. Alternatively, the distinction of the particles by assigning them to a system of states gives them the exclusion energy (ii).

See **7·1** § 36 (*a*) and **8·4** § 38.

48. **The special planoid.** The *special planoid* is defined (with relation to the standard uranoid N, R_0) by $N_1 = N$, $R_1 = R_0$; it is also defined to have the 'sub-threshold' Planck constant $\hbar_1 \equiv (\frac{3}{5})^{\frac{1}{2}}\hbar$ of § 42. Thus (contrast § 46) its uncertainty constant σ_1 is given by

$$\sigma_1^2 = R_1^2/5N_1 = R_0^2/5N = \tfrac{4}{5}\sigma^2, \tag{48·11}$$

and the weight constant of (38·72) is

$$\varpi_1 = \hbar_1/2\sigma_1 = (\tfrac{3}{4})^{\frac{1}{2}}\varpi. \tag{48·12}$$

Since quantum-specified lengths and masses are multiples of σ and ϖ, planoidal lengths and masses L_1 and M_1 differ from the uranoidal (or standard) L, M by

$$L = (\tfrac{5}{4})^{\frac{1}{2}} L_1, \quad M = (\tfrac{4}{3})^{\frac{1}{2}} M_1. \tag{48·2}$$

These results are needed in § 49. Rather than summarize the intricate discussion following (48·2) in **F**, we refer the reader to quotations in **8·4** § 37, which are on the whole clearer.

49. **The energy of two protons.** As neutral matter (conceived as equal distributions of protons and electrons) has no electric energy, the electric (Coulomb) energy of two protons is minus the proton-electron energy, and so is $+e^2/r$. But two protons in a neutral environment induce a charge $-2e$ in it. We avoid this by employing *quantum protons*, superposed on the environment without any compensatory induced charge (as with a 'rigid field'). Replacing the induction effect, an extra energy appears which is identified with the *non-*

Coulombian energy of the protons (cf. §5). We shall verify that the energy of a pair of 'quantum charges' is

$$E=-e^2/r \text{ (unlike)}, \quad E=e^2/r+B\delta(r') \text{ (like charges)}, \qquad (49\cdot1)$$

where $\delta(r)$ is Dirac's function:

$$\int\delta(r)\,dV=1, \quad \delta(r)=0 \quad \text{if} \quad r\neq0, \qquad (49\cdot2)$$

and r, r' are as in §5. Calculating in the special planoid, the energy of a *classical* proton (inducing $-e$) is $-\Omega$, where

$$\Omega=e^2[r^{-1}]=\tfrac{3}{2}e^2R_1^{-1}=\tfrac{3}{2}e^2R_0^{-1} \quad \text{by (47·1) and §48}$$
$$=(\tfrac{4}{3})^{\frac{1}{2}}.\tfrac{3}{2}e^2R_0^{-1}.$$

(In the last step we treat Ω as a quantum-specified energy, and apply (48·2) to convert back to normal uranoidal measure.) But by (49·1) the summed mutual energy of a *quantum* proton with the $\frac{1}{2}N$ electrons and $\frac{1}{2}N$ protons of the uranoid is

$$\tfrac{1}{2}NB\int\delta(r')\,dV/V=\tfrac{1}{2}NB/V.$$

Equating this to $-\Omega$ above, and using $V=\frac{4}{3}\pi R_0^3$, gives

$$B=-(\tfrac{4}{3})^{\frac{1}{2}}.3e^2\frac{4}{3}\frac{\pi}{N}R_0^2=-(\tfrac{4}{3})^{\frac{1}{2}}16\pi e^2\sigma^2 \qquad (49\cdot4)$$

by (3·8). This shows the 'quantum' formula (49·1) to be consistent with the 'classical', and also yields the constant B. If as in §5 we convert from r' (directly measured) to r (coordinate-difference distance), the non-Coulombian term in (49·1) becomes

$$B\delta(r')=-\left(\frac{16}{3\pi}\right)^{\frac{1}{2}}\frac{e^2}{\sigma}\mathrm{e}^{-r^2/4\sigma^2}. \qquad (49\cdot7)$$

A supplementary argument on the form $\delta(r')$ has been omitted (cf. **D** §28).

50. **Non-Coulombian energy.** Unlike the Coulomb energy, which is transition energy, the non-Coulombian energy $B\delta(r')$ is ranked as *initial* energy (an adjustment of initial energy which enables us to simplify a 4-particle system into a 2-particle system of two protons). Thus to adjust (49·7) to the proton of mass m_p, we multiply by m_p/m_0, because our treatment had been applied to a top (planoid) particle of mass m_0; the two initial energies (rest mass and non-Coulombian

energy) must be adjusted together by this factor. The non-Coulombian energy of two protons is thus

$$E = -A\,e^{-r^2/k^2}, \quad k = 2\sigma = R_0/\sqrt{N}, \quad A = \left(\frac{16}{3\pi}\right)^{\frac{1}{2}} \frac{m_p}{m_0} \frac{e^2}{\sigma}. \tag{50·2, 3, 4}$$

From formulae (40·8), (29·5) and (29·3) we find

$$\sigma = \frac{R_0}{2\sqrt{N}} = \frac{3}{136^2 . 137 . 16\pi\sqrt{5}} \frac{1}{\mathfrak{R}} = 9{\cdot}604 \times 10^{-14}\,\text{cm}. \tag{50·5}$$

in terms of the Rydberg constant $\mathfrak{R}$ (of the observational system). Hence

$$k = 1{\cdot}9208 \times 10^{-13}\,\text{cm.}, \quad A = 4{\cdot}2572 \times 10^{-5}\,\text{erg} = 52{\cdot}01 m_e c^2. \tag{50·6}$$

This is in good agreement with experiment.

There is very little non-Coulombian energy between pairs of electrons, and (theoretically) none between unlike charges.

Cf. **8·4** § 39. The amendment (3·8)′ would make (50·5) read

$$\sigma = R_0/3\sqrt{N} = \ldots 1/\mathfrak{R},$$

the numerical coefficient of $1/\mathfrak{R}$ being unaltered.

In his paper **2·1 α** (p. 10 above), Eddington obtained for $A/m_e c^2$ the value 39·2, in agreement with a current observational value. A tentative correction to 52, to agree with later experiments, was made in paper **2·1 ζ**.

51. The constant of gravitation. From the formulae (5·41), (40·8) and $hc/2\pi e^2 = 137$, $e/Mc = \mathfrak{F} = \beta^{\frac{1}{24}}\mathfrak{F}'$, we obtain

$$\kappa/\mathfrak{F}'^2 c^2 = \frac{136 . 137}{10} \left(\frac{9}{20}\right)^{\frac{1}{2}} \frac{\pi\beta^{\frac{1}{4}}}{\sqrt{N}}, \tag{51·4}$$

relating the constant of gravitation κ and the (observational) Faraday constant $\mathfrak{F}'$. Using the observed $\mathfrak{F}'$ and the theoretical

$$N = \tfrac{3}{2} . 136 . 2^{256}, \tag{51·3}$$

we have

$$\kappa = 6{\cdot}6665 \times 10^{-8} \quad (\kappa \text{ observed} = (6{\cdot}670 \pm 0{\cdot}005) \times 10^{-8}), \tag{51·6}$$

in good agreement.

The *force constant* $F = e^2/\kappa m_p m_e$—the ratio of electrical to gravitational force between proton and electron—can be written, using the (analytical) formula (5·41) for κ, (29·6), (40·8) for $m_p m_e$, and $hc/2\pi e^2 = 137$,

$$F = \frac{2}{3\pi\beta^2} \sqrt{(5N)}. \tag{51·7}$$

Thus F is a pure multiple of $\sqrt{N}$.

See **8·4** § 40. Examination of the genesis of (51·4) above will show the reader that the value of κ in (51·6) is consistent with the amended relation $\sigma = R_0/3\sqrt{N}$ of (3·8)′, provided the theoretical N ((51·3)) is multiplied by $\frac{9}{4}$. For in (51·4), $\kappa/\mathfrak{F}'^2c^2 \sim (\kappa M)\,M$; and when (3·8) is replaced by (3·8)′, $\kappa M \sim R_0/N$ (compare (5·41)) is unchanged, but $M \sim \sigma^{-1}$ by (39·9). Thus when $\sigma = R_0/2\sqrt{N}$ is changed to $\sigma = R_0/3\sqrt{N}$, an extra factor $\frac{3}{2}$ should appear on the right-hand side of (51·4); so to preserve the numerical value of the left-hand side, N in (51·4) must be multiplied by $\frac{9}{4}$. Similarly in (51·7), $F \sim \{(\kappa M)\,M\}^{-1}$, and the same change in N is required. This modified N is consistent with the comment (ii) after § 5 above. Compare also the remarks after § 40 and 50 concerning σ and $R_0/\sqrt{N}$.

52. Molar and nuclear constants. (This is solely a table of results. The reader is referred to **8·4** § 40 which gives in the table a few observed values and 'probable errors'. Nos. 17–22 are affected by the changes suggested on p. 18.)

CHAPTER 4

DRAFTS OF F I–V

(i) CHAPTER I OF DRAFTS B AND C, AND AN EARLY FRAGMENT

4·0. Preliminary note

Of the three drafts **B**, **C**, **H** (cf. **2·3**, **2·4**) covering the 'statistical theory', **B** and **C** will be treated together, chapter by chapter, in **4–7**; the final draft **H** is described in **8**. Prior attention is given to **C**, which is more polished than the earlier **B**, and presumably represents the stage reached when Eddington prepared to deliver his Dublin lectures.

Chapters I of **B** and **C** (**4·2**, **4·1** below) resemble each other and not **F** I. They begin (as **F** I does) with the uncertainty of the origin, but then continue with momentum distributions (as in **F** IV) without introducing the 'Bernoulli fluctuation' of **F** §3. They proceed from a flat background 'uranoid' to a curved closed space, and so to the origin of rest mass.

A very early fragmentary opening is given in **4·3**.

Note on presentation. A draft section resembling closely a section of **F** is represented merely by its number and title; it will then generally suffice the reader to refer to the summary of **F** in **3** and (for **F** VI onwards) **9** and **10**. If the draft section differs materially from **F**, or includes matter from various sections of **F**, these are noted and the draft may be summarized in standard type.

Passages quoted from drafts are in *small type*. All important variants are represented by such quotations, which sometimes cover whole sections. The reader interested in details of these variants may at times prefer to consult the full text of **F** rather than the summaries in **3**, **9**, **10**; but the equations of **F** referred to in discussing the drafts will be found in the summaries. *Italics* in drafts are Eddington's own.

4·1. Draft C *33* I. The Uncertainty of the Origin

1. **The conditions of observability:** close to **F** §1.
2. **The Gaussian distribution:** close to **F** §2.
3. **Relative distribution functions:** like **F** §37 to (37·5).

This radically different order of presentation corresponds to the Dublin lectures as delivered in 1942; the order in **F** (introducing the

Bernoulli fluctuation as §3, and long postponing momentum distributions) corresponds to the lectures as printed in **D** (**2·1** above). The summaries and quotations of the following sections of **C** make clear the original line of argument, and also contain important variants.

To introduce the method of handling these drafts, the gist of the present draft §3 is prefaced to a short quotation.

For the p.d.'s $f(x_0)$, $g(x)$, $h(\xi)$ of the physical origin (x_0, y_0, z_0) and of the geometrical (x, y, z) and physical coordinates ($\xi = x - x_0$, etc.) of a particle, the Fourier transforms satisfy

$$H(q) = 2\pi G(q)\, F(-q), \qquad (3{\cdot}4)\ [\mathbf{F}\,(37{\cdot}3)]$$

and for the 'Gaussian' distribution $f(x_0)$,

$$H(q) = \exp\left(-\tfrac{1}{2}\sigma^2 q^2\right) G(q). \qquad (3{\cdot}5)\ [\mathbf{F}\,(37{\cdot}5)]$$

The result (3·5) is important...effectively a relation between the distribution of geometrical momenta conjugate to x and the distribution of physical momenta conjugate to ξ, as will appear in the next section. Although...(3·4) enables us to treat a non-Gaussian distribution of the origin in the same way, the extension could scarcely be applied in practice; for we have seen that, except when $f(x_0)$ is Gaussian, no information as to its form can be obtained.

4. **Relative wave functions:** related to **F** §§36, 38, but making no use of a *Gaussian* wave packet for the particle coordinate. A 'two-particle transformation' (cf. **F** §§18, 26) is applied explicitly to the joint wave function of x (particle) and x_0 (origin), and the 'weight function' (the effect of origin uncertainty on observational momenta) is obtained.

In quantum theory a p.d. is generally specified by a wave function... the square root of a distribution function.... The momentum p_α ... is

$$p_\alpha = -i\hbar\, \partial/\partial x_\alpha \qquad (4{\cdot}1)$$

applied to the wave function, $\hbar$ being a natural constant.

If a particle has an exact momentum p conjugate to a coordinate x, the wave function must, by (4·1), contain a periodic factor $e^{ipx/\hbar}$. In general, however, p as well as x has a p.d.; and we have two forms of the wave function...the reciprocal Fourier integrals

$$f(x) = (2\pi\hbar)^{-\frac{1}{2}} \int_{-\infty}^{\infty} e^{ipx/\hbar} F(p)\, dp, \qquad (4{\cdot}21)$$

$$F(p) = (2\pi\hbar)^{-\frac{1}{2}} \int_{-\infty}^{\infty} e^{-ipx/\hbar} f(x)\, dx, \qquad (4{\cdot}22)$$

and the probabilities in the ranges dx, dp are respectively

$$2\pi\hbar\, |f(x)|^2\, dx, \quad 2\pi\hbar\, |F(p)|^2\, dp. \qquad (4{\cdot}23)$$

By (4·23) the distribution function of x or p is proportional to the square of the modulus of the corresponding wave function. Consequently the transformation of a wave function from the geometrical to the physical frame is not quite the same problem as the transformation of a distribution function treated in § 3. But the method is closely analogous. We use the notation as in § 3 except that f, g, h, F, G, H now denote wave functions instead of distribution functions.

The double wave function for the combined distribution of two uncorrelated variates x_0, x with conjugate momenta p_0, p is, by (4·21)

$$f(x_0)\,g(x)=(2\pi\hbar)^{-1}\iint e^{i(p_0x_0+px)/\hbar}\,F(p_0)\,G(p)\,dp_0\,dp. \tag{4·31}$$

We introduce new variables (the same result is obtained if the second variable is taken to be $\eta=(mx+m_0x_0)/(m+m_0)$, where m and m_0 are arbitrary constants)

$$\xi=x-x_0,\quad \eta=\tfrac{1}{2}(x+x_0).$$

By (4·1), their conjugate momenta are found to be

$$p_1=\tfrac{1}{2}(p-p_0),\quad p_2=p+p_0,$$

and it follows that

$$p_0x_0+px=p_1\xi+p_2\eta,\quad dp_0\,dp=dp_1\,dp_2.$$

Hence (4·31), transformed into a function of ξ, η, becomes

$$h(\xi,\eta)=(2\pi\hbar)^{-1}\iint e^{i(p_1\xi+p_2\eta)/\hbar}\,F(\tfrac{1}{2}p_2-p_1)\,G(\tfrac{1}{2}p_2+p_1)\,dp_1\,dp_2. \tag{4·32}$$

To obtain the distribution of ξ...we integrate (4·32) for all values of η. This gives the simple wave function $h(\xi)$ of the physical coordinate $\xi=x-x_0$. Thus

$$\int h(\xi,\eta)\,d\eta=h(\xi)=(2\pi\hbar)^{-\frac{1}{2}}\int \exp(ip_1\xi/\hbar)\,H(p_1)\,dp_1 \tag{4·33}$$

by the standard form (4·21). Comparing (4·33) with (4·32), we see that

$$H(p_1)=(2\pi\hbar)^{-\frac{1}{2}}\iint e^{ip_2\eta/\hbar}\,F(\tfrac{1}{2}p_2-p_1)\,G(\tfrac{1}{2}p_2+p_1)\,dp_2\,d\eta. \tag{4·34}$$

The waves extend over a large (nominally infinite) region, so that only the waves which cohere in phase over a large range of η make an effective contribution to $H(p_1)$. This restricts the p_2 integration to an infinitesimal range about $p_2=0$. The formal reduction can be made by using the δ function

$$\int_{-\infty}^{\infty} e^{ip_2\eta/\hbar}\,d\eta=2\pi\delta(p_2/\hbar).$$

Then (4·34) becomes

$$\begin{aligned}H(p_1)&=(2\pi\hbar)^{\frac{1}{2}}\int F(\tfrac{1}{2}p_2-p_1)\,G(\tfrac{1}{2}p_2+p_1)\,\delta(p_2/\hbar)\,d(p_2/\hbar)\\&=(2\pi\hbar)^{\frac{1}{2}}[F(\tfrac{1}{2}p_2-p_1)\,G(\tfrac{1}{2}p_2+p_1)]_{p_2/\hbar=0}\\&=(2\pi\hbar)^{\frac{1}{2}}F(-p_1)\,G(p_1),\end{aligned} \tag{4·4}$$

which can be compared with the result (3·4) for the reciprocals of distribution functions.

From (4·4) we obtain

$$|H(p)|^2\,dp = |G(p)|^2\,w(p)\,dp, \tag{4·51}$$

where
$$w(p) = 2\pi\hbar\,|F(-p)|^2, \tag{4·52}$$

or...*the distribution of geometrical momentum* $|G(p)|^2\,dp$ *is converted into the distribution of physical* $|H(p)|^2\,dp$ *by weighting the ranges* dp *with the weight function* $w(p)$.

By (4·23) $w(p)\,dp$ is the probability that the momentum of the physical origin is in the range $-p$ to $-(p+dp)$. Evidently the weight function can be explained dynamically as a factor depending on the recoil momentum $-p$ of the physical reference frame associated with the momentum p of the particle considered.

5. **The weight function:** related to **F** § 38.

When $f(x_0)$ is the wave (instead of the distribution) function the probability $(2\pi\sigma^2)^{-\frac{1}{2}}\,e^{-x_0^2/2\sigma^2}\,dx_0$ that the physical origin is in dx_0 must be set equal to $2\pi\hbar\{f(x_0)\}^2\,dx_0$ by (4·23); so that

$$f(x_0) = (2\pi\hbar)^{-\frac{1}{2}}(2\pi\sigma^2)^{-\frac{1}{4}}\,e^{-x_0^2/4\sigma^2}. \tag{5·11}$$

This gives by (4·22)

$$F(p) = (2\pi\hbar)^{-1}(8\pi\sigma^2)^{\frac{1}{4}}\,e^{-\sigma^2p^2/\hbar^2}. \tag{5·12}$$

Hence by (4·52)
$$w(p) = (2\sigma^2/\pi\hbar^2)^{\frac{1}{2}}\,e^{-2\sigma^2p^2/\hbar^2}. \tag{5·2}$$

Or setting
$$\varpi = \hbar/2\sigma, \tag{5·3}$$

the weight function has the standard Gaussian form

$$w(p) = (2\pi\varpi^2)^{-\frac{1}{2}}\,e^{-p^2/2\varpi^2}. \tag{5·4}$$

The total weight distributed by the weight function is unity.

It is instructive to compare this with the result for distribution functions in § 3. The factor $e^{-\frac{1}{2}\sigma^2q^2}$ in (3·5) can be interpreted as a weight function. It will agree with (5·2) if we set $q = 2p/\hbar$, so that the periodic factor in the distribution function is $e^{2ipx/\hbar}$ as compared with $e^{ipx/\hbar}$ in the wave function. This was to be expected, since the distribution function is effectively the square of the wave function. But whereas in the transformation of wave functions the total weight is automatically adjusted to unity, the total weight in the transformation of distribution functions is $(2\pi\varpi^2)^{\frac{1}{2}}$, which is not of the dimensions of a pure number.

The effect of the weight function is to reduce the frequency of very large momenta....

6. **Uranoids:** related to **F** §§ 7, 39. First (as **F** § 7) the need for a 'uranoid'—a simplified environment; then (as in part of **F** § 39) the calculation of energy in an 'infinite-temperature uranoid'—an infinite spread of geometrical momenta, controlled by the weight

function of §5 above. But the calculation is here for a sphere in *flat* space (foreshadowing the 'planoid' of **F** V), and the uncertainty constant σ corresponds to this case.

...we divide the universe into...*object-system* and its *environment*... the background on which the object-system is superposed...simple environments will be called *uranoids*....It appears that the commonly accepted formulae in quantum theory are not always consistent, some being valid in one form of uranoid and some in another. The uranoid, which we shall adopt as standard environment, will be found to validate the most essential part of quantum theory...especially the theory of eigenstates of atomic systems. Our standard uranoid will be a uniform and static distribution of matter; in molar relativity an 'Einstein universe',...in microscopic theory...a steady uniform p.d. of a very large number of particles.

Two lines of thought...led to a large number of particles as environment....In §2 it was a question of *metric*;...here, of *mechanical interaction*....The two aspects are not really distinct....In microscopic theory...the system of uranoidal particles determines the energy of the object-system by mechanical interaction, or...by fixing the uncertainty constant σ of the reference frame... and hence the weight function $w(p)$....

[*N.B.*] An Einstein uranoid is too difficult to begin with, since it involves curvature of space. We consider first (as environment...) n similar particles...in a sphere of volume $V=\frac{4}{3}\pi R^3$ in flat space. The spherical boundary is...a temporary substitute for...the forces in the curved Einstein space. Owing to symmetry the centroid has the s.d. in all directions

$$\sigma = R/\sqrt{(5n)}. \tag{6·1}$$

...Consider first a uranoid in which *the geometrical momenta have unlimited uniform p.d.*; i.e. the number of particles in any range is proportional to $dp_1dp_2dp_3$, $-\infty<p<\infty$. This is..commonly assigned in statistical mechanics as the *a priori* p.d....We are not concerned with its supposed *a priori* justification,..generally acknowledged to be obsolete; we employ the distribution merely as a standard of comparison. It is of more interest that, according to statistical mechanics, the momenta tend to this distribution as the temperature $T\to\infty$ and the Boltzmann factor $e^{-E/kT}\to 1$. We therefore distinguish this as the 'infinite-temperature uranoid'....

The corresponding distribution of physical momenta is $w\,dp_1dp_2dp_3$, where w is (5·5) [**F** (38·9)]. Hence the mean values for the physical distribution are

$$\overline{p_1^2}=\overline{p_2^2}=\overline{p_3^2}=\varpi^2=\hbar^2/4\sigma^2. \qquad (6·2)\ [\textbf{F}\ (39·1)]$$

The energy E of a particle is given by (..with $c=1$)

$$E^2=m^2+p_1^2+p_2^2+p_3^2, \tag{6·3}$$

where m is the rest mass. We omit m, because it is our purpose to explain how rest mass originates. 'Rest mass' is...concealed energy; and...the process of concealment will be fully exhibited. Accordingly, we adopt

$$E^2 = p_1^2 + p_2^2 + p_3^2, \tag{6·4}$$

except when the theory itself indicates...an extra term m^2.... By (6·2) the mean value is

$$\overline{E^2} = 3\varpi^2 = 3\hbar^2/4\sigma^2. \qquad (6·5)\ [\mathbf{F}\ (39·2)]$$

...$\overline{E^2}$ is the 'expectation value'...for each of the n particles....If, without altering V..., we vary the number, $\sigma \propto n^{-\frac{1}{2}}$, so $\overline{E^2} \propto n$ by (6·5). We therefore regard $\overline{E^2}$ as a mutual characteristic arising jointly from the particle considered and the n particles of the uranoid. The constant $\overline{E^2}/n$ represents the contribution arising from one pair of particles. As commonly happens when mutual properties of pairs are transferred to individuals as self properties, the contribution of each pair gets counted twice over when all the particles are considered in turn. A correcting factor $\frac{1}{2}$ must be introduced to eliminate this double reckoning.

The momentum formula $p_\alpha = -i\hbar\,\partial/\partial x_\alpha$ is primarily intended to be applied to small object-systems superposed on the standard environment. All the excess energy and momentum above that of the undisturbed uranoid is then attributed to the object-system; and we must accept it as a convention that the constant $\hbar$ is defined so as to give the momentum of a particle reckoned in this way. But when the formula intended for a particle added to a fixed uranoid is applied to the uranoid particles themselves, the correcting factor $\frac{1}{2}$ must be inserted. If, for example, we construct the uranoid by introducing particles one by one into the volume V, the $\overline{E^2}$ of each added particle will be proportional to the number of particles already present; so that the mean $\overline{E^2}$ of the uranoid particles is half the $\overline{E^2}$ of the last particle added, which is given by (6·5). Hence, for uranoid particles, we have in place of (6·2) and (6·5)

$$\overline{p_1^2} = \tfrac{1}{2}\varpi^2, \quad \overline{E^2} = \tfrac{3}{2}\varpi^2, \tag{6·6}$$

with $\varpi = \hbar/2\sigma$.

7. Spherical space: includes a calculation like **F** §6. Eddington has now to pass to a curved-space uranoid, without the previous discussion of length and curvature of **F** I. The main points are:

Quantum lengths are functions of σ, so that the number n of particles determines the measured radius $R = \sigma\sqrt{(5n)}$. But when n is increased, the flat-space sphere changes to a closed hypersphere of N particles, radius R_0. The s.d. of a centroid coordinate is now

$$\sigma = \tfrac{1}{2}R_0/\sqrt{N}. \tag{7·3}$$

This calculation is equivalent to orthogonal projection on a tangent flat space.

[*Begins*] In the last section the particles were in a sphere V. We now seek to remove this artificial constraint. If the particle density $s = n/V$ is fixed and the sphere continually enlarged, $\sigma \propto Rn^{-\frac{1}{2}} \propto R^{-\frac{1}{2}}s^{-\frac{1}{2}}$; so that

$\sigma \to 0$ as $R \to \infty$. But according to relativity the space is curved to correspond to..the matter..and the curvature sets a limit to the possible increase of R and decrease of σ. Even when n and V equal the number of particles and volume of the whole universe, σ is not negligible compared with nuclear dimensions, and ϖ is of the order of the masses of elementary particles. Since 1930 the cosmological constants have been sufficiently well determined from the recession of the spiral nebulae to make it clear that the scale of extension and mass of nuclei and atoms is fixed by the constants σ and ϖ derived from the universe treated as a whole.

We here present the inverse point of view. The uranoid of n particles in flat space with an artificial boundary may be called a 'laboratory', or more technically a 'planoid'. Subject to the condition that it is not unduly large or small, n can be chosen arbitrarily; $n = 10^{32}$ gives a laboratory of convenient size for ordinary experiments. Since the artificiality of the boundary involves a discontinuity, an observer in the laboratory must develop his physics entirely from measurements inside the laboratory. He cannot assign a radius R to the laboratory until he has provided himself with a standard of length in the laboratory. A suitable standard is the wave-length of the hydrogen line H_α. Considering a hydrogen atom in the laboratory, the 10^{32} particles form its physical environment and their centroid provides the origin of its physical reference frame. The theory developed later determines the linear constants of the hydrogen atom in terms of the uncertainty constant σ of the physical frame; in particular, the H_α wave-length is $6{\cdot}83 \times 10^8 \sigma$. By (6·1), $R = \sigma\sqrt{(5n)}$. We therefore know R in H_α units; and this can be converted into metres, if the metre is defined as 1,523,741 H_α units. (The observer has no access to the 'Paris metre'.... He must reconstruct the metre in his laboratory from its specification in terms of natural standards such as the H_α unit). For $n = 10^{32}$ the radius is $R = 21{\cdot}5$ metres.

The point..is that there is no question of adjusting the bounding sphere of the laboratory to a particular radial extension R necessary to give the proper uncertainty constant σ. A laboratory of 10^{32} particles will automatically turn out to have the radius of 21·5 metres *when measured by an observer within, using standards constructed in the laboratory*. If we try to double R, we double σ and so the extension of the H_α unit; so that R measured in H_α units, or equivalently in metres, is unaltered. We have therefore only to specify n, and the size of the laboratory will settle itself.

When we pass to astronomical and cosmical systems, we require a larger laboratory. This is provided by increasing n, R being proportional to $\sqrt{n}$. The different sized laboratories are freely interchangeable,...the same σ...we employ in cosmology the same ultimate standards as in atomic physics.

When at last n and R become so great that the curvature of space in the laboratory can no longer be neglected, the relation between them begins to be modified; but R (defined by great-circle measurements on the hypersphere) continues to increase with n until the space closes up as a complete hypersphere. This gives the greatest possible laboratory, namely the universe; and the corresponding volume V gives the volume of the region in which particles of the universe have a probability of occurring, namely the total volume of physical space.

We can therefore eliminate the artificial boundary, and consider a distribution of particles limited only by the finiteness of the space which is curved to correspond to the density and pressure of this distribution. Let R_0 be the radius of curvature of the hyperspherical space and N the total number of particles. The rectangular coordinates of a particle, referred to a geometrical origin at the centre of the hypersphere, satisfy

$$x_r^2+y_r^2+z_r^2+u_r^2=R_0^2, \tag{7·1}$$

so that for uniform p.d. over the hypersphere...

$$\overline{x_r^2}=\overline{y_r^2}=\overline{z_r^2}=\overline{u_r^2}=\tfrac{1}{4}R_0^2. \tag{7·2}$$

Hence the s.d. of a particle coordinate is $\frac{1}{2}R_0$, and the s.d. of a coordinate of the centroid of the N particles is

$$\sigma=\tfrac{1}{2}R_0/\sqrt{N}. \tag{7·3}$$

Then by (5·3)

$$\varpi=\hbar\sqrt{N}/R_0. \tag{7·4}$$

The corresponding formula for the planoid (laboratory) [*sic*] was (6·1)

$$\sigma=R/\sqrt{(5n)}, \tag{7·5}$$

so that the curvature does not make a great deal of difference; but, of course, R and R_0 are not strictly comparable.

We have here calculated the centroid in four dimensions, or equivalently we have used the *orthogonal projections* of the particles on a flat 3-space, in calculating the distribution of the x_0, y_0, z_0 of their centroid... This too naïve..?...another, e.g. stereographic, projection would give a different coefficient in (7·3). Orthogonal projection is correct, but we cannot yet..justify it. Our temporary difficulty is that space-curvature is a foreign conception..imported from..molar relativity. But in §[11: **5·1**], we shall derive space-curvature directly from the statistical properties of systems of particles, and (7·3) will then be obtained without any ambiguity.

Meanwhile...there is a dynamical reason for preferring orthogonal projection...Unless the projection is orthogonal, integrals of a.m. in the curved space will not correspond to integrals of a.m. of the projected particles in the flat space. But in elementary mechanics the reason for employing the centroid as origin is that the relative motion is then separable and has its own integrals of energy and a.m. We can see there would be an uncomfortable divergence between the dynamical and statistical theory if the one required us to calculate the centroid from the orthogonal and the other from the stereographic projection.

8. **The zero-temperature uranoid:** related to **F** §39. The change from infinite to zero temperature in the above *curved* uranoid 'creates' a proper density, and so gives particles rest mass. Henceforth we

take the zero-temperature uranoid, and the corresponding rest mass as standard.

Eddington begins with **F** (39·3–39·6). Then:

At zero temperature the particles are at rest, and ρ_0 is the sum of the rest masses in unit volume. Thus the particles, initially without rest mass have acquired rest mass through lowering the temperature. Considering any one particle, its rest mass is an invariant independent of the velocity; thus the rest mass has not been created by reducing the particle itself to rest, but by the change in its environment. This is easily explained by Newtonian theory. The object-particle is in the field of the rest of the universe; and, if ϕ is the gravitational potential, its rest energy or rest mass

$$m' = m(1-\phi) \qquad (8{\cdot}6)$$

includes its potential energy $-m\phi$ in that field. When the mass of the surrounding matter is decreased by lowering the temperature, ϕ is decreased; thus m' is increased. In this way the particle, whose rest energy m' was initially zero, acquires a positive rest energy through the change of the gravitational field emanating from its environment. For a rigorous calculation we employ Einsteinian theory...and the result is (8·5) [**F** (39·6)].

The term 'proper mass' is, I think, generally understood to refer to m in (8·6); to avoid ambiguity we shall generally call m the *mass constant*. It is the quantity ordinarily given...e.g. in tables. But the relation between m' and m is indeterminate until we specify the zero from which gravitational potential is measured. The condition 'ϕ vanishes at infinity' is, of course, inapplicable, since there is no infinity in the Einstein universe.

...In Einstein's theory there is no physical distinction between gravitation and inertia, but in practice we make an artificial distinction. The part of the field emanating from a standard uniform distribution of the matter of the universe is distinguished as the *inertial field*, and the field due to the irregularities superposed on the standard distribution is the *gravitational field*. In unifying physical theory the standard distribution which determines the inertial field is naturally taken to agree with the standard environment or uranoid adopted for the purposes of quantum theory. Thus in the adopted uranoid there is no gravitational (as distinct from inertial) field, and $m' = m$. If we change the adopted standard environment, as when a zero- is substituted for an infinite-temperature uranoid, we change the artificial partition of the rest mass m' into inertial energy (proper mass) m and gravitational energy $-m\phi$.

...the proper mass or mass constant, unlike the rest mass, does not depend on the *actual* temperature of the environment; it depends only on the temperature of the *ideal* environment adopted as standard uranoid. If the actual temperature were increased, m' would be decreased, but m would be unchanged unless the hotter surroundings are accepted as the standard environment. Proper mass..depends on how you define it; its apparent changeability appears only at this initial stage when we are feeling our way towards a settled definition. Henceforth the zero-temperature uranoid will be adopted as standard; so that proper mass etc....are understood to be measured in conditions equivalent to those of a zero-temperature uranoid.

The transformation from (8·41) to (8·42) [**F** (39·51, 39·52)] might be criticized on the ground that the supposed change of temperature of the universe would violate conservation of energy. But the criticism is inapplicable when it is realized that there is no question of identifying either uranoid with the actual universe—which is an 'expanding universe' very far from the static Einstein state.

9. **Primitive observables:** parts of **F** §§5, 11, 35; the opening is as **D** §4. The relevance to the present development lies mainly in the final passage; but the section is a valuable summary of the nature of 'observables'. The numeration (i), (ii), ... is mine. Begins:

The description of physical systems by p.d.'s requires certain precautions [(i), (ii), ... below] which...have no counterpart in the classical ...physics....

(i) The distinction between direct measurement, and measurement via the physical origin, of the separation of two particles shows the range of non-Coulombian forces to be $\sigma\sqrt{2}$. [A recession calculation was to be added to §7 above on this point.]

(ii) Ambiguity appears in the definition of the momentum vector $p_1, \ldots, p_4$ and proper mass m...Presumably p_1, p_2, p_3 are..measured directly; but doubt may arise whether p_4 stands for a fourth measured quantity of an analogous kind, or a quantity computed from other measures by the formula $p_4^2 = m^2 + p_1^2 + p_2^2 + p_3^2$. In the latter case, is m measured along with p_1, p_2, p_3, or is it a specific constant determined once for all for the type of particle considered? Occasions arise when we distinguish a *hamiltonian* p_4 computed from p_1, p_2, p_3, m from an *energy* p_4 measured directly; also we have to distinguish the mass m assigned to a particle as a *stabilized characteristic* when it is known to be of a particular type, from the *observable* m whose p.d. is found by measurements made on the particle itself.

(iii) Such ambiguities..occur unless we state explicitly the plan of measurement....The measurements will in the first instance furnish the p.d.'s of certain 'primitive observables'. Any additional observables must be defined as mathematical functions of these, not as quantities which would have been obtained by measurement if our programme of measurement had been differently designed. When variables are transformed, it is necessary to distinguish between a mathematical transformation and a physical transformation implying a change of primitive observables...[e.g. cartesian to polar coordinates].

(iv) If we attempt to lay down a uniform procedure of measurement ..comprehensive enough..for all problems,...the observables will in general not commute. It is therefore necessary to vary the plan of measurement to suit..the problem..; often the main step in..a quantum problem is the discovery of a set of observables which will commute in the particular..problem. Thus, although we follow certain principles in the choice of primitive observables, we cannot lay down definite rules to cover all cases.

In general theory a system is described by coordinates and momenta q_α, p_α. By using complex wave functions instead of real distribution functions our information about q_α and p_α is incorporated in a single expression which has two reciprocal forms..giving respectively the p.d.'s of q_α and p_α. Accordingly each pair is associated with one measurement....The plan..is then defined by the set of q_α (*or* p_α) supposed to have been measured.

...measurement disturbs the system...measured. It...does not necessarily involve active interference. But since the opportunity of acquiring information occurs only in special circumstances, when the system (spontaneously or through our interference) starts a chain of causation which reaches our sense organs, there is an automatic selection of disturbed conditions which biases the information in the same way as active interference....Thus description always relates to a system subjected to a certain amount of disturbance....For..standard description the normal (technically 'undisturbed') condition must be defined to be that in which the system is being subjected to *a conventional amount of observational probing*, namely, the measurement of a particular set of primitive observables....In different types of investigation and stages ..of the theory, the normal condition may be differently defined. The re-definition creates what is to all intent a new system..with different specific characteristics.

In the preceding investigations the probing has been limited to.. three coordinates ξ, η, ζ or their conjugate momenta. The conventions of language make it necessary to introduce a *conceptual carrier* of these variates..a 'particle'. A particle with no independent observable characteristics other than those here stated will be called a 'V_3 particle'.

10. **V_3 and V_4 particles:** related to **F** § 39 and the beginning of **F** § 40. Eddington extends the argument of §§ 6, 8 to calculate the mass of a (spinless) particle; the 'absolute determination' (with spin effects) of **F** § 40 is not reached.

If the mechanical characteristics of a particle are fully specified by an exact momentum vector p_α ($\alpha = 1, ..., 4$), its contribution to the energy tensor must have the form $\Delta T_{\alpha\beta} = Ap_\alpha p_\beta$, where A is a scalar coefficient. ...Correspondingly, if p_α has a p.d., the mean contribution..is

$$\Delta T_{\alpha\beta} = \overline{Ap_\alpha p_\beta}. \quad (10{\cdot}1)$$

For a particle of the infinite-temperature uranoid this gives

$$\Delta T_{11} = \tfrac{1}{2}A\varpi^2, \quad \Delta T_{44} = \tfrac{3}{2}A\varpi^2, \quad (10{\cdot}2)$$

by (6·6), since E is now denoted by p_4. Summing for the N particles, the total pressure (T_{11}) and density (T_{44}) are

$$P' = \tfrac{1}{2}NA\varpi^2, \quad \rho' = \tfrac{3}{2}NA\varpi^2. \quad (10{\cdot}3)$$

Now reduce the temperature to zero. Since the particles are reduced to rest, the momentum vector becomes $(0, 0, 0, m)$; and the total pressure and density are, by (10·1),

$$P=0, \quad \rho=NAm^2. \tag{10·4}$$

By (8·5) [F (39·6)], $\rho=\frac{4}{3}\rho'$; hence

$$m^2=2\varpi^2. \tag{10·5}$$

Then by (7·4)
$$m=\hbar\sqrt{(2N)}/R_0. \tag{10·6}$$

Comparing (8·42) [F (39·52)] and (10·4), we obtain

$$A=1/8\pi\kappa\hbar^2N^2. \tag{10·7}$$

The formula (10·6) will be further examined in §[28]. We consider certain consequences of the relation $\rho=\frac{4}{3}\rho'$, which implies that when rest energy is released a quarter of the energy disappears. It is easily verified from (8·1) and (8·2) [F (39·3)] that this remains true when only part of the rest energy is released,...e.g. in a nuclear transformation. The experimenter tells us there is no loss...but he does not check up the energy of the rest of the universe....The loss predicted is a diminution of the gravitational potential energy of the rest of the matter in the universe, the efficacy of the energy as a source of gravitation being affected by its change from a bound to a released form. Although there is no contradiction, it is desirable to transform our results to give recognition to the experimenter's point of view.

Since we have been investigating V_3 particles...we call m in (10·6) m_3. By definition the only observable mechanical characteristics of a V_3 are p_1, p_2, p_3. Consequently m_3 is an unobservable (a 'stabilized characteristic') introduced by our system of theoretical formulation. There can be no question of testing observationally whether energy is or is not lost in the conversion of the mass m_3 into thermal energy; for if the mass of a particle is measured it is *ipso facto* not a V_3. The additional probing... makes it a system with four..freedoms....We shall call a particle with four primitive observables p_1, p_2, p_3, m a 'V_4 particle'.

It is natural to couple with the transformation infinite- to zero-temperature uranoid a transformation of the uranoid particles from V_3's to V_4's, so as to allow observational measurement of the proper mass that has been introduced. It will be shown later that the mass of a V_4 is

$$m_4=\tfrac{3}{4}m_3=\tfrac{3}{4}\hbar\sqrt{(2N)}/R_0. \tag{10·8}$$

Thus the thermal energy which becomes multipied by $\frac{4}{3}$ when converted into proper mass, is reduced to its original amount when the proper mass is *measured*.

The difference between a V_3 and V_4 can also be regarded in the following way. Consider a particle with measured momentum p_1, p_2, p_3. It can only be considered in conjunction with an environment; and since this is in any case supposed to be a uniform static distribution, the only other variate involved is the temperature τ of the environment. The rest mass or inertial-gravitational energy due to the environment is a function of τ;

so that a measurement of m and of τ are equivalent. But we do not contemplate *measurement* of the temperature of the uranoid; we *prescribe* the temperature $\tau=0$ as standard condition in which the behaviour of the particle is to be theoretically investigated. The result of the investigation is that, corresponding to $\tau=0$, we obtain the mass m_3. This must equally be regarded as a prescribed characteristic of the V_3 particle, not involving any measurement and therefore not involving an uncertainty of the conjugated coordinate. If we substitute measurement for prescription, either of m or of τ, the three-dimensional p.d. over p_1, p_2, p_3 is extended into a four-dimensional p.d. over p_1, p_2, p_3, m (or τ). The mass m_4 refers to a particle with this additional degree of freedom....

4·2. Draft B *17* I. The Uncertainty of the Origin (originally 'The Uncertainty of the Reference Frame')

Drafts **A** and **B** alone have decimally numbered sections. At the end of **B** *17* there are variant (earlier) versions of some sections; these are juxtaposed here and lettered; for example, § 1·2 (*a*) is the version occurring in the main sequence of Eddington's draft and § 1·2 (*b*) was found at the end.

Draft **B** *17* resembles **C** *33* of **4·1**, but has an extensive account (§§ 1·6 (*a*), 1·8) of the origin of proper mass. There is an elaboration (§ 1·9 (*b*)) of 'four-dimensional theory', which shows the first form of the notion of 'stabilization' (**F** § 11). *In* **B**, *Planck's constant is called* '$2\pi\beta$', *as a quantity to be examined later; this must not be confused with the 'Bond factor'* $\beta=\frac{137}{136}$ *of* **F**.

1·1. **The conditions of observability:** like **C** § 1 (**F** § 1) above.

1·2 (*a*). **The Gaussian distribution:** like **C** § 2 (**F** § 2) above.

1·2 (*b*). **The centroid as physical origin:** an earlier version related to **F** §§ 1, 2. The following remarks are made on the centroid, and on correlations:

In treating..n similar particles it is usual to take the centroid as origin. The advantages..are well known in elementary mechanics. In microscopic physics..additional considerations make it especially necessary.

The advantage in classical mechanics is that motion relative to the centroid is 'internal', separable from the external, with its own independent integrals....Although..separation in relativity and quantum mechanics is rather more complex, there is sufficient resemblance to make it necessary to retain the centroid as origin of internal coordinates,.. to employ the analogues of the classical integrals of energy and a.m.

Another decisive consideration is the elimination of *correlations*. Suppose..we employ one of the particles ($r=1$) as physical origin. The relative positions are then specified by the coordinates $\xi_{1s}, \eta_{1s}, \zeta_{1s}$

$s=2, 3, \ldots, n$) of the other particles, since ξ_{st}...can be deduced from ξ_{1s} and ξ_{1t}. But when we are dealing with p.d.'s and expectation values, the p.d. of ξ_{st} cannot be deduced from the p.d.'s of ξ_{1s} and ξ_{1t}, nor the expectation value....

The failure is due to correlations of the distributions of the particles. The investigations of these is an important part of wave mechanics; the wave functions with..practical application are usually 'correlation wave functions'. But for that very reason we have to take care that the correlation data are systematically segregated from the general description of the density and extent of the system. Our plan should either give complete information as to correlations of pairs of particles or exclude it altogether. For complete information it would be necessary to employ the whole $\frac{1}{2}n(n-1)$..ξ_{st}. Since we do not at present aim at so much detail, we must avoid using any of the relative coordinates.

To avoid premature dabbling with correlations we adopt coordinates ξ_r, η_r, ζ_r measured from the centroid. The geometrical coordinate of the centroid is

$$x_0 = n^{-1} \sum_s x_s. \tag{1·21}$$

If the rth particle receives a displacement δx_r, the centroid receives a displacement $\delta x_0 = \delta x_r/n$, and

$$\delta\xi_r = \delta x_r - \delta x_0 = \frac{n-1}{n}\,\delta x_r.$$

Measures of ξ_r should therefore be multiplied by a scale factor $n/(n-1)$, so as to reduce displacements, velocities, etc., in relative coordinates to the equivalent scale in geometrical coordinates. It is convenient to include this correction in the definition of ξ_r, so that

$$\xi_r = (n-1)^{-1} \sum_s \xi_{sr}, \tag{1·22}$$

and ξ_r now represents the coordinate of the rth particle measured from an *uncorrelated* point in the p.d. of x_0.

Notwithstanding the definition (1·22) we may treat ξ_r as though it were furnished by a single direct measurement from the centroid to the particle; i.e. we may treat the centroid as equivalent to an actual particle for the purpose of providing a physical origin. This is justified because the uncertainty relations are the same for the mean of $(n-1)$ measures as for a single measure. The uncertainty of position is reduced in the ratio $\sqrt{(n-1)}$, but the $(n-1)$ random momentum kicks combine into a kick $\sqrt{(n-1)}$ times larger. Thus the product of the uncertainties of ξ_r and its conjugate momentum are the same as for a quantity, such as ξ_{rs}, which can be obtained by a single measure.

The distribution of the centroid x_0 follows as in **F** §2.

1·3. **Systems of description:** like **4·1** §9 (parts of **F** §§5, 10, 11, 35), adding a note on the importance of correlation wave functions in conjunction with uniform p.d.'s; cf. **F** §10. A remark on 'primitive observables' (cf. **4·1** §9 (iv)):

If we attempt to lay down a uniform plan of measurement,...the corresponding primitive observables will not in general commute. This means that we must either prescribe arbitrarily the order in which the measures are made, or must work with averages in which all permutations of the order of the measures have been equally combined. It is more practical to vary the plan of measurement to suit the..problem... discovering a set of commuting observables....

1·4. **Relative distribution functions:** close to **4·1** §3 (**F** §37).

1·5. **Application to wave functions:** calculation as **4·1** §§4, 5 (cf. **F** §§37, 38), but (i) β is used for $\hbar$, and (ii) the (recoil) momentum view appears only in a pencilled addition: 'The weight function represents the distribution of momenta of the physical origin.'

1·6 (*a*). **Uranoids:** like **4·1** §6 (**F** §§7, 39): the idea of 'uranoid', and the calculation of mean energy in flat space with 'proper mass' (the 'rest mass' of **4·1** §6) zero. The final discussion of 'double reckoning' (after (1·66) below) anticipates later work.

...current quantum theory postulates a uniform static uranoid.... We consider first (as environment) n V_3 particles..in a sphere radius R, volume V in flat space...[$\sigma = R/\surd(5n)$...(1·61)] at 'infinite temperature'...weight function

$$w = (2\pi\varpi^2)^{-\frac{3}{2}} \exp\left[-(p_1^2+p_2^2+p_3^2)/2\varpi^2\right], \tag{1·62}$$

where $\varpi = \beta/2\sigma$....For the physical distribution

$$\overline{p_1^2} = \overline{p_2^2} = \overline{p_3^2} = \varpi^2 = \beta^2/4\sigma^2 \quad [\beta \text{ for } \hbar]. \tag{1·63}$$

..For the S_3 [V_3] particles here considered m ('proper mass') = 0 and

$$\overline{E^2} = 3\varpi^2 = 3\beta^2/4\sigma^2. \tag{1·65}$$

We call $\overline{E^2}$ the *energy invariant*. In the more fundamental parts of physical theory it commonly plays the part which is subsequently transferred to the energy in the usual adaptation to a less relativistic outlook. For example, (1·62) has a resemblance to Boltzmann's formula, except that $E^2 = p_1^2+p_2^2+p_3^2$ has taken the place of E. This is not surprising when we recall that in relativity energy is not particularly important; the important quantity is energy density T_{44}, which, we shall find, is more nearly related to E^2 than to E.

The energy invariant is not intrinsic in the particle..[cf. **4·1** §6].. the contribution of each pair gets counted twice when we sum over all the uranoid particles considering each in turn as the object-particle. A factor $\frac{1}{2}$ must be introduced to correct this double reckoning; in place of (1·63, 1·65) the correct results are

$$\overline{p_1^2} = \tfrac{1}{2}\varpi^2, \quad \overline{E^2} = \tfrac{3}{2}\varpi^2 \quad \text{with} \quad \varpi = \beta/2\sigma. \tag{1·66}$$

This does not finally dispose of..double reckoning. The formula $p_\alpha = -i\beta\,\partial/\partial x_\alpha$ was primarily intended..for a small object-system..

an addition to the uranoid. The whole excess of energy and momentum ..is attributed to the object-system although it is actually a mutual property of the object-system and uranoid. We must therefore accept it as a convention that β is defined so that $-i\beta\,\partial/\partial x_\alpha$ is the *excess* momentum or energy due to the presence of the particle, and not the share of the excess properly attributable to the particle. (That this is the recognized convention is evident from the fact that, when e.m. terms are included, p_α is defined to be $-i\beta\,\partial/\partial x_\alpha + e\kappa_\alpha$, although $\Sigma e\kappa_\alpha$ taken over all the particles gives a double reckoning of the e.m. energy.) Applying this to the uranoid particles, the total energy of the uranoid will be counted twice over...This is not simply a reinstatement of the double reckoning that was eliminated in (1·66); a double reckoning of the energy E gives a four-fold reckoning of E^2.

The most convenient way of handling this further complication is to limit the summation to half the total number of particles when we calculate the total energy or mass of the uranoid. We therefore suppose the uranoid to consist of $\frac{1}{2}N$ 'external' and $\frac{1}{2}N$ 'internal' particles. Equation (1·66) applies to the external particles, it being understood that β and p_μ are defined to correspond to the whole additional momentum, as the recognized convention enjoins. Double reckoning is then avoided by assigning zero energy and momentum to the internal particles; dynamically they are merely passengers, but they are counted in the total number in determining the uncertainty of the centroid in (1·61) and in the next section. The term 'uranoid particle' will ordinarily.. mean external uranoid particle.

1·6 (*b*). **Uranoids:** an earlier fragment, containing this on 'weight factors' (cf. **F** §38):

Most writers have disregarded the distinction between the physical and the geometrical origin and treated observable momenta as though they were geometrical momenta. Current investigations therefore need ..the weight factor [(1·62) above] inserting...$\varpi\sqrt{2}$ is found later to be 136 times the mass m_e of an electron. The correction therefore becomes important for energies of the order $136 m_e c^2$. For example, it affects the calculation of the radiation of high-speed electrons passing through matter. A direct experimental determination of ϖ by this method will doubtless be possible in the future....

1·7. **Curvature of space:** close to **4·1** §7. The word 'laboratorioid' (cf. uranoid) is suggested—not approved! The final results are (corresponding to **4·1** (7·3, 7·4)):

$$\sigma = \tfrac{1}{2}R_0/\sqrt{N}, \tag{1·73}$$

$$\varpi = \beta\sqrt{N}/R_0. \tag{1·74}$$

1·8. **Proper mass:** much of **4·1** §§8, 10 (**F** §39)—the calculation of mass by lowering the temperature—with variants which are of interest in this central calculation.

...for a uniform static distribution (an Einstein universe, or a portion thereof)

$$8\pi\kappa P = -R^{-2} + \lambda, \quad 8\pi\kappa\rho = 3R^{-2} - \lambda. \quad (1{\cdot}81) \text{ [F } (39{\cdot}3)]$$

...The proper density (the invariant T of the energy tensor) is

$$\rho_0 = \rho - 3P = (3R^{-2} - 2\lambda)/4\pi\kappa. \quad (1{\cdot}82)$$

By (1·81),

$$R^{-2} = 4\pi\kappa(\rho + P), \quad \lambda = 4\pi\kappa(\rho + 3P). \quad (1{\cdot}83)$$

When the uniform distribution covers a limited region...with a boundary constraint (e.g. gas in vessel), ρ and P can be chosen arbitrarily, subject to $\rho > 3P$. Thus R (the local radius of curvature) and λ are disposable constants. When the uniform distribution is unbounded, R becomes a fixed constant $R_0 = 2\sigma\sqrt{N}$ by (1·73), but λ remains disposable. By varying λ we vary P/ρ_0, or equivalently the temperature.

The temperature increases with λ. As $\lambda \to \frac{3}{2}R_0^{-2}$, $\rho_0 \to 0$...similar to radiation, which is characterized by zero proper density. This is presumably the limiting condition which we should describe as infinite temperature. We must, however, make sure that it agrees with the definition of infinite temperature adopted in § 1·6. At low temperatures the heat energy (kinetic energy of the particles) of a monatomic gas has the density $\frac{3}{2}P$; but at very high temperatures the coefficient gradually changes from $\frac{3}{2}$ to 3, owing to the increase of mass with velocity. Thus $\rho \to 3P$, or $\rho/\rho_0 \to \infty$, gives the limit at which the whole energy is heat energy. In ordinary statistical mechanics infinite temperature corresponds to infinite energy, the ratio ρ/ρ_0 being made infinite by $\rho \to \infty$; and the momenta then have unlimited uniform distribution over all values of p_1, p_2, p_3, as stated in § 1·6. This is equivalent to setting $\varpi = 0$ in the weight function. But when the infinite temperature is applied to the whole uranoid which furnishes the physical reference frame, we have to proceed to the limit in such a way that the characteristic ϖ of the reference frame remains constant. Then ρ remains finite, and the ratio ρ/ρ_0 is made infinite by $\rho_0 \to 0$.

We assigned no proper mass to the particles in the infinite-temperature uranoid, on the ground that a fundamental theory ought to show the origin of proper mass...We now see this is consistent with ordinary statistical mechanics, the distribution being such that the proper energy is infinitesimal compared with the heat energy; and since the physical frame gives a finite reckoning of the heat energy, the proper energy is on the same reckoning zero.

Distinguishing values..in the infinite-temperature uranoid by an accent..,

$$\rho_0' = 0, \quad \lambda' = \tfrac{3}{2}R_0^{-2}, \quad 4\pi\kappa\rho' = \tfrac{3}{4}R_0^{-2}. \quad (1{\cdot}841)$$

Zero temperature is given by $P = 0$, so that $\lambda = R^{-2}$. Hence in the zero-temperature uranoid

$$\rho_0 = \rho, \quad \lambda = R_0^{-2}, \quad 4\pi\kappa\rho = R_0^{-2}. \quad (1{\cdot}842)$$

Hence

$$\rho_0 = \rho = \tfrac{4}{3}\rho'. \quad (1{\cdot}843)$$

The special value $\lambda = R_0^{-2}$, which applies to the zero-temperature uranoid, is called the *cosmical constant*.

If the mechanical characteristics of a particle...are specified by a p.d. of p_μ ($\mu = 1, \ldots, 4$), the mean contribution to the energy tensor is..

$$T_{\mu\nu} = A\overline{p_\mu p_\nu}. \tag{1·85}$$

..and in the infinite-temperature uranoid by (1·66)

$$T_{11} = \tfrac{1}{2}A\varpi^2, \quad T_{44} = \tfrac{3}{2}A\varpi^2, \tag{1·861}$$

since E is now denoted by p_4. Summing for the $\frac{1}{2}N$ external particles,

$$P' = \tfrac{1}{4}NA\varpi^2, \quad \rho' = \tfrac{3}{4}NA\varpi^2. \tag{1·862}$$

Hence by (1·841) $R_0^{-2} = 4\pi\kappa NA\varpi^2$; and, since $\varpi = \beta\sqrt{(N)}/R_0$ by (1·74),

$$A = 1/4\pi\kappa\beta^2 N^2. \tag{1·87}$$

Now reduce the temperature to zero..p_μ becomes $(0, 0, 0, m)$, where m is the proper mass of a particle. By (1·85), $T_{44} = Am^2$; and the whole density is

$$\tfrac{1}{2}NAm^2 = \rho = R_0^{-2}/4\pi\kappa$$

by (1·842). Eliminating A by (1·87), we obtain

$$m = \beta\sqrt{(2N)}/R_0. \tag{1·88}$$

This is the proper mass of an external particle of a zero-temperature uranoid, or as this is the standard environment..m is the mass of a uranoid particle.

...Proper mass depends not on the temperature of the surroundings or of the universe, but on the *ideal* surroundings adopted as uranoid. The actual rest energy...depends on the actual surroundings;..it consists of proper mass and gravitational energy combined....

We have assumed the contributions of the particles to the pressure and density are simply additive. The particles introduced in statistical investigations are necessarily of this type; but we can allow for non-linearity by including in the energy tensor an additional contribution representing interaction energy...this would be regarded here as a cosmical field energy....In passing from molar to microscopic theory, we replace the field description of the environment..by the uranoid description..as a p.d. of particles. We naturally employ particles which will make this replacement as simple as possible.

Our original V_3 particles possessed only three mechanical characteristics...In admitting the particle can contain a concealed energy, not determinable by observation of the three momenta, we have given it an additional characteristic, namely, proper mass. It must therefore be recognized that the transformation to a zero-temperature uranoid has introduced a new type of particle. This,..with mass (1·88), will be called a V_4 particle.

1·9 (*a*). **Object fields:** like **F** § 12; in parts identical with **C** III § 19 (**6·1**). To summarize:

Particle object-systems disturb the environment; we treat the disturbance as a ('complementary') object-field superposed on the standard uranoid. Current quantum theory postulates an unpolarizable uranoid; this compels us to use a uranoid of fictitious neutral particles, not of protons and electrons.

1·9 (*b*). **Four-dimensional theory:** much of this (early) section reappears in **B** II § 2·7 (*a*) (**5·2** below). To summarize:

Although p.d.'s cannot be extended over time as well as space coordinates, they can be extended to a 'phase coordinate', an internal time indicator corresponding to *proper* time; the conjugate variable is mass as rest energy. If we can measure this the particle becomes an S_4 (V_4 in later notation) instead of an S_3. The weight function is then

$$w = (2\pi^2\varpi^2)^{-2}\exp(-p_4^2/2\varpi^2), \quad p_4^2 = p_1^2 + p_2^2 + p_3^2 + m^2$$

(p_4 the 'hamiltonian'). Three-dimensional wave mechanics is recovered by 'stabilizing' m (a process discussed in this section) and the conjugate phase variate s_0. This restricts the p.d. to a hypersphere in (x, y, z, s_0) space—effectively this introduces space curvature without general relativity, and justifies the 'projection' technique of **B** § 1·7 [cf. **4·1** § 7] above for the centroid. There are modifications in the zero-temperature uranoid calculations for S_4 particles.

Notes for commentary: an odd sheet with this title appears in this MS., and contains these notes:

1. determination of p.d. of origin a matter of psychology rather than physics. Not my business to prove that quantum physicists are intelligent.

reality conditions. Merely part of the condition that the mathematics represents the particular system we are studying—generally the condition that the uranoid is neutral.

4·3. Draft *11* (An Early Fragment)

The following recently recovered fragment represents the opening of the work. Three later versions show drastic modification towards the form of § 1 of **F** and **B** and **C** above, suppressing the discussion of wave functions and of mutual and self properties, and concentrating on relativity of position. Relatedness is discussed in the present § 1·1 in wave-mechanical notions and in § 1·2 in terms of general relativity theory.

1·1. Wave functions

The most characteristic feature of wave mechanics is the use of a *product of two functions* to describe the observable properties of a system. This is evidently connected with the relativistic principle that an observable is a relation between two physical entities. Primarily the symbolism $\phi\psi$ in wave mechanics corresponds to the conception of a vehicle of observable properties in which two entities, represented by the factors ϕ and ψ, have an equal share. Or we may say the so-called physical entities—the relata which furnish the observable relations—have *potential* properties only, which must accordingly be specified mathematically by a symbol ψ with no observational interpretation; but two such entities in conjunction have *actual* properties and the product symbol $\phi\psi$ has an observational interpretation.

But alongside our recognition that an observable involves two physical entities we have to keep in mind the common outlook which assigns it to one of the two participants or else apportions it between them. Even when the relativity is obvious as in the case of relative velocity, one of the two bodies is said to 'have' the velocity, the other being regarded as a reference-body. Though we see through the common outlook we cannot disregard it, for it is the basis of most of the nomenclature of physics; and knowledge gained through deeper penetration must be translated in terms of the common outlook before its practical bearing can be appreciated. It would be idle to harp on the illegitimacy of this conceptual transfer of observables from the relations to the relata. The right course is to treat it as a recognized way of representing (or misrepresenting) our knowledge, and to examine its consequences.

We have therefore to provide for a transfer of the observable properties contained in $\phi\psi$ to ϕ and ψ separately. In general ϕ and ψ will be treated unsymmetrically, since they are cast for different roles in the resulting world picture. One of them, ϕ, will be called the *reference-body*, the other the *object-body*. In view of this distinction the product will be rewritten as $\phi^*\psi$, the reference factor being distinguished by an asterisk.

To fix ideas, consider particles. In transferring observable characteristics from the relations between particles to the particles themselves we necessarily modify the concept of a particle, since we attribute to the particle properties which according to the primary concept it could not possibly possess. This modified concept will be called the *secondary concept*, and the corresponding particle a *secondary particle*. Since the secondary object-particle is to have properties formally similar to those embodied in $\phi^*\psi$, it must be furnished with a product expression $\psi^*\psi$ from which its properties are derivable by the same mathematical procedure that applies to $\phi^*\psi$.

It is understood that ψ is the *complete* specification of the potential properties of the primary particle. Consequently ψ^* must be effectively a repetition of ψ. Either it is a simple repetition or a specification of the same potential properties in another code. In the usual form of wave mechanics the code is changed by substituting $-i$ for i; that is to say ψ^* is the complex conjugate of ψ. This has the effect of making the product $\psi^*\psi$, embodying the self properties of a particle, mathematically real—a condition which can be variously regarded as a useful simplification or a vexatious restriction. We shall accept it as a normal convention which can be relaxed when necessary. The restriction to mathematical reality cannot in any case apply to $\phi^*\psi$, which also represents physically real properties.

It is an essential part of the relativistic principle that the relations are between *physical* entities and not between an object-body and an abstract mathematical reference system such as a Galilean frame. The reference-body must be such that it might itself be studied as an object-body. Accordingly when the secondary concept of particles is employed, it will be assigned observable properties represented by $\phi^*\phi$.

To legitimatize the transfer of properties it is necessary that the self properties of the object and reference particles specified by $\psi^*\psi$ and $\phi^*\phi$ shall together be observationally equivalent to the mutual properties specified by $\phi^*\psi$. This condition of consistency is the source of some of the most fundamental formulae.

1·2. The fundamental tensor

Wave analysis bifurcates according as precedence is given to the positional or mechanical characteristics (coordinates or momenta) of the particles. The result, however, is two parallel, not two divergent, branches. There are reciprocal relations by which formulae can be transferred from one branch to the other. In customary wave mechanics the most elementary wave functions give precedence to mechanical characteristics; they represent systems with definite momentum and energy but entirely uncertain position. We shall begin in the same way, and examine the principles of the wave method of representing *mechanical* properties.

In general relativity these are specified by an *energy tensor* $T_{\mu\nu}$. Ostensibly this is referred to a geometrical coordinate frame, i.e. a fourfold set of numbered partitions imagined to extend through space and time. These partitions are unobservable, and a set of numbers or numerical functions referring to an unobservable frame would not in itself describe anything observable. But attached to the geometrical frame is a tensor function $g_{\mu\nu}$ called the *fundamental tensor*. Just as $T_{\mu\nu}$ is a form of description of the object body so $g_{\mu\nu}$ is a form of description of the physical system that is employed as reference-body. The geometrical reference frame is merely an intermediary, and the tensor method of description secures that it is eliminated in all observational measurements and relations, and in particular in the observable relations between the object-body specified by $T_{\mu\nu}$ and the physical reference-body specified by $g_{\mu\nu}$.

The tensor $g_{\mu\nu}$ was first introduced into relativity in connexion with positional characteristics. For that reason it is expressed in units of different physical dimensions from those employed in specifying mechanical characteristics. We have therefore to introduce a conversion constant $\lambda/8\pi\kappa$ to change it into units appropriate to the present discussion. Here κ is the constant of gravitation, and λ a constant so chosen that the tensor $\lambda g_{\mu\nu}/8\pi\kappa$ specifies the reference-body in precisely the same way that the tensor $T_{\mu\nu}$ specifies the object-body; i.e. $\lambda g_{\mu\nu}/8\pi\kappa$ is the energy tensor of the reference-body. In general relativity this is called the cosmical energy tensor; we therefore denote it by $(T_{\mu\nu})_c$, so that

$$8\pi\kappa(T_{\mu\nu})_c = \lambda g_{\mu\nu}. \tag{1·21}$$

By a well-known relativity formula we have for the energy tensor of the object-body

$$8\pi\kappa(T_{\mu\nu})_o = -\{G_{\mu\nu} - \tfrac{1}{2}g_{\mu\nu}(G - 2\lambda)\}. \tag{1·22}$$

Hence the total energy of the combination (which should be equal to the mutual energy tensor) is

$$8\pi\kappa T_{\mu\nu} = -\{G_{\mu\nu} - \tfrac{1}{2}g_{\mu\nu}G\}. \tag{1·23}$$

The problem of unification of relativity and wave mechanics now begins to take definite shape. We have an object and a reference-body specified in the one theory by energy tensors $(T_{\mu\nu})_o$ and $(T_{\mu\nu})_c$, and in the other theory analysed into 'particles' described by product expressions $\psi^*\psi$ and $\phi^*\phi$. In both theories the observable mechanical properties are relations between the object-body and the reference-body, and it is only by a transference of conception that such properties are attributed to the two bodies individually.

In making an analysis into particles we have to remember that in elementary wave mechanics physical systems are combined by *multiplication*. This is because the theory is concerned with probability; and since probabilities are combined by multiplication, there are obvious advantages in a formalism which reduces the combination of the systems and of their probability coefficients to one process. In conformity with this principle the primary particles have been specified by multiplicative quantities ϕ, ψ..A system of N object-particles is represented in elementary wave mechanics as a p.d. in $3N$ dimensions, i.e. actual space is treated as a multiplicative combination of N partial spaces, each containing one particle.

So far as metric is concerned each partial space is identical with ordinary space. The fundamental tensor $g_{\mu\nu}$ and the derivative curvature tensor $G_{\mu\nu}$ apply equally to the partial spaces and to actual space. But equations (1·21–1·23) must be modified, since a partial space contains only a fraction of the energy tensor which they determine. A commonly occurring case is when there are N similar particles with the same p.d. of position and momentum, e.g. a gas, so that each particle has $1/N$ of the whole energy tensor. There must then be a compensatory increase of the constant κ, i.e. the constant of gravitation applicable to a partial space is

$$\kappa' = N\kappa. \tag{1·24}$$

Substituting κ' for κ in (1·21–1·23) we obtain $(T_{\mu\nu})_c$, $(T_{\mu\nu})_o$, $T_{\mu\nu}$ for a partial space.

In introducing partial spaces elementary wave mechanics is merely carrying out the principle implied in § 1·1 that each object-particle ψ is to be provided with a reference-particle ϕ^*. Space as a *physical* frame of reference is to be identified with the reference-body specified by its $g_{\mu\nu}$. There is no need for more than one set of coordinate partitions, which are in any case unobservable; keeping the same geometrical framework, we obtain the required multiplicity of partial spaces by introducing N similar but independent reference-bodies. The distinction imposed by representing the particles of the object-system in different 'spaces' is then expressed equivalently and more intelligibly by connecting them with different reference-bodies.

The MS. ends here. At the end of the corresponding section of the one earlier draft this passage occurs:

The physical reference-body specified by the tensor $g_{\mu\nu}$ pervades all space and time. In the nineteenth century it was called the *aether*; nowadays it is known as the *field*. This rather ill-advised change of name was intended to emphasize its dissimilarity to matter; but in practice it has obscured the difference between the actual field-pervaded space of physics and the abstract spaces of pure geometry. It was much more necessary to emphasize its similarity to matter. The strictly mechanical properties (..omitting e.m. properties..) of both field and matter are described by second-rank tensors. Further progress, by abolishing the classical conception of particles and revealing 'wave properties' of matter comparable with those of the field, has bridged the gaps which remained.

Leaving aside e.m. characteristics, the field specified by $g_{\mu\nu}$ is called the *metrical field* or the *inertial field* according as we give precedence to positional or mechanical characteristics. The inertial field includes the gravitational field—a name referring more especially to its small deviations from uniformity. The mass (or energy) density of a distribution of matter is made up of one factor contributed by the distribution itself and one factor contributed by the inertial field, just as electrostatic energy eV is composed of factors contributed by the charged body and the field. The usual practice of attributing the mutual energy, whether inertial or electrostatic, wholly to the object-body is (as is well known) liable to lead to the energy being counted twice over; care must therefore be taken to make a proper apportionment between the body and the field.

CHAPTER 5

DRAFTS OF F I–V

(ii) CHAPTER II OF DRAFTS B AND C

5·0. Preliminary notes

After proceeding in Chapter I from the coordinate distributions of **F** §§1, 2 straight to the momentum distributions and mass calculations of **F** §§37–39, drafts **B** and **C** now cover in Chapter II the substance of **F** §§3–11, namely, the foundations of length measurement, curvature, 'pseudo-discrete states' and 'stabilization'. In **C** II (**5·1**), the first quotations here concern the dubious arguments of **F** §3, which are not materially strengthened. **C** §16 on pseudo-discrete wave functions is of interest; and the discussion in **C** §18 on Stabilization of Tensors is hardly represented in **F**.

The earlier draft **B** II (**5·2**) covers much the same ground, but progress after the first five sections is tentative. The earlier section **B** §1·9 (*b*) (**4·2** above) is here rewritten as **B** §2·7 (*a*), (*b*) to lead in two directions; the second introduces the 'widening' (phase) factor of **F** §24. There is an illuminating summary in **B** §2·9 of conceptions of the nature of degeneracy (multiplicity) factors.

5·1. Draft C *34* II. The Uncertainty of Scale

11. **The Bernoulli fluctuation:** close to **F** §3 and **D** §2. This begins:

In the last chapter we made use of a conception derived from molar relativity, the curvature or non-Euclidean metric of the space occupied by material systems. We shall now introduce curvature of space more directly by showing it arises out of the statistical fluctuations of a distribution of a large number of particles.

As in **F** (or **D**), the extraordinary (negative) fluctuation of $\zeta = \delta n/n_0$ (n the number of particles in volume V_0) is found to be Gaussian with s.d. $1/\sqrt{N}$. Then:

The fluctuation transforms an exact particle density s_0 into an uncertain density

$$s = s_0(1+\zeta). \qquad (11·41)$$

Instead of considering an uncertain number of particles n in a fixed volume V_0, we can consider an exact number of particles n_0 and transfer the uncertainty to the containing volume. The mathematical analysis for a system of n_0 particles occupying a volume V_0 is adapted to other values of V by a change of linear scale; thus the uncertainty is now contained in a linear scale factor $1+\epsilon$, defined by

$$V=V_0/(1+\epsilon)^3. \tag{11·42}$$

We have to find the connexion between ζ and ϵ. If we had merely to transform a distribution of discrete values of ζ into a distribution of discrete values of ϵ, the relation would be $n/n_0=V_0/V$, which yields $1+\zeta=(1+\epsilon)^3$. But in transforming a continuous distribution, discrete values of ζ and ϵ are replaced by constant ranges of ζ and ϵ (e.g. a range of one unit in the last decimal place retained), and we have to include a factor $d\epsilon/d\zeta$ transforming the constant ranges of ϵ into ranges which correspond to constant ranges of ζ. The relation is accordingly

$$1+\zeta=\text{const.}\times(1+\epsilon)^3\,d\epsilon/d\zeta,$$

which gives by integration

$$(1+\zeta)^2=(1+\epsilon)^4. \tag{11·43}$$

The argument then proceeds as in **F** (3·5–3·8), here (11·5–11·8). There is then a review of the approach to curvature and uncertainty in the present development.

There are alternative formulae for ds in spherical space according to the way in which r is defined. In (11·7), r, θ, ϕ are the ordinary polar coordinates of the orthogonal projection of the point in spherical space on the tangent flat space at the origin. These, and the equivalent rectangular coordinates, will be called *ortho-coordinates* of the point in spherical space.

To sum up: we began [in **C** I] with..particles..x_r, y_r, z_r in a flat space, and allowed for uncertainty of the origin but not..of scale. Knowing from relativity..the space must be curved, we later took x_r, y_r, z_r to be ortho-coordinates of particles in spherical space. On that basis we obtained (7·3) [**4·1**] for σ, but could give no decisive justification of the use of ortho-coordinates for the purpose. This difficulty has been met by deriving the curvature in a different way. We have found that the finiteness of the total number of particles introduces an extraordinary fluctuation equivalent to a (negative) uncertainty of scale. Scale uncertainty affects most strongly the positions of particles most distant from the origin; so that the relation between the physical and geometrical frames becomes non-uniform and also non-isotropic. Uniformity and isotropy are restored by changing to a representation in spherical space with x_r, y_r, z_r as ortho-coordinates, and the formula for σ is now obtained directly. The radius of the hypersphere is fixed by the limiting points where in the radial direction the negative uncertainty due to the scale just balances the positive uncertainty due to the physical origin, i.e. where $\sigma^2-r^2\sigma_\epsilon^2=0$.

12. **The standard of length:** close to **F** § 4.

...The standard of length must be *quantum-specified*....By establishing the identity of the quantum-specified metric with the σ-metric, we shall justify our earlier remark that the uncertainty (...the local uncertainty) of the physical reference frame 'puts the scale into' all that we construct in the physical frame....

13. **Non-uniform curvature of space:** like the second part of **F** §6, the first part having come earlier as **C** §7 [**4·1**].

The extraordinary fluctuation is best retained as an additional (scale) variate in wave mechanics, although its alternative representation by curvature provides, in the case of the uniform distribution, a link between wave mechanics and general relativity. The general-relativity representation of local irregularities by non-uniform curvature is not to be used in wave mechanics. *Two remarks*:

In wave mechanics generalized coordinates are decidedly objectionable, because it is no longer clear which variables represent the primitive observables.

It is the intermediate [uniform-curvature relativity] theory which links up with quantum theory—which, when fully developed, *is* in fact relativistic quantum theory.

14. **The extraneous standard:** covers much of **F** §8 (omitting 'scale-uncertainty'). The discussion is as in **2·1 S** §3.

We employ a 'natural' system of units such that

$$c=1, \quad 8\pi\kappa\hbar^2=1. \tag{14·1}$$

...The second is suggested by [**4·1**] (10·7)...We therefore retain one extraneous unit....

It may seem that we have now a redundancy of connexion between the various object-systems. But we are...unifying different branches of physics, in which what is ultimately the same connexion may appear in different guise. The physical frame with its uncertainty constant, the uranoid postulated as standard environment for each object in turn, and the extraneous standard are ways of expressing the relativistic principle that absolute scale is meaningless, and the scale of magnitude of the characteristics of a system necessarily involves reference to something outside itself. We could..avoid introducing an arbitrary extraneous standard as unit by comparing extensions in the object-system directly with the extensions σ or R_0 characteristic of the uranoid. But it is more straightforward to develop the theory in harmony with the practical outlook, which by employing an indifferent standard (quantum-specified to be available at any place and time) puts measurement of the uranoid on the same footing with measurement of an object-system. The extraneous standard is then an intermediary in a comparison of scale between an object-system and the uranoid just as it is an intermediary in the comparison of scale of one object-system with another.

15. **Scale-free physics:** close to **F** §9.

...Broadly speaking we only pass over into quantal physics [from scale-free quantum theory] when quantization of a.m. is introduced....

...a scale-free system..: if we specify the characteristics of a system numerically in terms of an extraneous standard, and consider the series of systems formed by varying the standard but keeping the numerical specification fixed, then if one system of the series is a physically possible system all are possible.

16. **Pseudo-discrete wave functions:** like the first part of **F** § 10. The long extracts throw light on the connexion with 'scale-free physics'.

Two kinds of wave function...in quantum theory. One..[e.g. hydrogen atom] specifies a concentrated..density....The 'normalized' density distribution $\rho_n(x, y, z)$ is such that the total mass is that of one particle; but the wave function may represent a density distribution $j\rho_n$. ..We call j the *occupation factor*...usually occupation by more than one particle is impossible.

The other kind..the 'infinite plane wave function' of elementary quantum theory....We..introduce an arbitrary normalization volume V_n,...unit occupation as 'one particle per volume V_n'. The normalized density ρ_n, corresponding to unit occupation, is then

$$\rho_n = mV_n^{-1}, \tag{16·1}$$

m the mass of a particle. The actual density can be any multiple $j\rho_n$, ..., j need not be less than 1.

In natural units (§ 14), V^{-1} is a mass. Let

$$V_n^{-1} = Cm', \tag{16·2}$$

where C is numerical...Then

$$\rho_n = Cmm'. \tag{16·3}$$

In later developments m' will appear as the mass of a 'comparison particle' which, besides fixing the normalization volume, has a general importance (§ 29, [**7·1**]).

Wave functions of this kind are *pseudo-discrete*, in contrast to the first kind which are discrete. A set of discrete eigenfunctions has the form $\psi(x, y, z; \alpha)$, α a set of parameters $\alpha_1, ..., \alpha_k$ defining constant characteristics of the eigenstate. When the eigenstates are continuous,...we may use *continuous wave functions* as in [**4·1**] § 4, treating the parameters (..usually the momenta) as well as the coordinates as continuous arguments of ψ....Alternatively, we divide the continuum of α into small ranges $\delta\alpha = \delta\alpha_1 ... \delta\alpha_k$ and represent the density associated with each range by a separate wave function ψ_α with an occupation factor j_α. These are the pseudo-discrete wave functions.

...continuous and pseudo-discrete wave functions belong to different methods of wave analysis. In (4·22) [**4·1**] the continuous p.d. over p was represented by a *wave function* $F(p)$; in the pseudo-discrete method we consider the *distribution function* of p, and represent it by occupation factors j_p of small ranges δp. In particular, for different values of p the wave represented by $F(p)$ are in definite phase relation and interfere systematically, whereas the waves represented by ψ_p for different values of p are in incoherent phase, and the corresponding densities are simply additive.... Quantum theory,.. for homogeneity, extends the representation of p.d.'s by wave functions to problems which.. individually are better treated by the ordinary statistical theory of distribution functions. We proceed the opposite way, using distribution functions as far as possible, and factorizing them into wave functions only when there is a definite object to be gained.

In unit occupation a pseudo-discrete wave function represents a uniform distribution of one particle per volume V over an extensive region... a single particle with uniform p.d. over a volume V.. would be represented by a wave packet (a continuous wave function) and its occupation of the volume would be only momentary, since a wave packet diffuses. Pseudo-discrete wave functions are not applicable to individual particles; they represent an indefinitely large number spread uniformly (apart from fluctuations) over an indefinitely large region. For brevity we often speak of the pseudo-discrete wave function of a particle; but.. the particle is.. a representative.. or.. *an unidentified particle in a large assemblage*.... Primarily the pseudo-discrete wave function represents a molar system.. but as an assemblage of particles... all having the same p.d. of characteristics....

Pseudo-discrete wave functions belong to scale-free physics. A set of wave functions ψ_α^v with occupation factors j_α^v represents a mixture in certain proportions of particles of various kinds v in various states α; but the absolute number of particles in any volume depends on the arbitrarily chosen normalization volume.... We can develop much of the wave mechanics of atomically constituted matter before introducing the characteristics which determine the actual size of the particles... we employ 'scale-free particles' whose masses are indeterminate... but.. mass ratios determinate. When finally we pass into quantal (or into cosmical) physics the scale becomes fixed; and the arbitrary normalization volume is replaced by a natural normalization volume V_0, or more conveniently the equivalent mass m' in (16·2) by a fixed mass m_0. The general results of scale-free physics have then a particular scale of application... in which the pseudo-discrete wave functions combine homogeneously with the discrete self-normalizing wave functions characteristic of quantal physics.

17. **Stabilized characteristics:** like **F** §11. To summarize: in an experiment on an electron the mass may be taken from a table of constants—it then has no uncertainty and is 'stabilized'. Stabilization can reduce V_4 to V_3 particles. The environment also may be highly 'stabilized'.

...In practice, if we had found a mass of, say $(8{\cdot}8 \pm 0{\cdot}4) \times 10^{-28}$ g., we should jump to the conclusion that the particle was an electron or positron, and substitute the accepted $9{\cdot}107 \times 10^{-28}$ g. for our own more uncertain determination... *stabilizing* the mass. When an observable is stabilized, its conjugate coordinate or momentum disappears from the analysis. As an observational determination of m becomes more exact, the conjugate coordinate s_0 becomes more uncertain, and in the limit the wave function contains a periodic factor $e^{ims_0/\hbar}$ representing uniform p.d. over all values of s_0; but when m is given an exact value by stabilization, the factor $e^{ims_0/\hbar}$ is dropped. We no longer regard s_0 as having classificatory significance, and adapt our system of wave analysis accordingly. ...*A V_3 particle is derived from a V_4 by taking the mass constant m to be stabilized instead of observable.*

We shall find...it important to define whether in a problem..a particular variate is an observable or a stabilized characteristic. The distinction is commonly overlooked because a stabilized characteristic is observable in the literal sense, i.e. it could be observed if we chose. But in quantum theory an 'observable' is effectively defined as a quantity to which the uncertainty principle applies, and does not include quantities furnished as free information.

It may be suggested that the admission of stabilized characteristics in current quantum theory...conflicts with its declared intentions; and that the right course is to amend the theory, replacing all such characteristics by genuine observables. But...this would not improve the efficiency of the theory. Stabilization is one of the most valuable features of the method of quantum theory, and is used in a great variety of ways....

18. **Stabilization of tensors:** represented in **F** by one paragraph of §11. First (as in **F**) a tensor may be 'stabilized' by conditions of symmetry or factorizability. Then:

Regarding a vector as a combination of magnitude and orientation, we can stabilize the magnitude without the orientation, or vice versa. This form is applied to the momentum $p_1, ..., p_4$ when we stabilize m.... To extend this to general tensors we resolve a tensor into two factors, analogous to magnitude and orientation, in the following way:

Associated with a tensor T there are one or more invariant functions $I_1, I_2, ...$ of its components. Selecting one, say I_1, we define a *unit tensor* to be a tensor with $I_1 = 1$. Then any tensor for which $I_1 \neq 0$ can be expressed as $T = \lambda T_0$, where T_0 is a unit tensor and λ is an invariant. We call λ the 'magnitude' and T_0 the 'orientation' of T.

The separation of an object-system from its environment is a matter not of observation but of convention. Physically they are one system, and there is no natural line of demarcation. Therefore, when we take them apart, the additional variates which define the boundary of separation are not really observables though they play the part of observables in our picture of the two systems as separated. These variates are accordingly stabilized characteristics. Each of the two partial systems must contain at least one genuine observable, since otherwise there would be no physical meaning in isolating it. The most elementary kind of separation is to include an observable magnitude in one system and an observable orientation in the other. This gives us a uranoid with only one genuine observable σ, and an object-system characterized by a tensor with observable orientation but stabilized magnitude.

5·2. Draft B *32* II. The Uncertainty of Scale

(See *Preliminary notes*, **5·0**.)

2·1. **The Bernoulli fluctuation:** close to **5·1** §11 (**F** §3). In the paragraph ending with (11·43) in **5·1** §11, there is this variation:

If we had merely to transform an exact value of ζ into an exact value of ϵ, the relation would be $n/n_0 = V_0/V$, which yields $1+\zeta=(1+\epsilon)^3$. But in transforming a continuous distribution exact values must be replaced by *constant* ranges of ζ and ϵ. We require a transformation which gives the same number of particles, not in constant, but in *corresponding* ranges of $d\zeta$, $d\epsilon$, and have therefore to include a factor $d\epsilon/d\zeta$....

2·2. **The standard of length:** mostly close to **5·1** §12 (**F** §4). The opening paragraphs (related to **F** §3) illuminate the 'σ-metric':

We set $x-\xi=x_0$, etc., as in (1·11), so that the s.d.'s in (2·162) [radial $\sigma(1-r^2/R_0^2)^{\frac{1}{2}}$, transverse σ; namely, **F** (3·6) with $\sigma/\sigma_\epsilon = R_0$] refer to the dispersion of x_0, y_0, z_0 about the mean value zero....This local uncertainty of the physical frame...has been derived as a combination of uncertainty of scale with the uncertainty of position of a distant origin; but x_0, y_0, z_0 can be interpreted more compactly as the coordinates of a local physical origin referred to a local geometrical origin. By (2·162) the dispersion of the local physical origin is unsymmetrical in the orthogonal coordinates employed. But the asymmetry could be removed by a local transformation of coordinates; and, when pictured in spherical space instead of in projection, it becomes symmetrical.

Independently of coordinate systems, the local s.d...defines..a unit for measuring lengths...the σ-metric....There is no metric [other than the quantum or σ-metric] that is universally defined. Accordingly in §2·1 we could not do otherwise than measure the length ds in a unit having a constant ratio to the local s.d. That yielded the formula $\sigma = R_0/2\sqrt{N}$, showing that space is a hypersphere of radius $R_0 = 2\sigma\sqrt{N}$....

2·3. **Non-uniform curvature of space:** close to **5·1** §13 (**F** §6, second part). The following 'aside' contains a forcible footnote on the '*classical electron*'.

...non-uniformly curved space has no *locus standi* in quantum theory [cf. **F** §6]....It may be helpful to recall that the local curvatures are introduced in general relativity, first by creating singularities in the metric which constitute pseudo-particles, and then averaging the pseudo-particles into continuous matter. I use the term 'pseudo-particles' for these ideal singularities to distinguish them from the 'particles' recognized in quantum theory, which result from an altogether different analysis of continuous matter. (*Footnote*. A great deal has been written about the 'classical electron', defined as the limit to which an actual electron tends if we make $h \to 0$ in the quantum equations.

It seems commonly supposed this limit would be a field-singularity of the type we have called a pseudo-particle. This is not true; for the limit is nonsensical, and there cannot be a classical electron—a fact which explains the failure of the many attempts to construct one. The reason is that $hc/2\pi e^2$ is identically 137; so that if $h \to 0$, $e \to 0$. The introduction of 'atomicity' into matter and radiation, which marks the transition from classical to quantum theory, is a single step; and to introduce atomicity of charge without atomicity of quanta naturally leads to glaring inconsistencies.) The local curvatures are..alien to..quantum theory...General relativity retains only one link with quantum theory, namely, that in empty (not necessarily uniform) space-time the invariant ds is measured with a quantum-specified standard.

2·4. **The extraneous standard:** covers **5·1** §§14, 15 (**F** §§8, 9). These quotations show the scope, and supplement later forms.

The 'natural system' of units..is such that

$$c = 1, \quad 8\pi\kappa\beta^2 = 1 \quad [\beta \text{ for } \hbar]. \tag{2·41}$$

...By eliminating superfluous standards we describe the internal structure of a system wholly in terms of angular variables or other numerical ratios....To fix the scale..we retain *one* extraneous unit....The single standard, external to all the systems, is the link which reconnects..so that we can put the results together in their proper scale relation.

In our theoretical investigations the connexion...is provided by the uranoid, which is postulated as the environment of each [object-system] in turn. The extraneous standard is the practical substitute for the uranoid....In the system of natural units we leave the extraneous standard entirely arbitrary. It suggests itself that the units would be still more 'natural' if we adopted as extraneous unit one of the constants σ, R_0 or ϖ of the uranoid or the mass of a proton, electron, etc. One reason for refraining is that by choosing an *important* entity as standard, we make it a thing apart....The..extraneous standard is to serve as intermediary equally in a comparison of scale of one object-system with another and in a comparison of scale of an object-system with the uranoid. We adopt then an arbitrary standard such as the centimetre....

...Scale-free, quantal, cosmical physics...

...the dimensions of certain quantities in natural units....

2·5. **Pseudo-discrete wave functions** (originally 'Occupation factors'): broadly like **5·1** §16 (**F** §10); the later part is quoted. The last paragraph is a survey of the present drafts. After **5·1** (16·3):

The concentrated type of wave function may be described as self-normalizing...the 'infinite' type can be normalized only by reference to an extraneously introduced volume V or mass m'....The first type is not scale-free, since the s.d. of the distribution is a fixed length independent of j [occupation]...the wave functions of scale-free physics are limited to the second type, where the vague extent of the distribution allows us to vary the linear scale freely.

Analytically..wave functions of the first type are *discrete*, of the second..*pseudo-discrete*. An eigenfunction has the form $\psi(x, y, z; \alpha)$,.. α a set $\alpha_1, \ldots, \alpha_k$ of parameters....When the eigenstates are continuous, we can use *continuous* wave functions such that $|\psi|^2 dx dy dz d\alpha$ is the probability in $dx dy dz$ corresponding to a range $d\alpha$....But it is more convenient to use pseudo-discrete wave functions, each corresponding to a small range $\delta\alpha$, which are treated as though they were discrete and determine a probability $|\psi|^2 dx dy dz$ in $dx dy dz$. Owing to the arbitrary choice of $\delta\alpha$, there is an indefiniteness in the absolute amount of probability associated with a pseudo-discrete wave function, which is removed by fixing a normalization volume. Choosing a normalization volume V is in fact a substitute for choosing a normalization range $\delta\alpha$. The number of dimensions k of the element $\delta\alpha$ is called the *multiplicity factor* of the pseudo-discrete state...it appears in many formulae in scale-free physics....

From a physical standpoint..self-normalizing wave functions refer to a single particle or system, scale-free wave functions to an *ensemble* of particles or systems....

...Primarily the pseudo-discrete wave function represents a molar system..as an assemblage of particles..and thereby takes the first step in the transition from molar to particle theory. The connexion of discrete wave functions with molar physics is not so simple. The process of quantization, which introduces them, will be assumed familiar; but ..another link..is less well known. Wave functions of simple particles, e.g. protons and electrons, do not admit of quantization, and we have to introduce correlation wave functions (which may or may not be quantized) before any examples of discrete wave functions occur. We must therefore, as an intermediate step, examine the connexion between pseudo-discrete simple and pseudo-discrete correlation wave functions.

We can now outline a chain of connexion which forms, as it were, the **trunk road of relativistic quantum theory.** The whole route will be traversed in a preliminary way in Chapter III. First, we must ascertain how far the characteristics of molar matter contained in its energy tensor are represented by pseudo-discrete (scale-free) wave functions. Secondly, we must find the relation between simple (or 'external') wave functions and correlation (or 'internal') wave functions. Thirdly, we must study the interaction of the latter with the aether, which results in radiation. Radiation brings us back again to molar physics, but at a different point from that at which we started.

2·6–2·9. **General Note.** Draft *32* appeared to end with §2·5; but the following partial draft was recognized to be a continuation. The plan is less coherent, but there are details of interest.

2·6 (originally 2·5). **Stabilized characteristics:** close to **5·1** §§17, 18. 'Stabilization' of mass and environment is as **5·1** §17. After the opening part of **5·1** §18 on tensors, there is this variant:

Associated with a tensor T there are..invariant functions $I_1, I_2, \ldots$ of the components, any one of which can be regarded as a measure of the magnitude of T. Selecting one, say I_1, we define a *normalized* tensor to be one with $I_1 = 1$. Then..

$$T = \lambda T_0, \tag{2·51}$$

where T_0 is normalized and λ is invariant. We call λ the *extensor* and T_0 the *rotor* of T. The usual stabilization..is of the extensor.

The standard uranoid results from..stabilizing the rotor....In field theory the environment is specified by the tensor $g_{\mu\nu}$. Setting $g_{\mu\nu} = \lambda(g_{\mu\nu})_0$ as in (2·51), so that

$$ds^2 = g_{\mu\nu}\, dx_\mu\, dx_\nu = \lambda(g_{\mu\nu})_0\, dx_\mu\, dx_\nu,$$

we give the rotor $(g_{\mu\nu})_0$ stabilized Galilean values $\delta_{\mu\nu}$. Then the extensor λ determines the extension of a coordinate mesh in terms of the extension of the standard of length. We do not stabilize this. Being a genuine observable, λ has an uncertain value, and consequently the physical measures contain an uncertain scale factor (compared with the geometrical coordinate frame), which has been investigated in § 2·1.

...Stabilization of identity of particles will be explained in §   .

2·7 (*a*) (originally 2·6). **The time-coordinate:** based on **4·2** § 1·9 (*b*) above. The effect of a quasi-time coordinate on the uranoidal constant A of **4·2** § 1·8 is discussed.

...A distribution function $f(x, y, z; t)$ gives the p.d. *over* x, y, z *at* the time t. If, however, a particle contains a time indicator, e.g. the phase θ of an internal vibration, we can consider the p.d. over θ along with the distribution over x, y, z at the moment t. This is an effective substitute, since it does in fact determine a p.d. of events over t, the 'event' being the passage of the particle through zero phase. But a time indicator carried by a particle registers *proper time*. Thus θ, though in a sense deputizing for t, has not the same relativistic relations as the coordinate t. We shall consider formally the theory which results from including such a phase coordinate.

The elementary wave function

$$a \exp\{i(p_1 x + p_2 y + p_3 z + p_4 t)/\beta\} \tag{2·61}$$

can also be written as $a\, e^{ims/\beta}$, where m is the proper mass and s the proper time recorded by a time indicator carried by a particle with momentum vector p_1, p_2, p_3, p_4. The coefficient a is not necessarily real; let it be $|a|\, e^{i\theta}$, and set $\theta = ms_0/\beta$. Then (2·61) becomes $|a| \exp\{im(s_0 + s)/\beta\}$, or

$$|a| \exp\{i(p_1 x + p_2 y + p_3 z + ms_0 + p_4 t)/\beta\}, \tag{2·62}$$

and s_0 is the initial phase at the proper time $s = 0$, expressed in a linear measure homologous with x, y, z.

The usual procedure is to treat m and s_0 as stabilized; but we now take them to be conjugate observables with p.d.'s satisfying the uncertainty relation. The phase s_0 must be measured from a physically defined zero; assigning the zero-point the same uncertainty σ as the physical origin of the space coordinates, we can extend (1·62) [**4·2**] symmetrically to the new dimension m, s_0. The weight function for the element $dp_1 dp_2 dp_3 dm$ is accordingly

$$w = (2\pi\varpi^2)^{-2} \exp\{-(p_1^2+p_2^2+p_3^2+m^2)/2\varpi^2\}$$
$$= (2\pi\varpi^2)^{-2} \exp(-p_4^2/2\varpi^2). \quad (2\cdot63)$$

In place of (1·65, 1·66) [**4·2**], we shall have for the particles of the infinite-temperature uranoid

$$\overline{E^2} = \overline{p_4^2} = 2\varpi^2 = \beta^2/2\sigma^2, \quad (2\cdot641)$$

$$\overline{p_1^2} = \tfrac{1}{2}\varpi^2 = \beta^2/8\sigma^2. \quad (2\cdot642)$$

The reduction to a zero-temperature uranoid, made as in §1·8, is found to involve a change of λ from $\frac{7}{5}R_0^{-2}$ to R_0^{-2}; and the constant A is found to be $\frac{4}{5}/8\pi\kappa\beta^2N^2$. In place of (1·88) we obtain

$$m_0^2 = \tfrac{5}{2}N\beta^2/R_0^2. \quad (2\cdot65)$$

..m_0 is not the mean value of m. The mass m_0 is still largely provided by the gravitational energy introduced by the reduction to zero temperature, though m now makes a contribution.

These differences may be attributed to the fact that, by altering the 'conventional..probing' to include the measurement of phase, we have substituted for the V_4 particles a new kind with different intrinsic properties (§1·3). We call the new particles V_5 particles. The difference between V_5 and V_4 particles is exhibited by the comparison

$$(m_0^2)_5 = \tfrac{5}{4}(m_0^2)_4, \quad A_5 = \tfrac{4}{5}A_4. \quad (2\cdot66)$$

By admitting a p.d. of s_0 we abrogate the convention of wave mechanics, that the phase of a wave function (at given x, y, z, t), though unknown, is discrete. We have done this to see how the convention has changed the course of development of the theory—not with any intention of permanently discarding it. Two conclusions are reached. First, the extension of wave mechanics to four dimensions is on altogether different lines from the extension of 3-space to 4-dimensional space-time. Secondly, it appears from (2·66) that the conventional limitation to three dimensions introduces a factor $\frac{4}{5}$ in certain fundamental constants. When this factor turns up rather oddly..later, we shall realize it is a consequence of employing stabilized instead of measured masses of the particles.

2·7 (*b*). **Four-dimensional theory:** this begins like §2·7 (*a*), but is confined to the theory of the internal phase θ, and the effect of its limitation as an angle to $(0, 2\pi)$; compare **F** §24. After the first two sentences quoted from §2·7 (*a*),

We have then a four-dimensional distribution $f(x, y, z, \theta; t)$. But a time indicator...registers proper time. Thus θ..has not the relativistic relations of the coordinate t.

An angular coordinate can be defined either as an angle θ' capable of all values from $-\infty$ to ∞, or as the principal value θ lying between $-\pi$

and π. The term *phase* implies the latter... When θ' is a time indicator, there can be no steady p.d. of θ'; but the distribution of phase θ can be statistically steady provided that it is uniform. Since we are at present occupied with equilibrium theory we adopt θ, not θ', as the extra coordinate... this introduces a distinction between the fourth coordinate and the flat rectangular coordinates, which would be represented geometrically by rolling the four-space into a cylinder with θ as the circumferential coordinate—thus making the points $\theta' = 0, 2\pi, 4\pi, \ldots$ actually coincident.

This restriction of angular variables to principal values is the basis of quantization, and also the key to the procedure of stabilization. It introduces a change in the basis of statistical enumeration by suppressing purely analytical distinctions between configurations which have no observational distinction. When the conjugate momentum p_θ becomes exact, the coordinate θ' becomes dispersed over an infinite range; but the infinitude no longer causes a breakdown of statistical enumeration, because the *physical* configurations depend only on θ, which becomes uniformly distributed over the finite range $-\pi$ to π. The difference between a stabilized and a highly accurate observed value of p_θ is that the former is treated as a constant having no conjugate coordinate, whereas the latter requires that we uniformly 'widen' the distribution (over the other coordinates) to a thickness 2π in an extra dimension. From the widened distribution we can pass continuously to the case in which p_θ is appreciably uncertain and the distribution in θ is correspondingly concentrated.

In measuring an object we employ a physical origin and an extraneous standard,... both subject to uncertainty... In § 2·1 we combined the two uncertainties, and thereby obtained a curvature of space. This establishes a connexion with general relativity; but it is not a convenient representation for comparison with wave mechanics which postulates flat space. If the scale uncertainty is not absorbed into the reference frame as curvature, we must provide for it in the object-system, i.e. the wave function $f(x, y, z; t)$ which represents the object-system on scale 1 must be repeated on all possible linear scales $1+\epsilon$, and the distribution function of ϵ is part of the specification of the object-system. This gives an extra dimension to the p.d.... so that there will be an additional coordinate and momentum.

The 'scale', defined less restrictedly, is an arbitrary power of the linear scale $1+\epsilon$, being measured by the ratio of an invariant of the system (the extensor of a tensor) to an extraneous standard of the same dimensions. Choosing the latter of dimensions of a.m. (in natural units $(\text{mass})^{\frac{2}{3}}$), the scale is specified by the invariant $m^{\frac{2}{3}}$ of the object-particle which is proportional to $(1+\epsilon)^{-2}$. The coordinate conjugate to the a.m. $m^{\frac{2}{3}}$ is an angle θ. The object-system is therefore represented in flat space by a wave function $f(x, y, z, \theta; t)$ giving a p.d. over four coordinates one of which is phase. It is evident.. that θ is separable from the other three coordinates.

The scale, conjugate to a separable phase, can now be stabilized, and we reach the usual three-dimensional wave mechanics of a particle with wave function $f(x, y, z; t)$ in flat space and stabilized mass-constant m. The result may be compared with that obtained by including the scale uncertainty as a curvature in the reference system, and then limiting the system to a region small enough to be treated as flat. In both cases the flat three-dimensional representation results from treating the scale uncertainty as negligibly small; but they differ significantly because we proceed to the limit in different ways.

2·9. **Degeneracy factors:** after an odd page from a missing §2·8 related to **F** §15, this section summarizes views of the multiplicity k; cf. **F** §§15, 16.

The number k of dimensions of the phase space over which the recognized states of the system are continuously distributed, will be called the *degeneracy factor*. Owing to the variety of methods..of physics, we often come across the same thing expressed in different ways, and we here review the various aspects in which the degeneracy factor presents itself.

(1) It represents a *marginal effect*, or 'law of diminishing returns'. Owing to the required adjustment of the self-consistent field the energy is not a linear function of the occupation factors. Thus the last element of probability dj to be added contributes to the energy in a proportion different from the mean. Disregarding..the difference of sign between $E_{\mu\nu}$ and $T_{\mu\nu}$, and regarding the continuous j as the limit of discrete j_r, the result (2·81) [?**F** (15·7)] is equivalent to saying that $T_{\mu\nu}$ is a homogeneous function of degree k of the j_r, so that the marginal effect is k times the mean effect.

(2) The marginal effect can also be regarded as an *exclusion effect*. When the eigenstates are discrete, late arrivals are forced up into higher levels by exclusion, so that their energy is greater than the mean of those already present. When the eigenstates are too close-packed to be treated separately, the exclusion effect is represented by increased energy of the added particles depending on the extent to which the block of states is already occupied. It is necessary to suppose that the blocks are already highly occupied in order that the changes of occupation which we study may be relatively small marginal changes. This is the origin of Dirac's picture of an infinity of occupied states below the threshold of apprehensible matter..often useful, but artificial, in that it represents $E_{\mu\nu}$ and $T_{\mu\nu}$ as having the same sign.

(3) The degeneracy factor measures the *recoil* of the rest of the universe, when the motion of the particle is changed. This is obvious in the case of the momentum components, since by ( ) [**F** (15·7)] a momentum density $E_{\mu 4}$ of the particle involves a momentum density $-(k+1)E_{\mu 4}$ in the field. This shows incidentally that the minus sign in ( ) is reasonable. The conception of recoil is extended to the other components by analogy....

(4) The degeneracy factor is a transformation factor arising from a change in the selection of primitive observables. In particular, it represents an economy in the use of the extraneous standard.

(5) It is simply a degeneracy factor of the type familiar in the more technical developments of quantum theory but not ordinarily considered in connexion with its foundations.

CHAPTER 6

DRAFTS OF F I–V

(iii) CHAPTER III OF DRAFTS B AND C

6·0. Preliminary notes

These draft Chapters III cover the main points of **F** II, namely, the division of total energy between particle and field, and the determination from this of the mass ratio of the proton and electron. We again begin with the more polished draft **C** and then proceed to **B**. The latter is represented in **6·2–6·4** by two complete versions, MSS. *26* and the earlier *24*, and by a substantial fragment *31*. The **B** chapters end with a discussion of *radiant energy* which is not in later versions.

6·1. Draft C *27* III. Multiplicity Factors

(Originally headed 'Fields and Particles', with a fresh heading 'Chapter IV Multiplicity Factors' at §25.) The content is most of **F** §§12–21, 40, i.e. multiplicity factors and the mass ratio, and also a continuation of the discussion in **C** I (**4·1** §10 above) of the mass of the primary (uranoid) particle.

19. **Object-fields:** like **F** §12.

...three-fold division of the universe into object-particles, field and uranoid....If the field is grouped with the uranoid as in general relativity ...the standard uranoid is distinguished as the inertial part, and the disturbance as the gravitational part of the unified inertial-gravitational field represented by $g_{\mu\nu}$. The other grouping gives an object-system and object-field with the standard uranoid as environment...corresponding to Newtonian and quantum theory....

We distinguish between the *complementary field* of the object-particles and an *extraneous field*....

20. **The rigid-field convention:** like **F** §§13, 14. The 'rigid' or 'self-consistent' field of quantum particles is discussed as in **F** §13, and the separation of field and particle energy for discrete states much as in **F** §14. Some amplifications and variants from this latter part:

...First..eigenstates discrete..eigenfunctions ψ_r..with occupation factors j_r..regarded as generalized coordinates. The total energy..H^0. Ordinarily we regard H^0 as the sum of contributions from the different states weighted according to their degree of occupation, so that

$$H^0 = \Sigma j_r H_r, \qquad (20{\cdot}1)$$

where H_r is the energy corresponding to unit occupation of the state ψ_r, or briefly the energy (hamiltonian) of the wave function ψ_r. This does not mean that H^0 is a linear function of the j_r; for the H_r must be calculated in the self-consistent field corresponding to the actual state of occupation, and will therefore themselves be functions of the j_r. This non-linearity, due to the continual readjustment of the self-consistent field and the eigenfunctions calculated in it, is familiar in the calculation of energy levels of complex atoms. The present theory, however, relates to molar (complementary) fields, and the parallel with intra-atomic fields should not be pressed too far.

Thus far we have not separated the energy of the particles from the energy of the field..we must introduce the stationary condition....Let

$$E_r = \partial H^0 / \partial j_r, \tag{20·2}$$

so

$$dH^0 = \Sigma E_r dj_r, \tag{20·3}$$

and let

$$E^0 = \Sigma j_r E_r. \tag{20·4}$$

Then the energy of a particle...is E_r, the total energy of particles..is E^0, and of field..

$$W^0 = H^0 - E^0, \tag{20·5}$$

so

$$dW^0 = -\Sigma j_r dE_r. \tag{20·6}$$

For large changes of occupation E_r and W^0 must be treated as functions of the j_r given by (20·2, 20·5). But, having calculated them for an initial j_r, we can treat them as constant for small changes dj_r, and dH^0 will then be given correctly...The energy E_r of a particle in an eigenstate is not the energy H_r of the eigenstate; the latter includes a share of the complementary field energy.

...Any additive characteristic H^0, e.g. momentum, pressure...must be apportioned..as

$$H^0 = \Sigma j_r E_r + W^0, \quad E_r = \partial H^0 / \partial j_r. \tag{20·7}$$

This allows an infinitesimal flexibility of occupation with E_r and W^0 kept constant....It follows that the complementary field has mechanical properties similar to those of the particle system, being in fact a dump for the residue of the energy, momentum, etc., which it is impossible or inexpedient to assign to the particles individually. 'Rigidity' means that the dump is not being added to or drawn from; so that, after being taken into account in the initial calculation of the framework of the eigenstates, the field is a passive element in the problem.

The argument proceeds as **F** § 14 to the end, with final comments:

Condition (1) [**F** § 14, end] points to the pseudo-discrete wave functions which represent unidentified systems in a large assemblage as the fertile field of application. They correspond to..continuous eigenstates (..next section). For intra-atomic and intra-nuclear fields, where the wave functions usually..cannot be occupied by more than one particle, the limitation to small changes remains; and the method, as it stands, is on that account infertile.

21. **The rigid field in scale-free systems:** like **F** § 15, stopping at (15·52). After an introduction as in **F**, the definitions are expanded:

. . .$j(X)\,d\tau$ is the probability associated with the states within $d\tau$. The consequent modifications of the formulae of § 20 are. . .

$$H^0 = \int H(X)\,j(X)\,d\tau, \tag{21·1}$$

where $H(X)$ is the energy[a] of the state X_α (distinguished from the energy $E(X)$ of a particle in that state). . .

$$E(X) = \eta H^0/\eta j, \tag{21·2}$$

$$E^0 = \int E(X)\,j(X)\,d\tau, \tag{21·3}$$

where η denotes hamiltonian differentiation. The meaning of (21·2) is that $E(X)$ is obtained by expressing the change of H^0 due to arbitrary small variations of the function $j(X)$ in the form

$$\delta H^0 = \int E(X)\,\delta\{j(X)\}\,d\tau, \tag{21·4}$$

which corresponds to (20·3). By (21·3)

$$\delta E^0 = \int E\,\delta j\,d\tau + \int j\,\delta(E\,d\tau) = \delta H^0 + \int j\,\delta(E\,d\tau)$$

by (21·4). Hence $$\delta W^0 = -\int j\,\delta(E\,d\tau), \tag{21·5}$$

corresponding to (20·6).

Let l be the dimension index of H^0. . .and so (by the equations) of $H(X)$, $E(X)$, E^0, W^0. Then the index of $E\,d\tau$ is $l+k$. Since the system is scale-free, we can take for the variation δ in (21·5) an infinitesimal change of scale $X_\alpha \to (1+\epsilon)\,X_\alpha$. Then $W^0 \to W^0(1+\epsilon)^l$ and $E\,d\tau \to E\,d\tau(1+\epsilon)^{l+k}$; so that $\delta W^0 = l\epsilon W^0$, $\delta(E\,d\tau) = (l+k)\,\epsilon E\,d\tau$. Hence by (21·5)

$$l\epsilon W^0 = -(l+k)\,\epsilon \int jE\,d\tau = -(l+k)\,\epsilon E^0$$

by (21·3). Hence $$W^0 = -\left(1+\frac{k}{l}\right)E^0, \quad H^0 = -\frac{k}{l}E^0. \tag{21·6}$$ [**F** (15·51)]

By. .dividing phase space into small numbered cells $(d\tau)_r$, we can replace the continuum of states by pseudo-discrete states (§ 16) and the continuous occupation function by pseudo-discrete occupation factors $j_r = j(X_r)\,(d\tau)_r$. Comparing (21·6) with (20·8) [**F** (14·6)] we see that:

The scale-free condition makes H^0 a homogeneous function of degree $-l/k$ of the pseudo-discrete occupation factors. [**F** (15·52)]

[a] In MS. **B** *26* § 3·2 (**6·2**) we have '$H(X)$ corresponds to unit occupation of the state X_α'.

We shall be chiefly concerned with the elementary case in which (initially) the whole probability is concentrated in one pseudo-discrete state. Its characteristics X_α are then..'almost exact'..e.g. almost exact rest. In almost exact conditions

$$H^0 \propto j^{-1/k}. \tag{21·7}$$

This shows the great importance of distinguishing between observables and stabilized characteristics. For if stabilizing conditions are imposed on the characteristics X_α, the number k of dimensions of the phase space ..is reduced.

22. **Partition of the energy tensor:** like **F** §15 (end) and §18 (beginning). This is clearer, less compressed, than **F** and is quoted almost complete.

Relativity mechanics is based on the energy tensor $T_{\mu\nu}$, which includes the density..T_{44}. In microscopic theory the molar density is..the sum of contributions from a great number of particles...Usually there is no information as to the location of any particular particle. Thus each of the particles into which a molar object is divided has a p.d. extending throughout the object, and contributes to the energy tensor at every part of it.

The energy tensor is a scale-free characteristic; for the density of a particle can be varied by varying the volume over which its p.d. extends. On the other hand, the mass of a particle is not scale-free; it has a fixed relation to σ. Thus[a] the density and energy tensor are the primary characteristics of a particle in scale-free physics...we shall..determine the ratio of the density of the positively..to the negatively charged material in molar hydrogen...by scale-free physics; this..is more commonly known as the ratio of the *masses* of the proton and electron.

...To include the later 'complete energy tensor', we define an 'energy tensor' as a tensor which includes the density....The vectors and tensors in the present theory are of the elementary type associated with orthogonal axes...Accordingly the components are of homogeneous physical dimensions. It is convenient to take the extraneous standard to be a density, so that the dimension index of $T_{\mu\nu}$ is 1.

The simplest system in scale-free physics has an energy tensor but no other characteristics...We call this a *bi-particle*..'the conceptual carrier of an element of the energy tensor' (...Since the state is pseudo-discrete, a bi-particle can (strictly) exist only as an unidentified member of a large assemblage of bi-particles).

Consider as object-system an assemblage of bi-particles together with the necessary complementary field. We apply §21 to apportion $T_{\mu\nu}$.. between particles and field. We put $H^0 = T_{\mu\nu}$. Also, since there are no other characteristics, we have to use $T_{\mu\nu}$ as the set of classificatory

[a] See the first quotations from **B** *26* § 3·3 in **6·2** below.

characteristics X_α. Since H^0 and X_α are now identical the dimension index l is 1; and (21·6) [**F** (15·51)] gives

$$T_{\mu\nu}=-kE_{\mu\nu},\quad W_{\mu\nu}=-(k+1)E_{\mu\nu}. \quad (22\cdot 1)\ [\mathbf{F}\ (15\cdot 7)]$$

..k, the number of independent components..is the *multiplicity factor*.

In an almost exact state...

$$T_{\mu\nu}=j^{-1/k}(T_{\mu\nu})_1 \quad (22\cdot 2)$$

by (21·7), $(T_{\mu\nu})_1$ being the value for unit occupation. By (22·1), $E_{\mu\nu}$ and $W_{\mu\nu}$ vary in the same way.

To apply rigid-field treatment, we..partition the initial $(T_{\mu\nu})_0$ into $(E_{\mu\nu})_0$ and $(W_{\mu\nu})_0$ by (22·1). (Previously T, E, W have been energy tensors of the whole assemblage, but we now define them as energy tensors per particle. They will be referred to as the *total*, *particle* and *field* energy tensors of a particle.) Since $W_{\mu\nu}$ is unaltered by transitions,

$$(T_{\mu\nu})_0=-k(E_{\mu\nu})_0\,(\text{initial}),\quad \delta T_{\mu\nu}=\delta E_{\mu\nu}\,(\text{transition energy}), \quad (22\cdot 3)\ [\mathbf{F}(15\cdot 8)]$$

so that for all states

$$T_{\mu\nu}=(T_{\mu\nu})_0+\delta T_{\mu\nu}=-k(E_{\mu\nu})_0+\delta E_{\mu\nu}. \quad (22\cdot 4)$$

In this..the particle is an unidentified member of a large assemblage, of which all but a small proportion are in the initial state.

The classification of states in phase space is determined by $T_{\mu\nu}$. An equivalent classification, generally more convenient, is given by

$$X_{\mu\nu}=-T_{\mu\nu}/k. \quad (22\cdot 5)\ [\mathbf{F}\ (15\cdot 91)]$$

..$X_{\mu\nu}$ the *generic* [in an earlier MS. *24*, **6·3** § 3·3, 'the *kinematic*'] *energy tensor*. It agrees with..$E_{\mu\nu}$ if the state which it represents is adopted as initial state; but if it is not the initial state,

$$X_{\mu\nu}=(X_{\mu\nu})_0+\delta X_{\mu\nu},\ E_{\mu\nu}=(X_{\mu\nu})_0-k\,\delta X_{\mu\nu}. \quad (22\cdot 6)$$

...$\delta X_{\mu\nu}$ is the change of energy we should calculate by ordinary mechanics, but the actual transition energy $\delta E_{\mu\nu}$ is $-k$ times as much. The tensor $X_{\mu\nu}$ (like $T_{\mu\nu}$) has the ordinary Lorentz-invariant properties, i.e. a change of velocity of the particle produces in it the same change as an opposite change of velocity of the origin of reference. It therefore varies in the way expected from the kinematical description of the system. This does not apply to $E_{\mu\nu}$ and $W_{\mu\nu}$, the latter being unaffected by the velocity of the particle but undergoing the usual transformation when the velocity of the origin is changed.

Looking at it from another point of view: when a transition to a different velocity occurs, relativity theory gives a change of the energy tensor of the particle with consequential changes of the tensor $g_{\mu\nu}$ of the gravitational field; quantum theory ignores the change of $g_{\mu\nu}$ but in compensation multiplies the change of energy tensor of the particle by $-k$. It is only in a scale-free distribution that the change of $g_{\mu\nu}$ can be compensated in so simple a way. It may be added that wave mechanics gives an elementary explanation of the fact that the energy and momentum of a particle do not correspond to its velocity in the classical way, namely, that the velocity corresponds to the group velocity of the waves. But since current wave mechanics takes account of the potential energy in the electric field only and ignores gravitational potential energy, it misses the difference we have here calculated.

[Cf. **F** § 18.] The treatment is simplified if we analyse systems in such a way that initial and transition energy are provided by different particles. We then distinguish:

Initial particles. . one (pseudo-discrete) state, usually almost exact rest.

Transition particles. . many possible states, the initial. . with zero-energy tensor.

This. . . is the normal procedure in classical mechanics and astronomy . . . where a system is replaced by an *external particle* moving with the centre of mass. . and *internal particles* describing relative orbits. . . . Taking the initial state to be that in which the particles are relatively at rest, the initial energy tensor is that of the external particle alone. The internal particles have. . . only transition energies.

We can therefore use the terms 'external' and 'internal' as alternative to 'initial' and 'transition' particles. By (22·6)

$$\left.\begin{aligned} E_{\mu\nu} &= X_{\mu\nu} && \text{external (initial) particles,} \\ E_{\mu\nu} &= -kX_{\mu\nu} && \text{internal (transition) particles.} \end{aligned}\right\} \qquad (22\cdot7)$$

23. **The inversion of energy:** the source of parts of **F** §§ 21, 18.

First, the method of field-particle partition is applied to a V_1 particle whose only observable is mass; secondly, the sign to be given to added ('transition') kinetic energy is carefully discussed.

. . . we consider V_1 particles. . as a link between classical (or relativity) and quantum particles.

In relativity, as in classical theory, matter is divided into ideal particles to which the uncertainty principle does not apply. These are included in the present treatment as a special case, namely, k [multiplicity] $= 1$. The number of degrees of freedom of the energy tensor is reduced to 1 by stabilizing its orientation (§ 18, **5·1**), leaving only its magnitude observable. . . [this stabilizes] the direction of the momentum vector in four dimensions, and hence the velocity. Thus the condition $k = 1$ implies the velocity is treated as free information, not involving uncertainty of position. This elimination of all observables except one is consistent with the elementary representation of the mechanical characteristics of a particle in relativity; the length m of the momentum vector is an 'absolute' quantity determinable by observation, but the orientation is unobservable, being referred to an unobservable space-time frame.

We therefore describe V_1 particles as semi-classical. Like classical particles they are exempt from the uncertainty relation. . . But as quantum particles they are superpositions on a rigid environment, unlike classical particles which disturb the environment. We shall see later that the formal equivalent of a classical particle in our theory would be a V_{-1}.

The mass of a body, as defined in molar physics, is not a net addition to the energy of the universe...elsewhere there is a decrease, the negative potential energy of surrounding objects..produced; this..constitutes the complementary field energy. We may distinguish the mass ordinarily defined as the E-mass, the negative energy of the surroundings as the W-mass, and their sum (the net addition to the energy of the universe) as the T-mass.

In classical theory the E- and W-energy are..inside and outside the body; but in microscopic theory this distinction becomes blurred by uncertainty of position, and disappears altogether in the uniform distributions of indefinite extent represented by pseudo-discrete wave functions. These distributions form a natural meeting-point of quantum and classical (or relativity) theory.

Setting $k=1$ in (22·1), we obtain for the density $T_{44}=-E_{44}$, $W_{44}=-2E_{44}$; so the relation between the three kinds of mass of a V_1 particle is

$$E=m, \quad W=-2m, \quad T=-m. \tag{23·1}$$

In quantum treatment (23·1) applies only to the adopted initial state; in classical treatment there is no distinction between initial and other states, and the relation $W=-2E$ (or differentially $\delta W=-2\delta E$) should apply to any steady distribution....The potential and kinetic energy of a steady gravitating system satisfy $V=-2K$. Since the kinetic energy is an addition δE to the particle energy, and the potential an addition δW to the field energy, the condition $\delta W=-2\delta E$ is confirmed.

In the rigid field, the basis of wave mechanics, we deal with net additions..to the universe; so the particle mass which concerns us is the T-mass, which is negative by (23·1)....We have to discover..how current wave mechanics...conceals this reversal of sign.

Consider V_1 particles in an initial state of rest, the rest mass (ordinary reckoning) being m_0. Let one acquire a velocity which according to classical or relativity theory would give it momentum p'_α $(\alpha=1, 2, 3)$. In § 22 we used 'generic' for quantities calculated by ordinary mechanics from the kinematical description of the system; thus p'_α is *generic momentum*, and if the velocity is not unduly large (*Footnote*: the formula $(m_0^2+\Sigma p'^2_\alpha)^{\frac{1}{2}}$ does not give any greater accuracy since it is valid only in the absence of a gravitational field—a condition evidently unsatisfied ..with the large potential energy of (23·1)) the generic energy is

$$X=m_0+(\Sigma p'^2_\alpha)/2m_0. \tag{23·2}$$

When the change is treated as a transition in a rigid field, the particle energy is

$$E=m_0-(\Sigma p'^2_\alpha)/2m_0, \tag{23·3}$$

by (22·6). To remove the incongruous negative sign, we introduce a quantum momentum

$$p_\alpha=ip'_\alpha, \tag{23·4}$$

so that

$$E=m_0+(\Sigma p_\alpha^2)/2m_0. \tag{23·5}$$

Then the quantum energy and momentum E, p_α are related in the same way as those of a classical particle.

Equation (23·4) shows the origin of the $\sqrt{-1}$ which appears so mysteriously in quantum theory, being introduced in the definition of the momentum operator $p_\alpha = -i\hbar\,\partial/\partial x_\alpha$. By (23·4) the generic momentum is $p'_\alpha = -\hbar\,\partial/\partial x_\alpha$. The $\sqrt{-1}$ is thus introduced as a means of compensating the reversal of sign of the energy.

From (22·6) the classically calculated energy X and the quantum E agree only if $k=-1$. But a V_{-1} is too formal an abstraction; it is better to connect quantum mechanics, not directly with classical mechanics, but with the inverted classical mechanics satisfied by the V_1 particles. The procedure of transforming the classical variables X, p'_α into the semi-classical E, p_α will be referred to as the *inversion of energy*, and will generally be taken for granted as a preliminary step when we treat the more typical quantum particles for which $k \neq 1$.

...The practical effect of the inversion is that we study systems whose states are such that the modified momenta are real so that the corresponding classical momenta are imaginary. Hence..the elementary constituents of the systems in quantum theory are represented classically by waves; and anything approximating to a classical particle can only be introduced by a complicated superposition of waves.

The momentum of a particle, unless otherwise stated, always refers to the quantum momentum p_α. The equation for the generic energy (23·2) is therefore more conveniently written as

$$X = m_0 - (\Sigma p_\alpha^2)/2m_0. \qquad (23{\cdot}6)$$

24. **Rigid coordinates:** like **F** §19. Begins:

We normally adopt the state of rest as the initial state..the only non-zero component of the energy tensor is the density E_{44}..accompanied by a field energy density $W_{44} = -(k+1)E_{44}$....Evidently the gravitational field which produces W_{44} has a constant potential. By ordinary standards the field is preposterously intense....Relativity theory rescues the [rigid-field] method from futility because it permits us to create a gravitational field of constant potential by a simple transformation of coordinates...

The rigid-coordinate formula **F** (19·1) is then established but not discussed.

25. **Standard particles and vector particles:** like **F** §17, with a clear statement of the approach to **F** (16·5), and (at the end) of the physical picture of the effects of 'multiplicity'. The first part is summarized.

'Ordinary' energy tensors and momentum vectors with 10, 4 independent components become 'complete' with 136, 10 when spin is taken in. The carrier of a 'complete' ($k=136$) tensor is called a *standard* (*bi-*) *particle*. By (22·1), in the initial state

$$T_{\mu\nu} = -136E_{\mu\nu}, \quad W_{\mu\nu} = -137E_{\mu\nu}. \qquad (25{\cdot}1)\ [\mathbf{F}\,(20{\cdot}1)]$$

An ordinary energy tensor with no internal stress is

$$T^{\mu\nu} = v^\mu \sqrt{\rho_0} \, . \, v^\nu \sqrt{\rho_0}, \qquad (25{\cdot}3) \ [\mathbf{F}\,(17{\cdot}2)]$$

and so has only four independent components (e.g. a classical or relativity particle). In quantum theory we have a choice between 'standard' ($k = 136$) and 'vector' particles, characterized by a complete momentum vector (or 'root vector' like a factor of (25·3)), and obtained by stabilizing a complete energy tensor to be an outer square; this reduces k for vector particles from 136 to 10.

Consider n particles per unit volume at rest, and let the total density T_{44} be given. By (22·1) the density apportioned to the particles is $\rho = E_{44} = -T_{44}/k$. The factor k depends on whether we adopt standard or vector particles, and the densities are in the ratio

$$\frac{\rho_v}{\rho_s} = \frac{k_s}{k_v} = \frac{136}{10}. \qquad (25{\cdot}4)$$

If m_0, M are the masses of standard and vector particle, $\rho_s = nm_0$, $\rho_v = nM$. Hence

$$M = \tfrac{136}{10} m_0. \qquad (25{\cdot}5)$$

Other modes of stabilization are sometimes introduced. For any two values k, k' of the multiplicity, the masses . .

$$m_k / m_{k'} = \rho_k / \rho_{k'} = k'/k. \qquad (25{\cdot}6)$$

The particles whose masses obey (25·6) are initial (external) particles... electrically neutral, since we have not yet treated the e.m. field. Evidently a distribution of electrons cannot be treated directly by this method... no steady distribution is possible.

The position contemplated is that we have to apply wave analysis to a density T_{44}. This is the *total* energy density, i.e. the net addition to an undisturbed background, . . the system analysed is superposed on the universe. The number of particles into which the system is to be divided is fixed by considerations outside scale-free physics; these do not matter at present, since we determine only relative masses. We can represent the distribution by alternative kinds of particle (. . one at a time) which differ only according to the amount of free information we accept. For example, it normally happens that in the molar energy tensor the spin components are negligible, and the body can quite well be represented by a distribution of spinless particles. In that case we incorporate in the definition of the kind of particle we are using free information that its spin is zero. The relative masses of these different kinds of particle are determined by (25·6).

The field energy will also vary with the degree of stabilization as represented by k. The more stabilization we introduce the greater is the particle energy and the less is the field energy. We may regard stabilization as a constraint, and the difference of field energy as the (negative) energy of the constraint.

There is a final passage like the end of **F** § 17.

26. **Transition particles:** like part of **F** § 18. This contains the crux of the mass-ratio calculation. Eddington's $p_1^2+p_2^2+p_3^2$ is written here as p^2.

Consider a standard particle of mass m_0 in an initial state of rest. If it makes a transition to a state of momentum p_1, p_2, p_3, the generic energy is

$$X = m_0 - p^2/2m_0 \tag{26·1}$$

by (23·6); by (22·6) the particle energy is

$$E = m_0 + 136p^2/2m_0 = m_0 + p^2/2\mu, \tag{26·2}$$

where

$$\mu = \tfrac{1}{136}m_0. \tag{26·3}$$

In accordance with § 22 we introduce separate carriers for the initial and transition energies, so that the standard particle is divided into an initial or external particle of energy E_e and a transition or internal particle of energy E_i, where

$$E_e = m_0, \quad E_i = p^2/2\mu. \qquad (26·41)\ [\mathbf{F}\ (18·33)]$$

Now stabilize the two particles as vector particles...no effect on transition energy, since the relation $\delta E = \delta T$ does not involve k. As already shown, the effect on the initial particle is to change the mass from m_0 to M. Thus for the two vector particles which replace the unspecialized element of energy tensor, we have

$$E_e = M, \quad E_i = p^2/2\mu, \qquad (26·42)\ [\mathbf{F}\ (18·4)]$$

so that the masses of the particles are

$$M = \tfrac{136}{10}m_0, \quad \mu = \tfrac{1}{136}m_0. \qquad (26·5)\ [\mathbf{F}\ (18·5)]$$

This development..applies particularly to a two-particle object-system...According to classical mechanics two 'actual' particles m, m' are equivalent to an external particle of mass M and an internal particle μ, where

$$M = m + m', \quad \mu = mm'/(m+m'). \qquad (26·6)\ [\mathbf{F}\ (18·1)]$$

...The masses m, m', M are rest energies; but the internal particle has zero energy when at rest, i.e. when $d\xi_\alpha/dt = 0$ ($\xi_\alpha = x'_\alpha - x_\alpha$), and its mass is therefore defined as the constant μ in the expression

$$E = p^2/2\mu, \tag{26·7}$$

giving the energy in terms of the momentum....To distinguish between mass as rest energy and mass defined by (26·7) we call the latter the *mass constant*...the mass in classical dynamics is the mass constant; the definition as rest energy was introduced by Einstein....That the classical result (26·6) applies to quantum particles will be proved in... by transforming a double wave function....

A final 'aside' on *bi-particles and two-sided surfaces*:

Although we adopt the description 'bi-particle' as a concession to the ordinary outlook,..the standard particle is essentially the simplest of all the carriers that exist in actuality or imagination. One might equally well attribute composite nature to a plane surface on the ground that it possesses two sides; but although it may be necessary to distinguish the two sides, it will scarcely be argued that a one-sided surface is a simpler conception. In the next section we shall introduce the 'elementary particles', namely, protons and electrons; the term implies only that they are normally the ultimate elements in wave analysis, not that they are conceptually simpler than the bi-particle.

27. **Protons and electrons:** close to part of **F** §18. The mass-ratio quadratic is now established in less than a MS. page.

The masses M, μ of the external and internal particles which form the standard bi-particles are given in (26·5). Their ratio is

$$\eta_1 = M/\mu = 136^2/10 = 1849{\cdot}6. \qquad (27{\cdot}1) \ [\mathbf{F}\ (18{\cdot}6)]$$

Presumably this combination of elementary external and internal particle is realized in the hydrogen atom. Accepting this... η_1 is a principal constant of the hydrogen atom....

...the masses m, m' of the proton and electron..from (26·6)..are the roots of

$$m^2 - Mm + M\mu = 0. \qquad (27{\cdot}2) \ [\mathbf{F}\ (18{\cdot}7)]$$

By (26·5) this becomes

$$10m^2 - 136mm_0 + m_0^2 = 0. \qquad (27{\cdot}3) \ [\mathbf{F}\ (18{\cdot}8)]$$

...the ratio

$$\eta_2 = m/m' = 1847{\cdot}60.... \qquad (27{\cdot}4) \ [\mathbf{F}\ (18{\cdot}9)]$$

The value can be found observationally with a probable error of about 1 in 8000, and the agreement is perfect. The quantity m/m' generally listed in tables (1836·6) is not directly comparable owing to differences of definition.

28. **The mass m_0**: related to **F** §40. Having found the mass *ratio*, Eddington now solves the complementary problem of the *absolute* mass of the standard particle. The main argument (concerning the Planck and gravitational constants) is as **F** §40, and the result (28·8) obtained is **F** (40·7). The following remarks concern the overall logical structure.

The basic formula is $m = \hbar\sqrt{(2N)}/R_0$ ((10·6) in **4·1** above), the mass of a V_3 particle. The consequent formula $m_4 = \frac{3}{4}\hbar\sqrt{(2N)}/R_0$ ((10·8) in **4·1**), the mass of a V_4, has now been justified by the multiplicity factor arguments of the present chapter. By the same arguments, the mass m_0 of a standard particle or V_{136} is $\frac{4}{136}m_4$. The appropriate Planck constant '$\hbar$' for (10·8) is related to the 'observational' $\hbar$ as in **F** §40. (The absence of a factor $\sqrt{2}$ from (10·6, 10·8) compared with **F** (39·9) is due to a shifted 'double reckoning', and the final results agree.) The final part is

...in ordinary units

$$m_0 = \frac{3}{4}\frac{h}{2\pi}\frac{(\sqrt{\frac{4}{5}N})}{R_0 c}. \qquad (28{\cdot}8)$$

This is the principal formula in the unification of gravitational with quantum theory...comparisons must be deferred until we have investigated a small correction arising from the electrical interaction of the particles which form the actual uranoid.

6·2. DRAFT B *26* III. MULTIPLICITY FACTORS

3·1. **The rigid field:** close to **6·1** §20 (**F** §§13, 14). 'Object fields' (**6·1** §19 or **F** §12) were discussed in **B** §1·9 (*a*) (**4·2** above) of this version.

3·2. **Scale-free systems:** close to **6·1** §21 (**F** §15, part). A final remark on *almost exact* states:

An almost exact state is represented by the same wave function as the corresponding exact state; but the energy and other mechanical characteristics are partitioned differently between the particles and the field, according to the multiplicity factor of the almost exact state.

3·3. **Allocation of the energy tensor:** mainly close to **6·1** §22 (**F** §§15, 18). The following excerpts should be read with **6·1**.

...The energy tensor is..scale free...the mass of a system is not.... In quantal physics the mass and momentum vector supplant the density and energy tensor as primary characteristics....This makes the connexion of quantal physics with molar relativity much less direct. We therefore begin with scale-free characteristics, and consider 'particles' which are effectively nothing more than small elements of an energy tensor, having no other distinguishing characteristics. It should be added that...we sometimes introduce a fictitious characteristic, an identification number (as a subscript); the consequences are treated in the theory of interchange.

...The meaning of *generic energy* is seen by considering a Lorentz transformation which changes the velocity of a particle from $\mathbf{v}$ to $\mathbf{v}'$. Then the change of $T_{\mu\nu}$ is given by the ordinary theory of Lorentz invariance; and $X_{\mu\nu}$, which is in a constant ratio to $T_{\mu\nu}$, is transformed in the same way. In other words, the dependence of $X_{\mu\nu}$ on $\mathbf{v}$ is just what we should expect if the particle were an ordinary relativity particle. But $E_{\mu\nu}$ is not Lorentz-invariant, since the assignment of energy to the particle depends on the field which is rigidly fixed. It follows...*the transition energy* $[\delta E_{\mu\nu}]$ *is* $-k$ *times the amount expected from the change in velocity*.

The deviation is explained by the fact that our expectation takes for granted a linear relation between the energy and the amount of matter present, whereas the actual dependence is non-linear. The marginal effect of additional energy is large for this reason. Or...the deviation arises from the distinction between wave and group velocity—applied here to scale-free wave functions.

'Initial particles' and 'transition particles' (here: 'or oscillators') follow as in **6·1** §22 to (22·7). A two-particle system is then treated as in **6·1** §26 (round (26·6, 26·7)). Finally, there is this note on *mass* (cf. **F** §18):

In classical mechanics the mass is defined as μ in the expression

$$\varpi_4 = (\varpi_1^2 + \varpi_2^2 + \varpi_3^2)/2\mu \tag{3·37}$$

for the energy ϖ_4 in momenta.... The distinction between a rest mass and a mass constant [i.e. μ in (3·37)] is important when we absorb a multiplicity factor into the mass. Thus if $\varpi_1, \ldots, \varpi_4$ are the generic momenta and energy of the internal particle, the true energy is by (3·35) [**6·1** (22·7)]

$$\varpi_4' = -k\varpi_4 = (\varpi_1^2 + \varpi_2^2 + \varpi_3^2)/2\mu', \tag{3·38}$$

where $$\mu' = -\mu/k, \tag{3·39}$$

so that, although the energy is $-k$ times the amount expected, the mass (i.e. mass constant) is $-1/k$ times the amount expected.

3·4. **Rigid coordinates:** like **6·1** §24 (**F** §19). The opening is an eloquent account of Eddington's conception of quantum theory:

Since the energy tensor gives a purely mechanical description of the system, §3·3 is not concerned with electromagnetism. Thus the self-consistent field introduced is a mechanical or inertial-gravitational field, and W_{44} is gravitational potential energy.

The self-consistent gravitational field $W_{\mu\nu}$ is the complementary object-field defined in [**4·2**] §1·9 (*a*), except that the particles are now in a special environment chosen to validate the rigid-field treatment, instead of in the standard uranoid. Evidently we must, before applying wave mechanics, put the system in the environment that method demands; it must be an environment whose energy is changed only by terms of the second order through the gravitational effects of small changes in the object-system. This environment turns out grotesquely different from the standard uranoid, which represents the condition in which measurements are normally supposed to be made. But that is the price to be paid for using a coordinate frame with rigidly fixed $g_{\mu\nu}$.

The question naturally arises: Of what conceivable use is a theory which investigates an atomic system in conditions absurdly remote from those of actual experiment? The answer to this question removes the mystery from quantum theory. Quantum dynamics is only fantastic because it is the dynamics of a fantastic problem. It is not the system of dynamics but the problem that strains our credulity. But it is not uncommon in physics to solve one problem and then by a transformation obtain the solution of another, e.g. conformal transformation in hydrodynamics. And so, having solved the comparatively easy problem of an atomic system in grotesquely unnatural conditions, we proceed by a mathematical transformation to derive the solution of the problem of an atomic system in natural conditions.

We shall find the transformation introduces the multiplicity factors k, including the minus sign attached to them in (3·31) [**F** (15·7)], etc. This reversal of sign of certain quadratic terms is responsible for the $\sqrt{-1}$ in corresponding linear terms, which is so common a feature of quantum formulae. In current theory these factors are absorbed into the masses and other constants in the equations, so the transformation is concealed. The current equations refer directly to the natural problem; but they appear as equations which, for unexplained reasons, just happen to fit, and the constants are empirical. The combination of relativity and quantum theory. . gives a rational derivation of the equations and constants. In this derivation the transformation of the elementary but artificial rigid-field problem into the natural problem of an atomic system in non-rigid environment is an important step.

We thus distinguish the 'rigid' and the 'natural' problem. The fantastically large field in the rigid problem is removed in the natural problem. The secret of the transformation is that a gravitational field can be removed by a change of coordinates. More strictly, we must distinguish between *reducible* and *irreducible* fields. If the self-consistent gravitational field corresponding to the initial state is irreducible, the rigid-field treatment is valid but infertile, because there is no way of transforming the rigid into a natural problem. We therefore restrict the choice of initial state by admitting only states which give a reducible field. Then the natural is obtained from the rigid problem simply by a transformation of coordinates.

The x, y, z, t accepted as rectangular coordinates and time in the rigid problem will be called *rigid coordinates*. Since a gravitational field is present their metric is not Galilean. To obtain the natural problem we have to transform to Galilean coordinates x', y', z', t'.

Formula **F** (19·1) is then established. In the course of this, it is observed that the velocity of a quantum particle can be defined as the *contravariant* $v^{\mu} = g^{\mu\alpha} p_{\alpha}/m$, where p_{α} is the *covariant* momentum $-i\hbar\partial/\partial x_{\alpha}$, so that velocity and momentum transform oppositely under **F** (19·1):

The elementary explanation of this difference. . . is that the velocity of the particle is the group velocity of the wave packet. But although current quantum theory contains the machinery for dealing with this difference, it does not apply it in practice, since it ignores gravitational fields and assumes the difference between group and wave velocity to depend wholly on the e.m. field.

The concluding passage looks forward to radiation theory:

It appears that no transformation satisfying (3·31) [**F** (15·7)] can be found unless $E_{\mu\nu}$ reduces to a single component E_{44} in some frame. . . Unless the molar ensemble particles in the initial state are relatively at rest. . the field is irreducible, and results have no practical value. . . . We are restricted to small changes. . e.g. for hydrogen the initial state of the molar system corresponds to a distribution of protons and electrons almost at rest. Our formulae cover transition to another state. . small velocities; and they also cover transition of a *small proportion* of the pairs of electrons and protons into hydrogen atoms in the ground state. This is sufficient to go on with. Later we consider an aggregation in which the majority of the particles form hydrogen atoms in the ground state. That will involve a somewhat different treatment explained in § 3·9.

3·5. **The inversion of mass:** related to **6·1** §23 and **F** §21. Eddington prefaces the main topic by this conspectus of quantum-classical relations, statistical ensembles, particles, field and radiation.

The problem of connecting quantum and molar-relativity theory is now beginning to take definite shape.... We prepared the ground by investigating the standard environment common to molar and microscopic object-systems. We have described this—the zero-temperature Einstein uranoid—both from the molar and the microscopic point of view, and have connected the constants used in the two descriptions; we have also described it from the practical view as supplying the solitary extraneous standard.... Into this environment the molar and the quantum physicist put objects usually of extremely different types, requiring different systems of description; so that before reaching observable phenomena the two theories have sprung apart. To make the desired connexion we must hold them together a little longer by introducing into the environment a simple type of object-system to which the methods of both theories are rigorously applicable.

We leave aside electrical phenomena for the present and consider only mechanics. Similar terms—mass..density—are used in molar and quantum theory, but...by different definitions. It is usual to justify these terms in quantum theory by an *analogy* between the dynamical equations.... Our plan is not so much to reunite the two theories, which have begun in different ways, as never to allow them to separate...in this chapter the mechanical terms applied to microscopic systems are not analogues but are *identical* with the terms in molar mechanics. It is the energy tensor $T_{\mu\nu}$ defined in molar theory, and determined by molar measurements, which we break up into contributions from a large number of particles. Thus the mechanical connexion is made definitive at the start.

The idea that quantum theory deals with single particles or small groups..whereas molar theory deals with large numbers is illusory. Both deal with averages of large numbers, but in forming the averages quantum theory takes note of correlations...By use of the concept of probability the analysis is applied ostensibly to one *representative system*, but the results are interpreted by substituting for the probabilities in the representative system frequencies of individual systems in a large ensemble. Primarily an ensemble is a purely conceptual association of systems, as when we imagine the same experiment repeated a large number of times. But molar objects commonly provide real aggregations of systems which can with sufficient approximation take the place of the conceptual ensemble.

Thus, if we introduce into the standard environment an object-system which is an ensemble of particles or simple systems, we satisfy both the molar and the quantum physicist. The latter is provided not only with the representative system which he ostensibly studies, but with the ensemble postulated in the interpretation of his results. But that is not all. The provision of a real, rather than a conceptual, ensemble is not merely a question of convenience. We have seen (§ 2·5) that a pseudo-discrete wave function represents an ensemble of particles. Now discrete wave functions occur only as correlation wave functions representing internal states of a system; the external particle must in any case be represented by a pseudo-discrete wave function and therefore as existing in a real ensemble. Except by the much more elaborate construction of a diffusing wave packet, quantum theory is unable to treat a particle or system singly. It is therefore just as insistent as molar theory that the object-system shall be a large collection of particles.

An internal wave function, if discrete, has an occupation factor j not greater than 1, and is said to represent one internal particle. But this means one internal particle in the representative system. The whole wave function is the product of external and internal wave functions, and the whole occupation factor the product of...factors. Thus the repetition of the external particle a great number of times in the ensemble carries with it a repetition of the internal particle associated with it.

The distinction between a *system* and a *real ensemble* is that there is no direct interaction between the particles or systems forming an ensemble. There may be a corporate interaction through the medium of a molar field. The particles of the ensemble collectively produce the molar field, which acts similarly on each individual particle. In the present method this corporate interaction is not left to be introduced as a correction in a second approximation. The energy tensor $T_{\mu\nu}$, with which we start, is not differentiated as between particles and field; and we proceed to determine the allocation between the particles themselves and the field representing their corporate interaction. Conformably with this approach the particles are represented, not by wave packets.., but by pseudo-discrete wave functions which explicitly represent them as constituents of an ensemble.

According to § 3·1 the energy E_r of a system in the rth state is a function of the occupation factors, i.e. of the frequencies of the states in the ensemble. It will perhaps be objected that the frequency of the light emitted in an atomic transition does not depend on the number of atoms in different states of excitation. But that only shows that the energy difference $\Delta E = h\nu$ referred to in the optical formulae is not the quantity $E_s - E_r$. The accepted radiation theory does in fact recognize that $h\nu$ is not the whole energy difference; there is, in addition, a perturbational energy depending on the j_r which is kept separate. We have not yet introduced this separation. At present it is sufficient to say that quantum physicists, having a special interest in spectral radiation, have developed this part of microscopic theory more intensively than the rest; and, realizing that it would be impossible to make headway with such problems under the restrictions of the rigid-field treatment, they have exempted the transverse part of the e.m. field (the part concerned in radiation) from the rigid-field condition...ceasing to regard it as field and treating it as particles (photons).

It would have saved much confusion if that overworked term 'field' could have been restricted entirely to molar fields. A molar field is the combined effect of many sources *in incoherent phase*, so that fluctuations are small and we employ expectation values. Molar potentials, being expectation values, are numerical—unlike the symbolic quantities.. in microscopic theory....The suppression of fluctuations introduces a clear distinction between the field of the particles as a causal agency and the particles themselves. But when there is no averaging, 'the effect of the field of a particle' is only a long-winded way of saying 'the effect of a particle', and there seems no reason for regarding the particle and its field as separate concepts. For this reason we shall not regard so-called intra-atomic and intra-nuclear fields as true fields....We recognize only the two pure concepts: (1) the molar concept of a 'true' field, and (2) the microscopic concept of direct particle interaction.

The *inversion of mass* is now discussed.

We have seen that eigenstates...are constructed in a rigid gravitational field..a 'rigid problem'. This is transformed into the 'natural problem' by identifying the time t of the rigid problem [**F** § 19] with $-1/k$ times the actual time t', and the energy p_4 with $-k$ times the actual energy p_4'. In particular, a particle of mass m initially at rest becomes in the rigid problem a particle of mass $-km$. The elementary explanation is that, by introducing the rigid field, a gravitational potential energy $-(k+1)m$ has been added. This is also shown in (3·31) [**F** (15·7)], where the gravitational potential energy is given separately as field energy, the energy density T_{44} associated with the particle in the rigid field being separated into a field energy density W_{44} and a proper particle energy density E_{44} which is the residue left after removing the field.

Attention must be called to the opposite sign of the rest mass in the two problems. Assume provisionally that the rest mass in the rigid problem is positive. Then transition energy (the same in the two problems) has the sign ordinarily expected; but the mass m of the particle in the natural problem is negative. This..disconcerting result..is explained by the fact that, in removing the field energy, we have reduced our 'particles' to fragments of energy superposed on the uniform uranoid. I think that is logically the correct conception of *pure* particle energy. In wave mechanics the energy represented by wave functions is a net addition to the undisturbed environment, and for consistency rest energy must be treated in the same way; and the same rule should apply whether the environment is the special one introduced in the rigid-field treatment or the standard environment postulated in the natural problem. Our calculations have been made on this basis. But the ordinary conception of a particle, which attributes positive rest energy, does not go so far in eliminating field energy. A molar object disturbs its environment gravitationally, so that the object-system includes a complementary field as described in § 1·9 (a) [**4·2**]. In the molar ensemble represented by a pseudo-discrete wave function, each particle will (from the ordinary point of view) have besides its true particle energy a proportionate share of the complementary field energy. *This addition reverses the sign of its rest energy.*

...The reversal is exact.. $-m$ for m. According to general relativity equality of energy and gravitational mass..holds only..in a steady state...then T [kinetic energy] is $-\frac{1}{2}V$, so that the whole energy Ω is $-T$. The gravitational mass of a steady energy distribution is therefore expressed indifferently as Ω or $-T$. The usual view is that Ω is acting as a positive source of gravitation, not that T is acting as a negative source.

Although this result does not refer directly to the energy concealed in the rest masses of the particles, we must extend it to rest energy because the methods make no discrimination between concealed and patent energy. A steady distribution of either kind of energy is represented by a pseudo-discrete wave function; and the separation of field and particle energy which can be verified for energy of visible motion must formally apply also to invisible energy. Thus the rest energy Ω of a static ensemble must be classified as a true particle energy $T = -\Omega$ together with a field energy $V = 2\Omega$; and the particles will be assigned a positive mass m or a negative mass $-m$ according as we do or do not count the field contribution as part of the particle mass.

Henceforth we..adopt the ordinary view which gives the particle positive mass, namely, $1/k$ of the mass in the rigid problem...(3·41) [i.e. $t' = -kt$ of **F** (19·1)] is then replaced by $t' = kt$....We shall refer to this adaptation to the ordinary point of view as the 'inversion of mass'.

3·6. **Standard particles and vector particles:** much of **6·1** §25 (**F** §17). The main result (3·64) is **6·1** (25·5), the mass ratio of vector and standard particles. Some different emphases are illustrated below; the final analogy is of interest.

...current theory describes phenomena as far as possible in terms of vector particles....There is no observational criterion for deciding whether a given molar system shall be represented as..standard or vector particles; for the internal stress can be represented as intrinsic in the individual particles..or as the resultant of their diverse motions. The assumption that the ultimate particles are free from internal stress is a conventional restriction....A pseudo-discrete state..such as almost exact rest, may be occupied either by a vector or a standard particle; the state is centred about a point of contact of the 10-space of the vector particle with the 136-space of the standard particle.

...Hence

$$M = \tfrac{136}{10} m_0 \quad (M \text{ vector, } m_0 \text{ standard particle}). \qquad (3·64)$$

...from the point of view of §3·4...the multiplicity factor only appears in the transformation of the rigid into the natural problem. In the rigid problem there is no distinction between the two kinds of particle; they are just particles of energy p_4 and mass m....The main purpose of the analysis..in the rigid field is to calculate transition energies $\delta T_{\mu\nu}$; and, since $\delta E_{\mu\nu} = \delta T_{\mu\nu}$, these apply directly to the particles irrespective of k. The multiplicity factor is involved..when we transform to the natural problem; for then we have to separate the initial p_4 into particle p_4' and field energy $p_4 - p_4'$ in order to remove the latter by transformation of the time coordinate. At that stage we decide whether the transition energies calculated shall apply to a natural problem relating to vector or standard particles. Both applications exist. But the vector particles to which the transition energies apply have mass $M = -\frac{1}{10}m$, and the standard particles $m_0 = -\frac{1}{136}m$; or, allowing for inversion of mass in the natural problem,

$$M = \tfrac{1}{10}m, \quad m_0 = \tfrac{1}{136}m. \tag{3·65}$$

...It will be shown...that a combination of *two* vector particles is sufficient to give an energy tensor with the full number (136) of independent components;..no need to attribute greater complexity to the standard particle...

This relation of the standard to the vector particle is the mechanical counterpart of the relation of the interval to the point-event. In the relativistic outlook the standard particle is, like the interval, the more primitive concept. The ordinary outlook takes the vector particle and the point-event as primitive concepts, and builds the standard particle by the association of two vector particles as it builds the interval by the association of two point-events.

3·7. **Mass ratio of the proton and electron:** related to **F** § 18. The derivation here is based on the foregoing development, invoking the simple 'inversion of mass' argument and not the quantum-classical momentum relation ($p_\alpha = ip_\alpha'$, **F** (18·31)) of later versions. I write p^2 for $p_1^2 + p_2^2 + p_3^2$.

...Consider a standard particle of mass m_0. By § 3·3, the ordinary Lorentz-invariant formula for the energy, $p_4 = \sqrt{(m_0^2 + p^2)}$, gives the *generic* energy (corresponding to $X_{\mu\nu}$)...we use the approximation

$$p_4 = m_0 + p^2/2m_0. \tag{3·72}$$

Taking the state of rest as initial, the first term represents initial, the second, transition energy. By (3·35) [**6·1** (22·7)] the second term has to be multiplied by -136 to give the true energy. Allowing for the inversion of mass, the minus sign is cancelled; so the true energy is

$$p_4 = m_0 + p^2/2\mu, \quad \mu = \tfrac{1}{136}m_0. \tag{3·73}$$

The two terms are now interpreted as energies of an external (initial) particle of rest mass m_0, and an internal (transition) particle of mass constant μ....

...next substitute vector for standard particles. This makes no difference to the transition energy, but the initial particle energy is increased from m_0 to M (§ 3·6). The true energy for vector particles is accordingly

$$p_4 = M + p^2/2\mu, \tag{3·74}$$

where

$$M = \tfrac{136}{10}m_0, \quad \mu = \tfrac{1}{136}m_0. \tag{3·75}$$

The mass ratio of the external and internal particles is therefore

$$\eta_1 = M/\mu = 136^2/10. \tag{3·76}$$

The mass quadratic follows as in **6·1** §27 or **F** §18:

$$m^2 - Mm + M\mu = 0. \tag{3·772}$$

By (3·75)

$$10m^2 - 136mm_0 + m_0^2 = 0. \tag{3·78}$$

3·8. **The fine-structure constant:** like part of **F** §20. The fine-structure constant is shown as a ratio of field and particle action; the purpose here is to lead into the following section on Radiation.

In..the relation between the field and particle energy of a standard particle,

$$W_{\mu\nu} = -137E_{\mu\nu}, \tag{3·81}$$

..137 is called the *fine-structure constant*. It is most conveniently described as the ratio of two natural units or 'atoms' of action..[cf. **F** 20].

We therefore identify the action unit for field energy with $\hbar$, for particle energy with e^2/c....The field unit is introduced as a coefficient in the definition of momentum in current quantum theory, $p_\mu = -i\hbar\,\partial/\partial x_\mu$. Evidently current formulae must be understood to refer to field momentum and energy...The quantum first appeared in the study, not of particles, but of radiation. The pre-eminence of the radiation field became further established when Heisenberg gave up the Bohr model of the interior of an atom, and proceeded on the principle rather too sweepingly expressed in the dictum that 'all we can really know about the interior of an atom is that certain kinds of radiation come out of it'.

This leads us to consider a modified form of the two-particle problem in which the two elementary vector particles are not in the standard environment but in a radiation field.

3·9. **Radiant energy:** this section is not in later MSS. nor in **F**. Its goal is a correction (3·96) to the mass quadratic, a correction which had been otherwise obtained in **RTPE**, and which is replaced by the more subtle $(\frac{137}{136})^{\frac{5}{6}}$ in **F** III. The topic of the section is the relation of radiant and particle energy, particularly for hydrogen.

When transitions occur the particle system emits or absorbs radiant energy. We..consider..a carrier of the radiant energy...long known as the aether...

Aether, like matter, can be treated molarly or microscopically...as a continuous medium..or a large number of potential carriers of energy ..generally called *oscillators*. We may distinguish *standard oscillators* which..carry unrestricted elements of energy tensor....

In ordinary relativity a molar treatment of radiation is combined with a molar treatment of matter. It might be expected that quantum theory would combine a photon treatment of radiation with a particle treatment of matter. This . . is not the standard procedure. The established practice makes the connexion of material and radiant energy a crossing-over point between microscopic and molar physics. Historically, quantum theory has been dominated by the 'equation of spectroscopy'

$$E_2 - E_1 = h\nu_{12}, \tag{3·91}$$

which connects the transition energy of a particle with the frequency of the waves which form a molar representation of the effect of a large number of transitions. The equation (3·91) therefore refers to a mixed treatment—microscopic matter interacting with molar aether. We call this combination . . a *mixed* or *spectroscopic* representation. Since the constant h of quantum theory is defined by (3·91) which refers to a mixed representation, we accept the mixed representation as the one normally contemplated in current theory. The theory would have been simpler if . . based on a consistently microscopic representation, . . with a modified Planck constant; but it would be inexpedient to make a change now.

Accordingly, the aether is to be treated as in the e.m. theory of radiation, as a system with an infinite number of d.f.'s. Its phase space is therefore ∞-dimensional, and the multiplicity factor of its states is infinite. Thus, for the part of the energy tensor corresponding to radiation, $E_{\mu\nu} = -T_{\mu\nu}/\infty$ and $W_{\mu\nu} = T_{\mu\nu}$. If the remainder corresponds to standard particles, we have in an initial state

$$\text{radiant energy (molar)} \qquad (T_{\mu\nu})_r = (W_{\mu\nu})_r, \tag{3·921}$$

$$\text{material energy (microscopic)} \qquad (T_{\mu\nu})_m = \tfrac{136}{137}(W_{\mu\nu})_m. \tag{3·922}$$

In a system of two or more particles, we generally recognize, in addition to the energy allocated to the particles individually, a mutual interaction or perturbation energy. If two particles are interacting so that energy passes from one to another, . . should the energy in transit be counted as belonging to the matter or aether? Is it emitted into the aether by one particle and absorbed thence by the other, or is it all the time a material energy belonging to the combined system but not to either particle separately? We show that for equations (3·92) it scarcely matters which view is adopted. The number of d.f.'s of a system of two standard particles is so great that there is little inaccuracy in treating it as infinite; for three or more . . . the error is entirely negligible. Thus in practice we can lump together all perturbation energy, i.e. all energy not assigned to the standard particles (or their stabilized substitutes) individually, as aetherial energy governed by (3·921).

Let H^0 be the energy-density of interaction of two standard particles. The classifying characteristics of the particles separately being $(T_{\mu\nu})_1$ and $(T_{\mu\nu})_2$, the characteristics X_α giving a complete classification of states of the combined system are of the form $(T_{\mu\nu})_1 (T_{\sigma\tau})_2$. Thus the number of dimensions of the phase space is of order 136^2. Taking account of 'dormant' characteristics of the two particles [F VI, VIII] and of their indistinguishability, the actual number of dimensions is 128×257. The dimension index of X_α being unity, the dimension index l of H^0 is $\frac{1}{2}$; hence by (3·26) [F (15·51)], $E^0 = -H^0/(256 \times 257)$. Thus the difference between $W_{\mu\nu}$ and $T_{\mu\nu}$ rapidly disappears as the carriage of the energy is shared by an increasing number of particles.

It should be remembered that a standard particle appears to us as a bi-particle (§ 3·6), which includes an internal particle as radiating unit. The 'interaction energy' which is grouped with the aetherial energy in (3·921) is the energy of perturbation of one radiating unit by another or energy in process of exchange between radiating units. There is no interaction energy of this kind in a hydrogen atom in an eigenstate, because the atom is derived from one standard bi-particle....(From the e.m. point of view the internal particle is a doublet..its electric energy.. governed by (3·922), not by (3·921) as energy of interaction.)

We now reconsider the problem of the hydrogen atom (§ 3·7) with the altered condition that its two particles (external and internal) are.. interacting with a third system, namely, the aether, instead of forming a solitary system in a rigid field. As usual...the atom is one of a large ensemble..actually present. The alteration makes an important difference because the system is now conservative, the energy released in transitions being transferred from the matter to the aether. (Formerly it simply disappeared, the 'transition' being a mental comparison of two prescribed states rather than a physical process.) If we adopt a microscopic representation of radiation by standard photons, (3·922) will apply both to the particle and the photon energy, and conservation of $T_{\mu\nu}$ will involve that of $W_{\mu\nu} = \frac{137}{136} T_{\mu\nu}$. The rigid behaviour of the field is no longer restricted to small changes of occupation. We have secured that $W_{\mu\nu}$ is constant for small changes from an initial state; but in this special problem the constancy of $T_{\mu\nu}$ (not hitherto assumed) extends the constancy of $W_{\mu\nu}$ to large changes. For example, the internal particles of all the atoms may make the transition from the zero energy to the ground state without altering $W_{\mu\nu}$, provided the emitted radiation is retained within the ensemble as a distribution of standard photons. The proviso is appropriate since the ensemble is supposed to have indefinitely great extent.

To conform to the spectroscopic representation we have to substitute a molar aether and replace the photons by energy of light waves. It is then no longer possible to conserve both $T_{\mu\nu}$ and $W_{\mu\nu}$. To satisfy (3·921, 3·922), we must either take

$$-(\delta W_{\mu\nu})_m = (\delta W_{\mu\nu})_r, \quad -(\delta T_{\mu\nu})_m = \tfrac{136}{137}(\delta T_{\mu\nu})_r, \tag{3·931}$$

or

$$-(\delta W_{\mu\nu})_m = \tfrac{137}{136}(\delta W_{\mu\nu})_r, \quad -(\delta T_{\mu\nu})_m = (\delta T_{\mu\nu})_r. \tag{3·932}$$

The choice depends on whether it is more important to keep the field energy constant, or to conserve the total energy $T_{\mu\nu}$. The former..is necessary...For a change of field energy means a changed disturbance of the environment of the system...According to the microscopic representation the environment is undisturbed by the transition; and we cannot admit that an environment undisturbed in microscopic representation is disturbed in spectroscopic representation. We therefore adopt (3·931), and admit a change of total energy of the system as the result of the transition. This does not conflict with conservation of energy;

it is simply an energy difference which occurs in adjusting two states of the same system (particle and aether) to an undisturbed environment.

Since (3·931) is correct, we have the important result

$$-(\delta E_{\mu\nu})_m = -(\delta T_{\mu\nu})_m = \tfrac{136}{137}(\delta T_{\mu\nu})_r, \qquad (3{\cdot}94)$$

giving the amount of radiant energy $(T_{\mu\nu})_r$ emitted in an atomic transition. In interpreting experiments physicists have taken for granted the relation is $-(\delta E_{\mu\nu})_m = (\delta T_{\mu\nu})_r$; therefore they habitually attribute to $\delta E_{\mu\nu}$ a value $\tfrac{137}{136}$ times too great. The ordinary theory of the eigenstates of a hydrogen atom determines the ratio of the transition energy between any two states to the mass μ of the internal particle; hence the value of μ currently determined from spectroscopic measurements is $\tfrac{137}{136}$ times too great. Denoting the currently determined mass of the internal particle by μ', we have

$$\mu' = \tfrac{137}{136}\mu. \qquad (3{\cdot}95)$$

This value is wrongly used in (3·772), so the current masses of the proton and electron should satisfy $m^2 - Mm + \tfrac{137}{136}M\mu = 0$. By (3·75) this becomes

$$10m^2 - 136mm_0 + \tfrac{137}{136}m_0^2 = 0. \qquad (3{\cdot}96)$$

The ratio of the roots is 1834·1, which agrees with the currently accepted mass ratio of proton and electron.

The roots of (3·96) will be called the *current masses* of the proton and electron..not physically significant, simply quantities arrived at by an erroneous method of treating observational data. The error mainly affects the electron: so that the current mass of the proton is practically correct, and the current mass of the electron is $\tfrac{137}{136}$ times the true value.

6·3. Draft B *24* III. Multiplicity Factors

This is an earlier version of **B** *26* above. A sheet attached to this MS. bears the words:

Memoranda. Mass inversion—origin of the $\sqrt{-1}$ in formulae.

$$\delta W_{\text{mat}} = -\delta W_{\text{aether}} \quad \text{when} \quad T_{\mu\nu} = \text{const.}$$

instead of $\quad \delta T_{\mu\nu} = \delta E_{\mu\nu} \quad$ with $\quad W_{\mu\nu}$ const.

3·1. **The rigid-field convention:** like **6·2** §3·1. Eddington here, by explicitly describing Hartree's 'self-consistent field' method, shows that it was in fact his own starting point (although his application is different).

...It will be useful to recall first a familiar use of the rigid-field treatment in connexion with atomic electric fields, because its limitations are there well understood. Since the approximation is inaccurate for large changes of occupation, it is the practice to proceed by a method of successive approximation called the method of *the self-consistent field.*

Starting with a prescribed field F, which we treat as rigid, we determine a set of eigenfunctions ψ_r $(r=1, 2, 3, \ldots)$. Having decided which of these are to be occupied, or more generally having assigned occupation factors j_r to each of the ψ_r, we calculate the field F' which would result from the distribution. We then vary the initial field F until by trial and error we find a 'self-consistent field' for the state of occupation, i.e. a field which satisfies $F'=F$. By the stationary condition, small changes of occupation are admissible without altering F; but any considerable change of the j_r involves an entire recalculation of the self-consistent field, with corresponding changes of the ψ_r. Thus the energies H_r of the eigenfunctions are functions of the j_r; and the total energy of the system,

$$H^0 = \Sigma j_r H_r, \tag{3·11}$$

is a non-linear. . function of the j_r.

The foregoing is a familiar, but rather unfavourable, illustration of a method which can be applied much more rigorously to molar fields. Any field treatment of the interaction of the particles in an atom is rather a makeshift. . . . In the following the field is. . . molar. . . especially the inertial-gravitational field. . . .

3·2. **Scale-free systems:** close to **6·2** §3·2. At the end, a precise definition:

A system is in a *pseudo-discrete* state when the whole probability is concentrated in a small wave packet at a point in phase space. The characteristics are then '*almost exact*'. . . (e.g. rest). The number of dimensions k of the infinitesimal spread of the wave packet is called the *multiplicity* of the state. It determines apportionment of energy, etc., between particles and field. . . . In other respects, the pseudo-discrete state can be treated as if it were discrete; it can, for example, be represented by discrete wave functions.

3·3. **Partition of the energy tensor:** close to **6·2** § 3·3.

3·4. **Rigid coordinates: 6·2** § 3·4 is a somewhat fuller account.

3·5. **The fine-structure constant** (cf. **F** § 20): the 'standard particle' (one with unspecialized complete energy tensor) is introduced; the ratio 137 of its field to its particle energy is the 'fine-structure constant'. The discussion is then as in **6·2** §3·8 (omitting the last sentence).

3·6. **Vector particles: 6·2** § 3·6 is developed from this. The last part is:

A vector particle [mass M] conforms to the traditional conception. . . standard particles would generally be described as analytical fictions. Thus M is a mass ordinarily recognized. But to obtain the vector particle we have to. . . [stabilize the energy tensor. In researches on the unstabilized energy tensor] vector particles of mass M are to be replaced (one for one) by standard particles of mass $m_0 = 10M/136$. The change of mass is due to the fact that we alter the potential of the field, so that there is a change in the gravitational potential energy included in the rest mass; or equivalently we change the scale of the time coordinate (§ 3·4), so that there is a reciprocal change in the reckoning of the energy conjugate to it. The field introduced..is the '*stabilizing field*'—the physical cause of the stabilizing constraint—so that its removal releases the vector particles into standard particles..and the wave packet..which was dispersing in 10 is allowed to disperse in 136 dimensions.

The relation (3·63) $[\rho_v/\rho_s = \frac{136}{10}]$ applies only to initial energy density. Thus in (3·64) $[M = \frac{136}{10} m_0]$, M and m_0 are the masses of the external (initial) elementary particles of the vector and standard types. The apparent mass-constant μ' of an internal particle is not altered by changing it to a vector particle, because the transition energy $\delta T_{\mu\nu}$ forms an addition $\delta E_{\mu\nu}$ to the particle energy independent of k.

3·7. **Mass ratio of the proton and electron:** as **6·2** § 3·7, followed by the *summary* and note on current theory quoted here; a brief discussion of the role of $\sqrt{-1}$ in quantal momenta is not quoted.

The steps in the derivation of M, μ may be summarized:

(1) We consider a standard particle of mass m_0 in field-free conditions, regarding it as a coalescence of an external particle of rest mass m_0 and an internal of mass constant m_0.

(2) We introduce the self-consistent rigid gravitational field..wave functions..quantum conditions....

(3) The particles are transformed to vector particles by stabilizing conditions. This involves a change of the rigid field....

(5) We remove the rigid field so as to obtain external and internal vector particles in the usual field-free conditions.

By § 3·4 [the 'rigid' transformation, **F** § 19], the space coordinates and conjugate p_1, p_2, p_3 are unchanged throughout these transformations. ..Thus particle density is unchanged; this..enables us to substitute energy and momentum per particle for energy..per unit volume, so that although only scale-free physics is employed, the results can be expressed in terms of the momentum vector. The invariance of x_α and p_α secures that the angular momenta are unchanged. Accordingly the quantization rules, the wave functions (excluding periodic time factors) and most of the ordinary theory developed for a rigid field can be carried over unchanged into the natural system. We can therefore see why current quantum theory is to a large extent successful although it ignores the self-consistent gravitational field. The effect is manifested mainly in the values of the masses; and the quantum physicist is content to accept empirical determinations of these....

3·8. **Radiant energy:** this unfinished section differs in emphasis and detail from the later **6·2** § 3·9 quoted above: the 'Memoranda' at the head of **6·3** (q.v.) bear on the rewriting.

6·4. Draft B *31* III. Degeneracy Factors

This fragment, the earliest on this part of the theory, consists of §§3·1–3·3 much as in **6·2**, **6·3** above, and an odd page on the fine-structure constant. 'Degeneracy factors' (cf. **2·1** δ) are the 'multiplicity factors' k.

3·3. **Partition of the energy tensor:** part of this is quoted, as the earliest approach to the fundamental equation **F** (16·5).

...for a particle in scale-free physics...

$$T_{\mu\nu} = -kE_{\mu\nu}, \quad W_{\mu\nu} = -(k+1)E_{\mu\nu}, \tag{3·31}$$

where the number k is now the number of independent components of the energy tensor....As an application, we consider the physical transformation (..one to one) of one kind of particle A into another kind B ..the distinction being that one has a more restricted type of energy tensor...phase spaces..dimensions k_a, k_b.

Consider a uniform p.d. of A-particles superposed on the standard uranoid. The object-system then consists of these with their self field, the equivalent of the disturbance..produced by their gravitational action. The energy or mass density of the particles is

$$\rho_a = (E_{44})_a = -T_{44}/k_a \tag{3·321}$$

by (3·31), T_{44} being the total energy-density of the object system. Now let the transformation occur...T_{44} is conserved, but differently distributed...the energy density of the B-particles being

$$\rho_b = (T_{44})_b = -T_{44}/k_b. \tag{3·322}$$

Since each A transforms into one B the ratio of the densities is the same as the ratio of the masses. Hence

$$m_a/m_b = \rho_a/\rho_b = k_b/k_a. \tag{3·33}$$

Thus for particles connected by this kind of transformation *the masses are inversely proportional to the number of independent components of the energy tensor.*

The simplest examples..are the transformation of a neutron into a hydrogen atom and of a mesotron into an electron....

It is worth while to scrutinize the physical principles....The particles (say, a hydrogen atom and a neutron) are not observed in a standard environment. But a physicist..would for the experiment provide an environment suited to his purposes; and similarly for our 'theoretical experiment' we provide a suitable environment, namely, the standard uranoid. In these conditions the energy $(E_{\mu\nu})_a$ of the hydrogen atom occurs in conjunction with a total energy $-k_a(E_{\mu\nu})_a$. The determination of the total energy is important as a step in the experiment; but since it includes a mutual gravitational energy of the atom and its surroundings, it has no importance outside the experiment. The value of the theoretical experiment, like that of the practical experiment, is that it determines a ratio m_a/m_b which is unaffected by the environment, and therefore applies in actual observational conditions. Since this mass of a particle belongs to quantal physics, this is not the place to consider its invariance. The problem we have treated is, accepting the conclusion of quantal physics that the hydrogen atom and neutron have constant proper masses (or a constant ratio of proper masses) however they are circumstanced, to design and carry out a theoretical experiment for determining the actual value of the ratio.

The remarkable simplicity of the formula (3·31), which replaces what would ordinarily be an exceedingly troublesome calculation of self-consistent fields, results from the scale-free condition.... The present type of investigation is a continuation of Dalton's scale-free atomic theory.

CHAPTER 7

DRAFTS OF F I–V

(iv) CHAPTER IV OF C AND V OF B

These are the last extant chapters of the early draft **C** and the earlier **B**. In the scheme of these drafts, electrical theory (Chapter V, or **F** III) was placed after the gravitational treatment of exclusion and interchange (Chapter IV, or **F** IV, V).

7·1. Draft C *25* IV. Exclusion and Interchange

This interesting MS. passes from a discussion of comparison particles and scale (**F** §§ 22–24) to the 'exclusion principle' method of calculating m_0, and thence to some of the planoidal theory of **F** V. There is a 'planoidal' calculation on Saturn in the last section.

29. **The phase coordinate:** related mainly to **F** § 24. The order of argument, however, is the reverse of **F** §§ 23, 24; here we pass from the concept of measurement based on scale uncertainty to that of a 'comparison particle', which is required for the following sections.

...Instead of taking the scale uncertainty into account...by curvature, we shall include it..as an additional variate....Primarily the scale uncertainty is a fluctuation...of the standard....But the effect is indistinguishable from an opposite fluctuation...of the object-system. The scale has two aspects,..as *reference*- and *object-scale*. If considered as a reference-scale it is part of the environment; as object-scale, part of the object-system....An object-system which includes an object-scale as an additional variate will be called *perfect*....

The treatment of scale as *angular* momentum follows as in **F** § 24. Turning to the treatment of scale as *linear* momentum:

The object-scale is then a mass or momentum; so that the reference scale or extraneous standard, whose fluctuations are opposite to those of the object-scale, is the reciprocal of a mass or equivalently a volume. We have already introduced (**5·1** § 16) a normalization volume V and an associated mass m' which have to be extraneously defined. These now can be identified with the reference-scale and object-scale. It is convenient to introduce a particle called the *comparison particle* as carrier of the mass m'. Then, by our previous definition, *a perfect object-system consists of an ordinary object-system together with a comparison particle.*

For example, a perfect electron is a bi-particle—electron and comparison particle...Analytically, a simple wave function

$$\psi = a \exp\{i(p_1 x + p_2 y + p_3 z)/\hbar + i\alpha\},$$

where a is real, describes a distribution over x, y, z but not over..the initial phase α... But if ψ is treated as a double wave function representing a bi-particle, so that $\psi = \psi_1 \psi_2$, where [β denoting $\hbar$]

$$\psi_1 = a_1 \exp\{i(p_1 x + p_2 y + p_3 z)/\beta\}, \quad \psi_2 = a_2 e^{i\alpha} = a_2 e^{im's'/\beta},$$

the restriction is removed, and ψ_2 gives the p.d. over the coordinate s' conjugate to the momentum m' of the second particle.

'Perfecting the object-system' and 'de-stabilizing the scale' are two aspects of the same process, by which the representation is changed from curved to flat space. Ideally the introduction of an extra pair of variates m', s' should not affect our conception of the system as a single entity. But the common outlook, and the practice of wave mechanics, makes it necessary to dissect the perfect object-system into an imperfect object-system and a comparison particle, much as it dissects space-time into separate continua of space and time. The analogy is, in fact, so close that the coordinate of the comparison particle (the linear phase s') is very commonly mistaken for the time coordinate t.

However complicated the object-system, only one extraneous standard is used; so only one pair of variates—a scale and phase—is added in perfecting the object-system. There is therefore only one comparison particle in an object-system, and its multiplicity $k = 1$....

30. **Mutual and self energy:** contains part of **F** § 22. First, the mutual energy tensor (of object and comparison particle) is introduced via the density of a pseudo-discrete wave function:

In III (**6·1**) the extraneous standard (or reference scale) was..a density (§ 22)....But Chapter III was concerned with scale-free theory....We pass over into quantal theory by introducing an extraneous standard of physical dimensions different from the energy tensor. The object-scale is now taken as the mass m' of a comparison particle and the reference scale is a corresponding volume V' given by $V'^{-1} = Cm'$. In scale-free physics the normalization volume V is chosen arbitrarily, but we now impose a scale by employing the natural normalization volume V' corresponding to the mass of a comparison particle. Then if ρ is the normalization density (density of unit occupation) of a pseudo-discrete wave function representing particles of mass m at rest,

$$V'^{-1} = Cm', \quad \rho = mV'^{-1} = Cmm', \tag{30·1}$$

...C a pure number. In (30·1) both particles are..at rest. More generally, ρ is the component M_{44} of the energy tensor for unit occupation given by

$$M_{\mu\nu} = \tfrac{1}{2}C(p_\mu p'_\nu + p'_\mu p_\nu). \tag{30·2}$$ [**F** (22·1)]

...Since the comparison particle has multiplicity 1,..[p'_ν] is stabilized except as regards magnitude....

The interpretation of the orientation of p'_μ:..in p.d.'s, time is differentiated from space coordinates..we must select a particular section of space-time as space frame....Coordinates and momenta in this space are the primitive observables, and the uncertainty of the physical origin is limited to directions in it. The natural normalization volume V' is constructed in this space. Its reciprocal is a 4-vector $(V'^{-1})_\mu$ in the corresponding time direction; and this also is the direction of the momentum $p'_\mu = C(V'^{-1})_\mu$...(30·2) holds in any coordinates; transformation..will not alter the primitive observables, which remain distinguished as measurements in the space orthogonal to p'_μ.

The argument of **F** §22 follows as far as reformulating the mass quadratic, and discussing the multiplicative nature of probabilities; this ends:

> Actually wave mechanics has developed a general theory (second quantization) of the transformation of multiplicative into additive wave functions; but this begins with the elementary particles, whereas the foregoing discussion leads up to them.

31. **Elision of comparison particles:** contains part of **F** §22. The formulae **F** (22·61, 22·62) are given for the mass of the internal and external particle, and it is observed that the 'comparison particle' (representing scale) of an object-system (in particular, the hydrogen atom) is attached to the *internal* particle, the external particle remaining scale-free, and having a pseudo-discrete wave function. (In parentheses, internal particles have 'correlation wave functions' (**F** §10) but, beyond hydrogen

> an atomic or nuclear system is a multiple correlation...we have to do the best we can with two-particle correlations, picking out the strongest ...and devising a technique of 'coupling' to cement them. These two-particle correlations are represented by internal particles; and since the mathematical procedure treats them as separate systems perturbing one another, a comparison particle is attached to each.

—Thus the ideal of having only one comparison particle for a general object-system is not in practice possible.) Continuing the main argument:

> When we put together a distribution of external and one of internal particles to form hydrogen, we adjust the former (which innately is scale-free) to a particular scale, so that there is one external particle to each internal. Accordingly, *in combination*, we recognize a scale of structure of the external distribution which has been borrowed from the internal; i.e. we give the external particle a mass M, as well as the scale-free characteristic of density....The ratio M/μ is primarily a [scale-free] density ratio,...turned into a mass ratio by one-one coupling of external and internal particles.

The formulae **F** (22·7) are collected, with the comment that only particles having comparison particles are electrical, as though the comparison particle gave them the closer kind of attachment to the environment which we interpret as electrical binding.... Since we wish to proceed with pure mechanics,...we accept this hint to confine attention to *external* particles in the rest of this chapter.

32. **Exclusion:** the substance of **F** §§41, 43 (omitting the final result (43·7)); close also to **D** §25. From the introduction:

...We have hitherto supposed there is no scatter of the rest energies of similar particles; so that in an assemblage at zero temperature the energies are all equal. But exclusion forces the energies to scatter....Let μ_0 be the [external particle] mass-constant, so that the energy of a particle (classical) is $\mu_0 + p^2/2\mu_0$.... If the assemblage is at zero temperature, the 'exclusion energy' is $E = p^2/2\mu_0$, since p_1, p_2, p_3 would be zero if there were no exclusion. We have to distribute $p_1, \ldots$ so that ΣE is a minimum.

The calculation follows as in **F** §41, and then passes to **F** §43. In the present argument, this 'exclusion' result for m_0 is compared with the 'gravitational' result of **6·1** §28. The correction **F** (42·2), from the appropriate Planck constant h_1 to the standard h (a correction applying to both **6·1** §28 and to §32 here) is made in §34 below.

33. **The negative energy levels:** like the first part of **F** §42. The concept of the standard (uranoid) particles filling a set of negative energy levels, below the zero level of energy $\mathfrak{E}$, is compared with Dirac's scheme. In the present theory, an object-particle is formed by excitation upwards from the $\mathfrak{E}$-level, leaving a hole at that level; this hole constitutes the comparison particle.

The mass m_0, which is equally the mass of a standard and of a comparison particle, is doubly identified with $\mathfrak{E}$: first by the identification of the comparison particle with a hole at the $\mathfrak{E}$ level, and secondly because the top exclusion energy $\mathfrak{E}$ (now identified as that of a standard particle) is in our ordinary outlook the energy of a particle at zero temperature in the standard uranoid.

34. **The factor $\frac{3}{5}$:** like the latter part of **F** §42. The object is to show that the Planck constant $\bar{h}$ (called h_1 in **F** §§42, 43) in exclusion theory is $(\frac{3}{5})^{\frac{1}{2}}$ times the normal h; this result is used to complete the calculation of m_0 (or $\mathfrak{E}$). In the present version, both the modified 'gravitational' constant $\bar{\kappa}$ and $\bar{h}$ are found, so that **F** (8·1) can be used merely as a check.

...a microscopic object-system is formed by exciting particles from the top level. But if we consider a large object-system, say a planet, it is necessary to go deeper to obtain (locally) sufficient particles. Thus the particles of a large concentrated object have, on the average, lower energy....This applies primarily to the comparison particles or holes which form a deep cavity in the uranoid...but the reduction is passed on to the object-particles whose masses have a fixed ratio to their comparison particles. This reduction is the negative gravitational energy of the planet, now appearing as a negative exclusion energy....

The exclusion treatment is..ill-adapted to..ordinary gravitational problems..; but an important result is obtained by applying it to the uniform distribution which forms the uranoid. By (32·3) [**F** (41·23)] the mean energy $\bar{E}$...is $\frac{3}{5}\mathfrak{E}$. Thus if M is the mass of a solitary hydrogen atom, the mass of the $\frac{1}{2}N$ atoms forming the static zero-temperature universe is $\frac{1}{2}N\bar{M}$, where

$$\bar{M} = \tfrac{3}{5}M. \tag{34·1}$$

The mass ordinarily recognized is M; for the reduction..in a large assemblage is treated separately as gravitational energy and not as a modification of M. The constant M applies also in Einstein's theory, where the reduction is eliminated by transformation to a non-Euclidean metric. Both Newtonian and Einstein theory effectively transfer every particle to the top level, so far as particle energy is concerned, and therefore give an energy and energy density of the universe $\frac{5}{3}$ times that given by the exclusion calculation. Nominally the difference is recoverable in Newtonian theory;..the potential due to the galaxies..would yield the energy $-\frac{2}{5}M$. In practice this correction is ignored; and, indeed, the Newtonian approximation is inadequate for the problem. In the relativity equations (8·1) [**F** (39·3)], actually used for the cosmical problem, the correction is absorbed in the metric..so that it is included in the scale relation between the energy tensor and the curvature tensor of space. To exhibit it, we write (8·42) [**F** (39·52)]

$$4\pi\kappa\rho = R_0^{-2} = 4\pi\bar{\kappa}\bar{\rho}, \tag{34·2}$$

where $\bar{\rho}$ is the density corresponding to $\bar{M}$, and $\bar{\kappa}$ is the corresponding constant of gravitation or, more appropriately, of *exclusion*. Since $\bar{\rho} = \frac{3}{5}\rho$,

$$\bar{\kappa} = \tfrac{5}{3}\kappa. \tag{34·3}$$

It appears then that the energy scale in the sub-threshold theory of the packed energy levels below $\mathfrak{E}$ is $\frac{3}{5}$ of the energy scale we subsequently adopt in the super-threshold theory of object-systems constructed above $\mathfrak{E}$....Consider a super-threshold process. By quantum theory we can calculate the change δM of the energy of a hydrogen atom excited from one level to another. The quantum constant h must be such as to give this energy on the same scale of reckoning as M. From the Newtonian view, M and δM are masses at the same gravitational potential; the latter has therefore a gravitational energy $-\frac{2}{5}\delta M$, which reduces the net energy to $\frac{3}{5}\delta M$. But in § 32 it is assumed the quantum calculation gives the net energy of each particle; the constant which gives the net energy must be distinguished as $\bar{h}$. Since $E \propto p^2 \propto h^2$, $\bar{h}^2$ must in super-threshold theory be increased to

$$h^2 = \tfrac{5}{3}\bar{h}^2 \tag{34·4}$$

in order to give the gross energy which is there employed.

By (34·3, 34·4) $\bar{h}^2\bar{\kappa} = h^2\kappa$, as required by (14·1) [**F** (8·1)].

Since $\mathfrak{E}$ has been identified with m_0, by (32·9) [**F** (43·6)] and (34·3)

$$m_0^2 = \mathfrak{E}^2 = \frac{3}{5} \cdot \frac{3}{4} \left(\frac{h}{2\pi R_0}\right)^2 N, \qquad (34 \cdot 4)'$$

which agrees with (28·8) [**6·1**]. Thus the exclusion theory gives the same masses of the particles as the gravitational theory.

Note. The rest of this MS. is an incomplete version of 'planoidal theory' (**F** V).

(§35): an odd page from a §35 contains formulae **F** (46·1, 46·2), (47·1–47·3) and ends (cf. **F** (47·4)):

Denoting the masses of the internal particle and the external V_3 particle by μ and M_3, we have $\mu = \frac{1}{136} m_0$, $M_3 = \frac{136}{3} m_0$. Hence

$$M_3 = \Sigma E_s = \Sigma p_s^2 / 2\mu. \qquad (35 \cdot 7)$$

36 (*a*). **Interaction of V_{10} particles:** like **F** §47, with extensions. It is first argued (as in the latter part of **F** §47) that the result (35·7) refers to particles for which the 'observational probing' consists of measures of three momenta or their conjugate coordinates, i.e. to V_3 particles.

The result is then transformed for V_{10} particles (extracules); the argument has points of interest, linked with the exclusion theory:

The V_3 particles are fictions.... Two criticisms of the foregoing analysis. First, it is adopted to the view that the particle under consideration is superposed on the rest of the universe.... But when we apply the representation of M_3 by half-quanta, it makes it necessary to introduce two independent half-quanta of a.m. for each pair of particles; having used the original half-quantum to form the mass M_3 of the first particle considered, we cannot use it again to form the mass M_3 of the second particle. ... But there is only one internal particle associated with a pair of particles... and only one half-quantum can be supplied. Revising the calculation so that each particle is alloted half the energy of their common internal particle, the resulting mass is $\frac{1}{2}M_3$.

The mass M of an external V_{10} particle is.. $3M_3/10$ by (25·6) [**6·1**]. Hence

$$\tfrac{1}{2}M_3 = \tfrac{5}{3}M. \qquad (36 \cdot 2)$$

This is reduced to M if the half-quantum is defined to be $\bar{h}/4\pi$ or $\frac{1}{2}\hbar$, as in the exclusion theory of mass, instead of $\frac{1}{2}\hbar$.

We can accordingly adopt (35·8) [? a formula related to **F** (47·4, 47·5)] so as to give a direct representation of the energy of a V_{10} particle, namely,

$$E_{10} = \{\tfrac{1}{2}(\bar{p}_1^2 + \bar{p}_2^2 + \ldots + \bar{p}_{\frac{1}{2}N}^2) + p_x^2 + p_y^2 + p_z^2)\}/2\mu. \qquad (36 \cdot 3)$$

Here $\bar{p}_s$ is the linear momentum for a.m. $\frac{1}{2}\bar{\hbar}$ about the sth planoid particle, and the factor $\frac{1}{2}$ occurs because the resulting energy is equally divided between the two particles. The observed momenta p_x... are formed with a constant $h = \bar{h}\sqrt{(\frac{5}{3})}$.. because the wave functions from which they are derived postulate a rigid field; and, of the energy $p_x^2/2\mu$, $\frac{2}{5}$ is a spurious addition due to our neglect of the negative gravitational energy, or due to our treatment in exclusion theory of every object as a top particle.

In Chapter III [**6·1**] the rigid-field condition was introduced in a purely analytical form. The exclusion principle provides a physical picture of a rigid environment.... In the general analytical treatment the rigid environment can be chosen to suit the multiplicity of the particles. The physical picture is inflexible...a rigid environment adapted only to particles of multiplicity 10. The fact that actual particles are complete vector particles of multiplicity 10 is therefore intimately bound up with the exclusion principle. To put it another way, the exclusion principle.. is physically associated with the fact that particles possess a spin 6-vector. Finally..apart from the exclusion principle there is no rule for determining the number of particles into which a given object is to be divided; the gravitating particles previously considered were scale-free. Thus the spin momentum may be regarded as the physical cause of the breaking up of continuous matter into discrete particles.

...The concepts of quantum,..modelled on classical dynamics, lead us to consider particles with three observable momenta. Initially therefore we deal with V_3 particles. But when we assign excluding properties, we transfer attention to top-level particles in a rigid exclusion field. This introduces a factor $\frac{5}{3}$ in the energies, increased to $\frac{10}{3}$ when we allow for double reckoning of mutual energy. This factor is absorbed by changing the multiplicity from 3 to 10; and it is therefore the V_{10} particles which appear in quantum dynamics as the precise analogues of the particles of classical dynamics.

We therefore distinguish *general* and *special* rigid-field theory,...in the latter..the exclusion field..the self-consistent gravitational field for particles of multiplicity 10, is the counterpart of the Galilean inertial field in molar theory....As we are approaching the more familiar and specialized applications of quantum theory, we shall..generally be concerned with the special field.

38. **Interchange:** a fragment related to **F** § 25, connecting notions of extra-spatial a.m. with indistinguishability.

When the half-quantum of a.m. is in the plane containing the phase direction s, the relative radial coordinate r_{12} of the two particles is coupled in rotation with the relative phase coordinate $s_{12}=s_2-s_1$; and the a.m. gives a circulation round the circle $r_{12}^2+s_{12}^2=\text{constant}$. The coordinate conjugate to the a.m. J is $\theta_{12}=\tan^{-1}(s_{12}/r_{12})$.

When J is given the exact (stabilized) value $\frac{1}{2}\hbar$, the p.d. over θ_{12} becomes uniform; and..θ_{12} ceases to appear as a classifying variate in our p.d.'s. But..the configurations $r_{12}=\rho$, $\theta_{12}=0$ and $r_{12}=-\rho$, $\theta_{12}=\pi$ are identical, so that the dropping of θ implies that r_{12} is not distinguished from $-r_{12}$.... In three-dimensional mechanics the distinction is retained; and the half-quantum circulation gives a continual flow of probability from $r_{12}=\rho$ to $-\rho$, and from $-\rho$ to ρ—in short, it continually interchanges the positions of the two particles.... We restrict the term 'interchange' to this extra-spatial interchange.... The extra-spatial transfer of probability...is a common concept in quantum theory...change by *transition* is as normal..as by *motion* is in classical theory.

37 (*a*). **The Newtonian potential:** this calculation, exemplifying planoidal treatment, of *the potential of Saturn's ring*, is quoted entire; it is not represented in other drafts or in **F**.

In the last two sections we have found a representation of inertia by half-quanta of a.m., which gives the same masses of the particles as the curvature and exclusion representations. A uniform static environment was postulated, and it is only in this case that the agreement of the three theories is established. As explained in § 13 [**5·1**], the uniform static distribution is the only common meeting ground of molar relativity and quantum theory; their methods and approximations diverge when there is non-uniformity, and we are not particularly concerned as to whether there is any corresponding half-quantum interpretation of the ordinary gravitational energy of molar bodies. It is, however, instructive to consider a rather simple case of non-uniformity.

As an example of a non-uniform but steady distribution we may consider Saturn and its ring, idealized as two bodies S_1, S_2 with constant separation a and masses M_1, M_2, the latter consisting of $\frac{1}{2}n$ hydrogen atoms of mass M. As standard environment we take a planoid of $\frac{1}{2}N_1$ hydrogen atoms, whose radius R_1 agrees with the radius R_0 of the uranoid, so that $N_1 = \frac{4}{5}N$ by (35·1) [**F** (46·1)]. The mean value [**F** (47·1)] of $1/r$ for the planoid is $3/2R_1$; but S_1 is in a modified environment with $\frac{1}{2}n$ particles at a distance a, which in standard conditions would have been spread over the planoid with a very much greater mean distance. Distinguishing the modified environment by an accent, we have[a]

$$\tfrac{1}{2}N_1[1/r'] = \tfrac{1}{2}N_1[1/r] + \tfrac{1}{2}n(1/a) = \tfrac{3}{4}(N_1/R_1) + \tfrac{1}{2}(n/a). \qquad (37·1)$$

By (28·22) [**F** (5·41)] $\kappa M = \pi R_0/N = \frac{4}{5}\pi R_1/N_1,$ (37·2)

so that by (34·3) [**F** § 42]

$$R_1/N_1 = \tfrac{3}{4}\bar{\kappa}M/\pi \quad [\bar{\kappa} = \kappa_1 \text{ in } \mathbf{F}]. \qquad (37·3)$$

By (37·1, 37·3) $[1/r'] = [1/r](1+\alpha),$ (37·4)

where $\alpha = \pi^{-1}\bar{\kappa}M_2/a.$ (37·5)

Since the energy is proportional to the square of the momentum in (35·8), we suppose tentatively that the mass M_1 is increased in the ratio $(1+\alpha)^2$ by the modification of the environment. Since α is small, the increase is

$$2\alpha M_1 = \frac{2}{\pi}\frac{\bar{\kappa}M_1 M_2}{a}. \qquad (37·6)$$

This is an increase of quantum energy which, owing to the inversion of mass, appears as a decrease of classical energy. Except for the factor $2/\pi$, it agrees with the potential energy of Saturn due to the ring.

[a] The *square* brackets denote an average as in **F** §47.

The use of the mean value $[1/r']$ implies that the field can be described by simple additive Newtonian potentials. In assuming this holds not only for the small gravitational field due to the ring but also for the large inertial field which gives the whole rest mass M_1, we are evidently straining the Newtonian approximation too far. It is therefore not surprising that the result is in error by a numerical factor. The point to which we wish to call attention is that this method gives substantially the right result, whereas the method which at first sight seems more obvious gives a totally wrong result. The latter method is to insert terms corresponding to the $\frac{1}{2}n$ particles of the ring in (36·3); the additional energy is then

$$\frac{n}{8\mu}\left(\frac{\hbar}{2a}\right)^2 \tag{37·7}$$

per particle of M_1. Not only does this give the wrong law of potential (varying as a^{-2}), but the order of magnitude is much too large.

The a.m. representation of inertia must therefore be coupled with a condition, which admits the treatment (37·6) but excludes the treatment (37·7). The condition is that non-uniformities are caused by modifying the distribution function applying to all the $\frac{1}{2}N_1$ particles, not by assigning a restricted distribution in space to a small number of them. Thus Saturn's ring is a region in which each of the $\frac{1}{2}N_1$ particles has a probability n/N_1 of occurring; so that the expectation number of particles in the ring is $\frac{1}{2}n$, but the particles are not identified.

37 (*b*), 36 (*b*). **The Newtonian potential:** two fragments.

7·2. Draft B *19* V. Electric Charge

There is no extant Chapter IV of Draft **B**. The following Chapter V has been assigned to **B**; it is roughly analogous to **F** III §§ 26–28 and 33, but is aimed more at the wave equation and the Coulomb energy. Considerable use is made here of the quadratic wave equation.

5·1. **The two-particle transformation:** like **F** § 26. The two-particle energy formula **F** (26·22) is regarded as applying to a *quantal energy tensor* defined *ad hoc* by

$$T_{\alpha\gamma} = \frac{p_\alpha p_\gamma}{m} = -\frac{\beta^2}{m}\frac{\partial^2}{\partial x_\alpha \partial x_\gamma} \quad [\beta \text{ denotes } \hbar \text{ in Draft } \mathbf{B}]. \tag{5·141}$$

For the *wave operator* $\Box = g^{\alpha\gamma}\partial^2/\partial x_\alpha \partial x_\gamma$, by **F** (26·14),

$$\Box_x/m + \Box_{x'}/m' = \Box_X/M + \Box_\xi/\mu. \tag{5·143}$$

The hamiltonians h, h', H are as in **F** (26·42), η as in (26·44), so that

$$h + h' = H + (\eta - \mu). \tag{5·152}$$

The exclusion principle is unchanged by the two-particle transformation [cf. **F** §41]. The normalization volumes for m, m', M, μ are taken equal as in **F**. The last paragraph is:

When we apply these formulae to transform two ordinary particles into an external and internal particle,...these are carriers respectively of the initial and transition energies; so that they are not identified with the particles M, μ unless the particle M is in the initial state of rest. Thus the space-time frame must be chosen so that

$$P_1, P_2, P_3 = 0 \quad \text{[external particle]}. \tag{5·19}$$

In the applications which follow it is understood that this is fulfilled.

5·2. **The electrical stabilization:** the argument here on the effect of the 'electrical degree of freedom' plays a similar role to the later concept (**F** §§27, ..) of 'freeing the intracule'. The quoted version is condensed.

The transformation in §5·1 applies primarily to a perfect bi-particle ($k = 137$), since $x'_\alpha - x_\alpha$ is distinguished from $x_\alpha - x'_\alpha$. But in practical application we need to couple with it an electrical stabilization (§4·6) [not extant] reducing k to 136, in order that ξ_α may be the reversible internal coordinate whose steady distributions constitute the eigenstates of the system.

Since our previous work has been based on the standard bi-particle ($k = 136$), we proceed conversely and consider the de-stabilization which converts the standard into the perfect bi-particle...The mass of a perfect bi-particle at rest is $136m_0/137$. But..we give a different.. interpretation to electrical stabilization. The essence of the change of k from 136 to 137 is that 'rest' is differently defined; part of what was formerly counted as interchange of indistinguishable states is now recognized as motion, and has to be eliminated before the perfect bi-particle is considered to be at rest. In this view the change $\frac{1}{137}m_0$ in the rest mass is caused by the removal of the energy of concealed motion. Since quantum energy has the opposite sign, eliminating the concealed motion..makes the E-energy of the perfect bi-particle greater than that of the standard bi-particle ($-\frac{136}{137}m_0 > -m_0$). We have therefore to give the standard bi-particle a positive transition energy $\frac{1}{137}m_0$ in converting it into a perfect bi-particle at rest....

In replacing standard..by perfect bi-particles we allow...a *change of reckoning* of energy—because a form of energy unprovided for in the purely mechanical theory is brought into account. The changed definition of 'rest' gives a new picture of the state of motion and we change the reckoning of energy to suit.

The state of rest..of the standard bi-particle is adopted as the initial state for both types..so both have initial energy m_0. In this state the perfect bi-particle has an energy of motion $\frac{1}{137}m_0$; and in a state with transition energy E_t it has energy of motion $E'_t = E_t + \frac{1}{137}m_0$. In electrical de-stabilization we change the zero reckoning of transition energy by

$\frac{1}{137}m_0$ so as to make the transition energy E'_t instead of E_t and therefore equal to the energy of motion.

The de-stabilization then increases the total energy (apart from transition energy) from $136m_0$ to $137m_0$, and the field energy from $137m_0$ to $138m_0$. The extra m_0..identified as the energy of the self-consistent electric field corresponding to the [new] initial state....

We consider a perfect bi-particle with wave function $\Psi(x_\alpha, x'_\alpha)$, and take m, m' to be the *current masses* of the electron and proton...so to conform with our regular notation we must rewrite (5·13) [**F** (26·21)] as

$$p_\alpha p_\beta/m + p'_\alpha p'_\beta/m' = P_\alpha P_\beta/M + \varpi_\alpha \varpi_\beta/\mu', \tag{5·21}$$

where $\mu' = \frac{137}{136}\mu$. The r.h.s. operates on the same wave function Ψ. When a standard bi-particle is substituted on the right a different wave function $\Phi(x_\alpha, \xi_\alpha)$ is involved, and the operator is changed..to give the same result on Φ as the old operator on Ψ. The required change of operator ...: incorporate in the r.h.s. (a) the change $\frac{1}{137}m_0$ in reckoning of transition energy and (b) the interchange energy and momentum given by the half-quantum rule, which is additional to the momenta P_α, ϖ_α (operating on Φ) but is included in p_α, p'_α, ... The net result [of (a): (b) is deferred] is that the transition energy derived from Φ is $\frac{1}{137}m_0$ less than that derived from Ψ; so that this must be added on the r.h.s. of (5·21) when the two-particle transformation includes electrical stabilization.

We first apply this result to the hamiltonians [in §5·1]. Setting $p^2 = p_1^2 + p_2^2 + p_3^2$, etc., by (5·21)

$$m + p^2/2m + m' + p'^2/2m' \approx M + P^2/2M + \tfrac{1}{137}m_0 + \varpi^2/2\mu'. \tag{5·22}$$

..$\approx$ denotes 'operational equality', i.e. equality of the results of applying to the respective operands. Since $\mu = \frac{1}{137}m_0 = \frac{136}{137}\mu'$, the last two terms give

$$\tfrac{136}{137}(\mu + \varpi^2/\mu) = \tfrac{136}{137}\eta.$$

Thus (5·22) reduces to

$$h + h' = H + \tfrac{136}{137}\eta. \tag{5·23}$$

We have seen (§3·8) [i.e. **6·2** §3·9] that current theory, by assuming the radiated is equal to the transition energy, obtains the latter too large by $\frac{137}{136}$, and that it is from these enlarged values that the current masses are derived. If the enlarged value $\eta' = \frac{137}{136}\eta$ is employed, (5·23) becomes

$$h + h' = H + (\tfrac{136}{137})^2 \eta'. \tag{5·24}$$

Our second application is to the wave operators $\Box$...a wide new field. The three vector particles represented in ordinary space-time should have rest masses equal to their mass constants, so that

$$p_4^2 - p^2 = m^2, \quad p_4'^2 - p'^2 = m'^2, \quad P_4^2 - P^2 = M^2. \tag{5·25}$$

Multiplying (5·21) by $\delta^{\alpha\beta}$ (the Galilean $g^{\alpha\beta}$), and inserting the extra energy $\frac{1}{137}m_0$ on the right, we obtain by (5·25)

$$m^2/m + m'^2/m' \approx M^2/M + (\varpi_4^2 - \varpi^2)/\mu' + \tfrac{1}{137}m_0,$$

so that

$$\varpi_4^2 - \varpi^2 = -\tfrac{1}{137}\mu' m_0 = -\mu^2. \tag{5·26}$$

Thus the internal particle has an imaginary (space-like) momentum vector of length $i\mu$.

Alternatively, if we assign space-like momentum vectors to the three particles in ordinary space, so that

$$p_4^2 - p^2 = -m^2, \quad p_4'^2 - p'^2 = -m'^2, \quad P_4^2 - P^2 = -M^2, \tag{5·25a}$$

we obtain $\quad m^2/m + m'^2/m' \approx M^2/M - (\varpi_4^2 - \varpi^2)/\mu' + \frac{1}{137}m_0,$

so that $$\varpi_4^2 - \varpi^2 = \mu^2. \tag{5·26a}$$

Particles with momentum vectors p_α and ip_α will be called *antithetic*. We see x- and ξ-space-time have antithetic character, particles in the one being antithetic to those in the other...[cf. **6·2** §3·4]. Owing to the inversion of mass, the transition momenta p_1, .. (now called ϖ_1, ..) are taken in quantum theory to be i times the classical momenta...The formal expression of antithetic character of coordinates is postponed until we develop wave-tensor calculus....

Choosing the equations (5·24) and (5·25a), and inserting the operational values of the momenta, we obtain the four 'wave equations'

$$\left.\begin{array}{ll}(-\beta^2\square_x - m^2)\,\Psi = 0, & (-\beta^2\square_{x'} - m'^2)\,\Psi = 0,\\ (-\beta^2\square_X - M^2)\,\Phi = 0, & (-\beta^2\square_\xi - \mu^2)\,\Phi = 0.\end{array}\right\} \tag{5·27}$$

These are regarded as standard although..not self-consistent; the function Φ satisfying the last two represents a combination of the external with the antithetic internal particle...Although m and m' are current masses, M and μ are standard masses. If we had taken m, m' to be standard masses, (5·26) would have been $\varpi_4^2 - \varpi^2 = -\frac{1}{137}\mu m_0$, and the fourth equation in (5·27) would become

$$(-\beta_s^2\square_\xi - \mu^2)\,\Phi = 0, \quad \beta_s = (\tfrac{137}{136})^{\frac{1}{2}}\beta. \tag{5·28}$$

5·3. The time coordinate: this has ideas in common with **F** §35 (and **4·2** §1·9 (b)). The quotations are mainly relevant to the nature of the internal phase coordinate.

...A distribution function $f(x, y, z; t)$ represents the p.d. *over* x, y, z *at* time t. If, however, the object-system contains a time indicator, e.g. the phase θ of an internal rotation or vibration, we can consider a p.d. over θ....We have then..$f(x, y, z, \theta; t)$. But a time indicator carried by a particle registers proper time. Thus θ, though in a sense deputizing for t, has not the relativistic properties of the coordinate t. Moreover, a 'time indicator' is a space-like coordinate, e.g. the position angle of the minute hand of a watch.

Time appears in two aspects—as *measure* and as *epoch*. Time as a measure can be replaced by the spatial coordinate of a time indicator. In a hydrogen atom at rest the time T of the external particle can be regarded as the reading of a time indicator contained in the atom, and the energy M conjugate to T is then the energy of this internal indicator. By this substitution M and T become space- instead of time-like; this is the anomaly discovered in the last section, namely, the space-time in which external characteristics are represented is antithetic to that of the internal characteristics....

It has already been found necessary to introduce a phase θ, or in linear measure s, as a fourth coordinate in §4·1, though without any suggestion it should act as time indicator. The principal application has been in... interchange, the interchange momentum being in the direction of the relative phase coordinate $s_{12}=s_2-s_1$. For uniformity with (5·11) [**F** (26·12)] we shall denote the phase coordinate of the internal particle by $\sigma=s'-s$.

The condition that σ be a simple time indicator is that the internal particle have a constant σ-momentum independent of its coordinates ξ_1, ξ_2, ξ_3. Adopting a linear measure of phase such that s is homologous with $x_1, \ldots, t$, we have simply

$$\sigma=i\tau=i(t'-t). \qquad (5\cdot31)$$

Thus in a two-particle system *the linear interchange coordinate can be represented as an imaginary time difference introduced in combining the two particles.*

Since σ is real by definition, there is no real time difference; i.e. a 'system' consists of particles contemplated simultaneously so far as real time is concerned. Simultaneity is here defined by the frame in which the external particle is at rest, since the state of rest is the initial state which determines the amount of transition energy and momentum allotted to the internal particle.

From the physical view an interchange coordinate is more intelligible than imaginary time, and we should naturally use σ rather than τ in the wave equation. But the standard wave equation (5·27) for Φ refers to the antithetic particle, so that τ there represents the real interchange coordinate, the τ of the antithetic particle being the σ of the actual particle.

The identification of $i\tau$ with σ depends on the σ- or τ-momentum being constant. Accordingly, in determining eigenstates we couple with the wave equation the condition

$$-i\beta\,\partial\Phi/\partial\tau=\epsilon\Phi, \qquad (5\cdot32)$$

where ϵ is a numerical constant; i.e. we consider it to be part of the definition of an eigenstate that it shall have a numerical internal energy $\varpi_4=\epsilon$. By (5·32) Φ will contain a factor $e^{i\epsilon\tau/\beta}$. This applies to the antithetic particle; in the wave function of the actual particle the factor would be $e^{i\epsilon\sigma/\beta}$. The common impression that the hydrogen atom wave function contains a factor $e^{i\epsilon\tau/\beta}$..has no justification, and leads to quite erroneous attempts to develop the relativistic theory of the atom....

5·4. **Quadratic and linear energy:** this continues the approach to the wave equation (§5·5), ending with 'interchange a.m.' The earlier part on 'quadratic and linear energy' is relevant to the discussion of *super-dense matter* in **F** §§44, 45.

We have two expressions for the energy..the 'quadratic hamiltonian'

$$h = m + (p_1^2 + p_2^2 + p_3^2)/2m. \qquad (5\cdot411)$$

The four-dimensional treatment provides an alternative

$$p_4 = (m^2 + p_1^2 + p_2^2 + p_3^2)^{\frac{1}{2}}. \qquad (5\cdot412)$$

In VI we...express $\sqrt{(-p_1^2 - p_2^2 - p_3^2 + p_4^2)}$...as

$$m = E_1 p_1 + E_2 p_2 + E_3 p_3 - E_4 i p_4, \qquad (5\cdot421)$$

and..derive $$p_4 = -i(E_4 m + E_{14} p_1 + E_{24} p_2 + E_{34} p_3). \qquad (5\cdot422)$$

This is the 'linear hamiltonian'.

...h is the energy for a particle characterized by a root vector, p_4 for a particle characterized by a true momentum vector.

When the multiplicity is reduced from 136 to 10 by the condition that the energy tensor is the square of a vector, the sign of the vector is ambiguous. The same pseudo-discrete state (as classified by the energy tensor) is therefore represented indifferently by p_α and $-p_\alpha$. The state defined by a root vector $\pm p_\alpha$ may therefore be regarded as a superposition of two substates (with equal occupation in a steady distribution) defined by true momentum vectors p_α and $-p_\alpha$. The 'velocity' of the particle has definite magnitude but ambiguous sign $\pm v$, and does not imply any progressive displacement. Without altering the multiplicity we can reclassify the states so as to recognize the distinction (not indicated in the energy tensor) between p_α and $-p_\alpha$; i.e. the unambiguous momentum vector is substituted for the root vector as the classifying characteristic of the states. The corresponding particles have a single-valued progressive velocity v.

The most familiar distinction between (5·411, 5·412) is that the latter takes account of 'change of mass with velocity'. But change of mass assumes a one-way velocity, and is clearly inapplicable to a two-way velocity $\pm v$. A distribution of root-vector particles with velocity $\pm v$ is not equivalent to a distribution of true vector particles half with velocity v and half with $-v$; for the latter form a rapidly dispersing system. They can be converted into a static system by continual interchange of momentum between the groups,...or of position. In either case the energy (5·412) is modified by...interchange energy. In the root vector particles the energy required to maintain equilibrium is embodied in the particles themselves; and the continual interchange of $+v$ and $-v$ appears in the attribution to the particle of both velocities with equal probability.

The wave functions corresponding to (5·411) are *standing waves*,.. to (5·412) *progressive waves*....

Plane progressive waves are not important in equilibrium theory.... But a one-way circulatory motion is compatible with equilibrium; and the practical application of the linear hamiltonian is to systems in which there is a steady circulation. In the theory of the atom we are concerned with states of one-way rotation, corresponding to progressive waves along an angular coordinate; and the interchange circulation is also one-way. Thus the internal states of an atom have to be specified by true momentum vectors, as distinguished from root vectors; and the linear hamiltonian, in the rationalized form (5·422), is the basis of the theory of atomic eigenstates.

Corresponding to the quadratic and linear hamiltonians..[we have] Schrödinger and Dirac wave functions...a Dirac wave function introduces a half-quantum of 'spin' a.m. in each plane in which the system is relativistically rotatable with respect to its environment. This is a consequence of the indeterminacy of orientation in such a plane. Besides the spin momentum in spatial planes there arises in the same way a spin momentum of the internal particles in the plane of τ (or σ) and the relative radius vector $\rho = (\xi_1^2 + \xi_2^2 + \xi_3^2)^{\frac{1}{2}}$, since the quasi-Lorentz invariance of the wave equation makes this a plane of relativistic rotation. This constitutes the interchange a.m.—the correction (*b*) in §5·2, which has still to be introduced into the wave equation. There is one important difference; the interchange a.m. is a whole quantum. The difference arises because a spatial rotation of the axes affects x_1, x_2, x_3 and ξ_1, ξ_2, ξ_3 equally, so that the wave functions ψ_e and ψ_i of the external and internal particles are transformed together. A half-quantum of spin momentum is added to each particle; but in practice we recognize only the half-quantum associated with ψ_i, since we are not interested in the wave representation of the external particle. On the other hand the $\tau\rho$ or $\sigma\rho$ transformation affects only the coordinates of the internal particle.... Rotatability in this plane gives the system a whole quantum of spin momentum as before; but this is now wholly associated with the internal particle instead of being divided between the two particles.

5·5. **The Coulomb energy:** one basis of **F** §33. This is the crux of the present exposition.

The alteration of the rigid field when the multiplicity is changed by stabilization does not affect p_1, p_2, p_3, but by (4·761) p_4 varies as k. The expressions

$$\eta = \mu + (\varpi_1^2 + \varpi_2^2 + \varpi_3^2)/2\mu, \quad \epsilon = (\mu^2 + \varpi_1^2 + \varpi_2^2 + \varpi_3^2)^{\frac{1}{2}}, \tag{5·51}$$

giving the quadratic and linear energy of an internal particle in terms of the spatial momenta and mass constant, are unaltered by change of the rigid field; and they are valid in any field since the energy of an internal particle, being wholly transition energy, is independent of the field. But we must not put $\varpi_4 = \epsilon$ unless the field is Galilean. For a general field the relation is

$$\epsilon = (\omega_4 \varpi^4)^{\frac{1}{2}} = \varpi_4/k = -(i\beta/k)\,\partial/\partial\tau_r, \tag{5·52}$$

where τ_r is used to denote the τ-coordinate in the rigid field considered. Thus (5·421) adapted to an internal particle in rigid coordinates becomes

$$\left\{-i\beta\left(E_1\frac{\partial}{\partial\xi_1} + E_2\frac{\partial}{\partial\xi_2} + E_3\frac{\partial}{\partial\xi_3} - \frac{i}{k}E_4\frac{\partial}{\partial\tau_r}\right) - \mu\right\}\Phi = 0. \tag{5·53}$$

We have to insert the interchange momentum in this equation. The corrections (*a*) and (*b*) in § 5·2 appear at the same stage of transformation; so that the interchange momentum is to be inserted at the same time as the correction $\frac{1}{137}m_0$, namely, when the system is in the rigid field corresponding to $k = 137$. We have seen in § 5·4 that it amounts to one quantum β ($=h/2\pi$) of a.m. This will give a linear momentum β/ρ in the direction of the phase coordinate σ of the actual internal particle, or of the coordinate τ of the antithetic particle employed in Φ. This momentum has therefore to be added to the component $-i\beta\,\partial/\partial\tau_r$ in that direction explicitly shown in Φ. Considered as classical energy it is positive, since it represents a motion of the particles concealed by the indistinguishability of interchanged states; it will therefore be a negative quantum energy.

Thus we incorporate the interchange energy in (5·53) by changing $-i\beta\,\partial/\partial\tau_r$ to $-i\beta\,\partial/\partial\tau_r - \beta/\rho$; or equivalently, by changing

$$-i\beta\frac{\partial}{\partial\tau} \quad \text{to} \quad -i\beta\frac{\partial}{\partial\tau} - \frac{\beta}{137}\frac{1}{\rho}, \tag{5·54}$$

since $k = 137$. Thus, reverting to Galilean coordinates, the completed wave equation of the internal particle is

$$\left\{-i\beta\left(E_1\frac{\partial}{\partial\xi_1} + E_2\frac{\partial}{\partial\xi_2} + E_3\frac{\partial}{\partial\xi_3}\right) - iE_4\left(-i\beta\frac{\partial}{\partial\tau} - \frac{\beta}{137}\frac{1}{\rho}\right) - \mu\right\}\Phi = 0. \tag{5·55}$$

With the usual $h/2\pi$ for the momentum constant β, and the factor c for ordinary units, this becomes

$$\left\{-\frac{ih}{2\pi}\left(E_1\frac{\partial}{\partial\xi_1} + E_2\frac{\partial}{\partial\xi_2} + E_3\frac{\partial}{\partial\xi_3}\right) - \frac{iE_4}{c}\left(-\frac{ih}{2\pi}\frac{\partial}{\partial\tau} - \frac{e^2}{\rho}\right) - \mu\right\}\Phi = 0, \tag{5·56}$$

where

$$e^2 = \frac{1}{137}\frac{hc}{2\pi}. \tag{5·57}$$

Except that current theory confuses the internal particle with the electron, (5·56) is the recognized wave equation of the hydrogen atom, $-e^2/\rho$ being the Coulomb energy...This establishes the identity of the Coulomb energy with the interchange energy. This identity,..pointed out in 1928, may be regarded as the distinctive feature of the unified theory set forth in this book. It seemed clear at the time that those who chose to disregard it were following the wrong track in their attempts to relate quantum to relativity theory; so that from that date the author's researches have been uninfluenced by what is often called the 'accepted' relativistic quantum theory—accepted in the sense that it is considered important to discover why it breaks down.

By (5·57) the fine-structure constant $hc/2\pi e^2$ is precisely 137.

The complete theory..of hydrogen must take in the interaction.. with the radiation field; in other respects (5·56) is exact, i.e. the Coulomb term $-e^2/\rho$ is a complete expression of the interaction of a proton and electron....

5·6. **Pairing:** an involved argument (discarded in later versions), based on the ambiguity of sign in (5·412) above as affecting the internal phase space, and using 'planoids'. The conclusion concerns

the physical interpretation of the mass $\frac{1}{137}m_0$ introduced by electrical stabilization. A standard bi-particle ($k=136$) in an assemblage in an initial state of rest in the self-consistent gravitational field has a mass m_0. When the additional d.f. corresponding to interchange is admitted, electrical energy is added; and by (5·64) the Coulomb energy in this initial state amounts to a mass $-\frac{1}{137}m_0$. Thus the bi-particle, which has now become a perfect bi-particle ($k=137$), has a total mass $\frac{136}{137}m_0$. This agrees with the ordinary rule that the mass is inversely proportional to k. In forming the wave equation (5·56), which applies to any steady state of the bi-particle, we first add $\frac{1}{137}m_0$ in order to remove the Coulomb energy $-\frac{1}{137}m_0$ present in the initial state, and then insert the Coulomb energy $-e^2/\rho$ present in the state considered.

5·7 (*a*). (No title): a fragment on the charge induced by a proton in a planoid.

5·7 (*b*). **The electromagnetic potential:** a short fragment.

CHAPTER 8

DRAFTS OF F I–V

(v) CHAPTERS I–IV OF DRAFT H

8·0. Preliminary note

In Draft **H** we meet the plan, and much detail, of **F** in its final form. Quotations from **H** do not show many variants in the concepts, but illuminate the argument on some points. The main quotations (**8·3**) are concerned with the electrical theory of **F** III. Draft **H** IV became **F** IV and V; so in **H** I–IV we review the whole of the statistical theory.

8·1. Draft H *13* I. The Uncertainty of the Reference Frame

This is close to **F** I.

1. **The uncertainty of the origin:** close to **F** § 1.

2. **The Gaussian distribution:** close to **F** § 2, with slight variants as in **D** § 1.

3. **The Bernoulli fluctuation:** close to **F** § 3, omitting the final paragraphs of the full text.

4. **The standard of length:** close to **F** § 4, omitting discussion of the constancy of the velocity of light.

5. **Range of nuclear forces...:** like **F** § 5.

6. **Non-uniform curvature:** close to **F** § 6. A remark:

...we are not attempting here to give an independent determination of the values of σ and σ_ϵ already found in § 3. What we have here found is that a naïve interpretation of the four-dimensional picture, rather unexpectedly, turns out to be valid; so that the uncertainties are not distorted in passing from three to four dimensions.

7. **Uranoids:** like **F** § 7. Two amplifications, (i) on gravitation, (ii) on using the whole uranoid, occur here typifying Eddington's views:

...Einstein's theory removed the absolute distinction between gravitation and inertia; in practice we still separate them conventionally, and for example, distinguish the gravitational energy $-m_e\phi$ of an electron in a field of $g_{\mu\nu}$ from its inertial energy m_ec^2. The distinction implies that in the standard conditions in which m_e is supposed to be measured the inertial-gravitational field is wholly inertial. Thus..the standard uranoid provides the inertial part, the deviation of the actual environment provides the gravitational part....

...A complete uranoid is considered...even if effectively we are concerned with no more than a cubic mm....this extrapolation is modest compared with non-relativistic quantum theory which employs *infinite* plane waves for the same purpose.

8. **The extraneous standard:** close to **F** § 8. A footnote on scale:

The actual scatter of our measures is reduced by assigning part of it to the fluctuation of the scale-unit used in the measurements. This agrees with the condition that the extraordinary fluctuation combines negatively (§ 3).

9. **Scale-free physics:** close to **F** § 9.

10. **Pseudo-discrete distributions:** like **F** § 10, with amplifications. On 'pseudo-discrete':

A set of distribution functions $f_\alpha(x, y, z)$ merges into a continuous $f(x, y, z, \alpha)$ when α becomes continuous; but this does not apply to wave functions. This is because a continuum is not the limit of a discrete set of eigenstates. (*Proc. Roy. Soc.* A, **174**, 28, 1940. This has also been stressed by [H. A. C. Dobbs], *Phil. Mag.* [(7), **34**, 651, 723] 1943, in a critical examination of Dirac's Representation Theory.) For example, consider the inclination of the plane of a.m. of an electron in an atom. The inclination, referred to a controlling plane xy, has a discrete set of eigenvalues, which become closer together as the a.m. increases, and may be considered to merge into a continuum as the a.m. tends to infinity. This is an example of 'continuity by close-packing'. But actual continuity of inclinations comes about in a different way through the absence of any controlling plane xy; then there is a set of eigen-inclinations, more or less numerous, for *every* possible plane. This is an example of 'continuity by symmetric degeneracy'. The latter continuity corresponds to a much higher order of infinity—the infinity of points on a line instead of the infinity of rational numbers. The difference which directly affects wave analysis is that in continuity by close-packing the contiguous eigenstates are orthogonal, whereas in continuity by degeneracy they are not orthogonal.

When the eigenstates are continuous, their separate occupation factors $j_1, j_2, \ldots$ are replaced by a continuous occupation function $j(\alpha)$ of the parameter, or set of parameters, α defining the states, $j(\alpha)\,d\alpha$ being the probability of the states comprised in $d\alpha$. But, since the wave functions cannot be made continuous functions of α, we divide the domain of α into arbitrary small ranges $\delta\alpha_r$ which are treated as infinitesimal. A finite amount of probability concentrated in one of these ranges constitutes a pseudo-discrete state.

A final paragraph is a first discussion of the relation of electrics and mechanics:

It is worth while setting down...a general idea of the relation of scale-free mechanical to scale-fixed electrical theory....In some sense the electrical characteristics form the square root of the mechanical. Signs of this..appear..; the scale-fixed characteristic is a first-rank tensor (momentum vector), the scale-free a second rank (energy tensor). The ambiguity of sign of a square root gives us positive and negative charges—a distinction with no parallel in mechanics. Dirac found it necessary to take the square root of the mechanical (quadratic) hamiltonian to obtain the hamiltonian of a fixed-scale particle (electron). In wave analysis the ultimate element is a 'pure' state characterized by an idempotent vector; this enables electrical and mechanical characteristics to be united in one vector without infringing the condition that the mechanical characteristics belong to a vector which is the square of that providing the electrical characteristics.

11. **Stabilization:** like **F** § 11. One sentence is:

We shall find that the number of dimensions of the p.d. enters as a coefficient into the leading formulae of quantum mechanics....

In **F**, Eddington replaces 'quantum mechanics' by 'fundamental theory'; this term occurs also in **F** § 39, and the related passage in **B** (p. 66).

8·2. Draft H *10* II. Multiplicity Factors

This is close to **F** II, leaving **F** §§ 22, 23 to the next chapter. The mass quadratic is reached most expeditiously here from the partition of the rigid field; other views of partition, such as the 'top particle', 'rigid coordinates' and 'mass inversion', are given afterwards.

12. **Object-fields:** like **F** § 12. One amplification on 'field':

In our microscopic theory..'field energy'..will refer to the complementary field. Extraneous fields are molar..usually specified by their potentials which are connected..with their energy tensors. Interactions between the particles of a microscopic object-system are commonly described as fields—intra-atomic fields—..this..misleading; ..the energy of this interaction forms part of the particle energy tensor and, like any other particle energy, has its own complementary field energy representing the disturbance which it creates in the environment.... 'Field'..as we use it, implies some..molar averaging....Field variates are thereby distinguished from particle variates in which there is no smoothing of fluctuations by averaging.

13. **The rigid-field convention:** like **F** § 13. A final paragraph on the 'self-consistent field':

Self-consistent fields..familiar through Hartree's calculations of eigenfunctions..in atoms. We use the same conception..with a very different kind of field; and the analogy must not be pressed too closely. Our analysis is..for the complementary (..gravitational) field arising from interaction of the microscopic system with its environment...doubtful if this has any useful application to so-called intra-atomic fields....

14. **Separation of particle and field energy:** close to **F** § 14.

15. **Application to scale-free systems:** close to **F** § 15; but the last paragraph is omitted, and there is an account of 'initial' and 'transition' particles (cf. **F** § 18).

...The generic energy tensor $X_{\mu\nu}$ gives the particle energy expected.. [kinematically], disregarding the rigid field. Special relativity does not apply in a gravitational field, and the actual particle energy $E_{\mu\nu}$ is not Lorentz-covariant. The meaning of (15·92) is..the transition energy is $-k$ times the amount expected. It is understood that we are dealing with transitions of unidentified members of a large assemblage of which the great majority are in the initial state; thus for practical purposes it is necessary to choose an initial state which is likely to occur in nature.

The theory is simplified if...the initial and transition energy are associated with different particles. In that case we distinguish *Initial*.. and *Transition particles*...[**F** § 18]. The classical procedure is to replace the system by *external*..and *internal particles*...the rest energies of the internal are zero....These correspond to initial and transition particles. The effect of the rigid field is to give an unexpected factor $-k$ in the internal energy tensor....

Note. **F** § 16 is represented by part of § 18 below.

16. **Standard carriers:** close to **F** § 17.

17. **Mass ratio of the proton and electron:** mainly in **F** § 18. The section begins at **F** (18·21); a standard carrier of mass m_0 makes a transition to a momentum state with the particle energy (with quantal reckoning of momenta) **F** (18·32). That the carrier is 'composite' is emphasized

by the fact that the rest energy m_0 does not agree with the 'mass constant' $\mu = \frac{1}{136} m_0$.

The carrier is divided into two carriers (**F** (18·33)) of E_e, E_i, which are stabilized to be vector particles ($k = 10$) *after* the separation. The *effect of multiplicity changes* (i.e. of stabilization) *on mass* has now to be established. As the 'top particle' method (**F** § 16) has not yet been given in this version, the method is essentially the alternative one as at the end of **F** § 18 (full text); but this derivation of the fundamental (17·42) [**F** (16·5)] is the most simple and general and deserves quotation.

Consider a distribution of s particles per unit volume at almost exact rest, and let the total energy density T_{44} be given. By (15·7) [as in **F**] the density apportioned to the particles is $\rho = E_{44} = -T_{44}/k$. Thus if k is changed by stabilization...$\rho_1/\rho_2 = k_2/k_1$. If s remains unaltered, the masses of the particles...are $m_1 = \rho_1/s$, $m_2 = \rho_2/s$; so that

$$m_1/m_2 = k_2/k_1. \qquad (17\cdot42)$$

This applies to initial energy...changes E_e from m_0 to $\frac{136}{10}m_0$. The transition energy E_i is wholly particle energy, and is unaffected....

The results **F** (18·4–18·9) then follow.

18. **The fine-structure constant: F** §§ 16, 20. 'Multiplicity factors' may be grasped through the method of 'rigid coordinates' (§ 19 below) or *the principle of the top particle*. The latter is described much as in **F** § 16 up to (16·3); then the argument proceeds:

The important formula (17·42) was obtained by changing k..with the total, i.e. the mean, energy tensor unchanged..so the molar characteristics and curvature...are constant. The change is only in the top density and mass of a top particle..[as k^{-1}]. This..is the mass of a quantum particle superposed without disturbance....Denoting the top density by ρ, we may either give the mean environment particle a density $-k\rho$ or, if we treat all the particles as having the same mass and density as the object [top] particle, we must introduce a field density $-(k+1)\rho$ per particle.

In this we postulate an environment composed of the same kind of particles as the object system....

We must always begin by dividing the molar distribution into standard particles. The relation (15·7)..is then

$$W_{\mu\nu} = -137E_{\mu\nu}. \tag{18·4}$$

The coefficient 137 is called the *fine-structure constant.*

The full argument of **F** § 20 on the fine-structure constant then follows.

19. **Rigid coordinates:** close to **F** § 19.

20. **Unsteady states:** the calculation is as in **F** IV § 34 (see note in **3·4**). The only result used immediately (§ 21; cf. **6·1** § 23) is the correct steady-state formula $2K+V=0$ for a gravitational system.

21. **The inversion of energy: F** § 21 is slightly expanded from this. A note on $\sqrt{-1}$:

..we write $m-\frac{1}{2}mv^2$ [the whole energy in the rigid field in inverted reckoning] as $m+\frac{1}{2}m(iv)^2$, thus attributing velocity iv. Correspondingly the particle in the rigid field has momentum i times the classical, as noted in (17·31) [**F** (18·31)]. This momentum in the rigid field is the 'quantum momentum' represented by the operator $-i\hbar\partial/\partial x$...the classical is.. $-\hbar\partial/\partial x$....

This..the source of $\sqrt{-1}$..in quantum formulae. It appears because they apply to the rigid field, which introduces inversion of energy.... Practical quantum problems [with real momenta] are so different from classical that we cannot compare them unless we use the formal extrapolation of classical quantities obtained by introducing $\sqrt{-1}$.

8·3. Draft H *9* III. Electrical Theory

This begins with **F** §§22, 23, and then follows **F** III fairly closely; variants are quoted because of the difficulty of **F** III. The chief amplifications concern the use of molar electrodynamic control (§27), the nature of the 'observational system' (§28), and two-particle hamiltonians (§30). The two-particle transformation comes at the end of this chapter, not early as in **F**.

22. **Mutual and self energy:** close to **F** § 22.

23. **Comparison particles:** close to **F** § 23 in the description of a *perfect* object-system. The discussion of *unit* occupation is amplified:

In natural units [**F** §8] a momentum vector is a volume-reciprocal, and the comparison particle can equally well be regarded as carrying a standard volume V_0 whose reciprocal is

$$V_0^{-1} = m_0. \qquad (23\cdot1)$$

We can adopt this as a *natural normalization volume*. Applying it first to the distribution of comparison particles, the density is $A_0 m_0^2$ by (22·52). The number of particles in unit volume is therefore $A_0 m_0$, and..in V_0 is A_0 by (23·1). Thus A_0 is the occupation factor.

For the object-particles the density is[a] $A_0 m^2$, so that the number of particles in volume[b] V_0 is[c] $A_0 m V_0 = A_0 m/m_0$. Thus the occupation factor is $A_0 m/m_0$. Thus there is not a (1, 1) correspondence of object and comparison particles in the distributions treated in §22. The theory did not assume such a correspondence, since it was concerned with comparison particles as carriers of a standard of density, namely, $m_0^2 = m_0/V_0$, rather than as carriers of a standard of mass m_0.

The self energy tensor $Ap_\mu p_\nu$ accordingly corresponds to an occupation factor $A_0 m/m_0$. Except in the fundamental investigations on which we are here engaged, we usually require the energy tensor for unit occupation. This is $(Am_0/A_0)\,p_\mu p_\nu/m$; or, omitting the multiplicity factor $k = A/A_0$, the particle self energy tensor for unit occupation is

$$E_{\mu\nu} = \frac{1}{V_0}\frac{p_\mu p_\nu}{m}. \qquad (23\cdot2)\ [\mathbf{F}\ (23\cdot1)]$$

For a particle at rest this reduces to the density $E_{44} = m/V_0 = mm_0$.

24. **The phase coordinate:** like **F** § 24 in main outline. This excerpt shows a varied approach to 'scale'.

Primarily the scale uncertainty is the fluctuation of magnitude of the extraneous standard. If the scale is taken to be a momentum of dimensions homogeneous with ordinary linear momenta, it is implied that the extraneous standard is a mass. We have seen the mass standard is embodied in a comparison particle; so the scale and phase are primarily

[a] The MS. has $A_0 m_0^2$. [b] MS. V. [c] MS. $A_0 m_0 V$.

variates of the comparison particle. But, as explained in § 23, we bring the comparison particle into the system in order to form a perfect object-system or object-particle. Analytically this means the scale and phase are transferred to the object as extra variates.

Thus perfecting the object-system means de-stabilizing the scale momentum. This is the procedure to be adopted when the scale fluctuation is no longer taken into account as curvature.

A footnote on *molar* scales is of historical interest:

..in pure self-contained molar theory..the standard of length.. must be a directed radius of curvature..in empty space the radius of spherical curvature of the three-dimensional section of space-time. Calling..this the molar metric, Einstein's law $G_{\mu\nu} = \lambda g_{\mu\nu}$ is a tautology. ...An experimental test of the law of gravitation is a test that the σ-metric to which practical measurements are referred agrees with the molar metric to which the theoretical law is referred. It is not difficult to see that the connexion between particle and field theory will ensure that the two metrics are equivalent; and as early as 1921 (*Phil. Mag.* **42**, 800; **43**, 174) it had become clear that Einstein's law was directly deducible from the principles of measurement in this way.

25. **Interchange:** close to **F** § 25. Three details:
On measurables:

...recognized physical quantities are defined by measurements of such a nature that the quadruple p.d. has a specially simple form, and effectively degenerates into a double or even a single distribution.

On the permutation or interchange coordinate:

The permutation coordinate, like the phase, may be expressed either in linear or in angular measure....The interchange dimension is the phase dimension of relative space....In general the complete equation of continuity of flow [spatial *and* extra-spatial or interchange] admits of no simple solution; and the interchange term has to be treated wholly or partly by perturbation methods as a perturbing energy inducing transitions between steady states.

On electrical neutrality:

Interchange does not apply to two external particles even when they are indistinguishable; for they have no comparison particles to interchange. Thus there is no Coulomb force...they are electrically neutral. This explains our comment on (22·7) [**3·2** above], namely it is the particles which have comparison particles attached that show distinctively electrical characteristics.

26. **Hydrocules:** the substance of **F** § 27 (save the penultimate paragraph of the full text), and § 28, with a first sentence:

For brevity, we call the external and internal particles of hydrogen atoms *extracules* and *intracules*.

To summarize: so far we have had ('system A') a uranoid of 'standard particles' (V_{136}'s) of mass m_0, which as object-particles could acquire transition (bound intracule) energy $p^2/2\mu$. We now consider a uranoid in which the transition energy of a standard particle is $\mu+p^2/2\mu$ (a 'free intracule'), so that the rest energy is $m_0+\mu$. The background particles here have no transition energy, and so are reckoned still to have mass m_0; we call these particles '*hydrocules*', and the uranoid with this reckoning of energy 'system B'. In passing from A to B, the measured mass of a standard (as object-) particle changes from m_0 to $m_0+\mu$, and so the systems will be related by the rule that, in passing from A to B, *measured densities* are multiplied by $(m_0+\mu)/m_0=\frac{137}{136}=\beta$. The 'hydrocules' of B, if considered to be present in system A, will there have mass $m_0'=m_0/\beta$.

The more straightforward argument of **F** § 28 then follows.

27. **The β-factors:** like **F** §§ 29, 30. First, the full argument of **F** § 29. The main argument of **F** § 30 follows, introducing the system B' determined by the invariance of classical electrodynamic action. The diagram in **F** does not appear in this MS., but the corresponding description of the relations of the systems is quoted here.

...in $A\to B$, (a) densities are multiplied by β, (b) lengths and times by $\beta^{-\frac{1}{6}}$....Rule (a) is determined by molar considerations (gravitational equilibrium..) and, independently of (b), makes κ invariant. Rule (b) is determined by $8\pi\kappa\hbar^2=1$, and makes $\hbar$ (and therefore e)...and the action in any quantum process invariant. Classical electrodynamics is derived as a limit $\hbar\to 0$, the number of quantum-specified processes being countless, and (b) has then no useful purpose. Let us try replacing it in classical electrodynamics by the condition that action is invariant in the processes specified by classical methods. Since classical action has dimensions ρL^3T, the new rule (b') to be coupled with (a) is: lengths and times are multiplied by $\beta^{-\frac{1}{4}}$. Calling the resultant system of measurement B', we have

In passing from B' to B lengths and times are multiplied by $\beta^{\frac{1}{12}}$, densities being unchanged.

To fix ideas consider the action E_1E_2/c obtained by multiplying the Coulomb energy of molar charges E_1, E_2 by the time-equivalent of their separation. All measurement depends on a quantum-specified standard, which in its most primitive form is the quantum of action $\hbar$ introduced by quantization of the a.m. of a free intracule, or e^2/c which [equals] $\frac{1}{137}\hbar$. The measurement of E_1E_2/c should therefore be regarded as the simplest of all molar measurements, since the quantum-specified standard can be used directly in its primitive form. Then $E_1E_2/c=n_1n_2e^2/c$, where n_1, n_2 are the numbers of elementary charges in the molar charges. But classical electrodynamics goes a much more roundabout way to reach the quantum-specified standard. The e.m. units are defined in terms of molar mechanical units; these have to be connected with the microscopic mechanical units; and these again have to be connected with the primitive quantum constant of the free intracule. Let us trace a complete circuit:

(1) Connect molar e.m. action with the microscopic e.m. action e^2/c. (Carried out in system B, because e^2/c is defined by a free intracule.)

(2) Connect microscopic e.m. action (carried by the intracule) with microscopic mechanical action (carried by the extracule). (Transform from B to A.)

(3) Connect microscopic mechanical action with molar mechanical action. (Carried out in system A, because the analysis of the molar energy tensor into elements gives standard particles.)

(4) Connect molar mechanical action with molar e.m. action by classical electrodynamics. (Transform from A to B'.)

It is understood that in (1) and (3) the connexion is that the molar action is simply the sum of the microscopic elements. There is nowhere any change of measure of action, since it is invariant in the measure transformations in (2) and (4). Yet we end in a different measure system B' from the system B in which we started. Thus the 'classical' e.m. action is not to be identified with the 'fundamental' e.m. action $n_1 n_2 e^2/c$ obtained by summing the microscopic elements. It can only be given the value $n_1 n_2 e^2/c$ by changing the measure system.

Consider the effect of the transformation from B to B' on a hydrogen atom in a definite quantum state...the various energies...transform like M....We have seen that M, and the energies which have a fixed ratio to it, are strictly densities prematurely converted into mass by application of the normalization volume determined at a later stage. Since densities are unchanged in $B \to B'$, these energies are unchanged. Hence

In passing from B to B′ the Coulomb energy of a quantum-specified system is unchanged.

It follows that the Coulomb action $n_1 n_2 e^2/c$ will be multiplied by $\beta^{-\frac{1}{12}}$ in $B \to B'$. This transformation is necessary to make it agree with the classical action; so that if e' is the classical measure of the elementary charge,

$$n_1 n_2 e'^2/c = \beta^{-\frac{1}{12}} n_1 n_2 e^2/c.$$

Since molar charges are measured in accordance with classical e.m. theory, the charge defined by molar control is e'. We have therefore.. [**F** (30·5)], giving the molarly controlled constants in terms of the actual constants. [In **F**: 'actual quantum constants'.]

The numerical results follow as in **F** § 30.

28. **The observational system:** this ends with the substance of of **F** § 31 on 'secondary anchors'. This is preceded by a discussion of the place of the 'observational system' in the present theory: its identification with System A of **F** § 27 (the standard-particle hydrogen uranoid) is for the present considered to be justified by the agreement (**F** (30·6), here (27·7)) of the theoretical and observational mass ratio, rather than by epistemology. Much of this discussion is quoted.

The agreement shown in (27·7) leaves no doubt that the observational system is A. We now consider the theoretical aspect of this identification.

Although separation of mechanical and electrical energies is important ..for analytical treatment, they are not separated in anything that the observer handles—or supposes he handles. The elementary particles, protons and electrons, carry both electrical and mechanical characteristics; the aether is a carrier of electrical and gravitational waves. Thus the identification of the observational system with system A is in accordance with natural expectation. On the other hand, the difference between A and B lies wholly in the standard environment postulated; and it may be urged that a uniform hydrogen uranoid is in any case so different from actual environments that it can make little difference whether the hydrogen is deprived of its electrical energy or not. This brings us to the crucial point. If the environment is purely mechanical there can be no e.m. waves in the environment, and it is meaningless to determine the Rydberg constant of waves which cannot exist. We can transform $\mathfrak{R}$, like any other length-reciprocal, from A to B conformably with the change of extraneous standard; but as a length-reciprocal with a direct observational interpretation in spectroscopy, it has no existence in B. By accepting (27·1) [**F** (29·1)] as the formula for the experimentally determined Rydberg constant, we tie the resulting value of m_p/m_e to a system which is certainly not B. The transformation from B to the observational system corresponds to the recognition in classical theory of an e.m. aether as part of the environment of every object-system.

We can now give a formal definition of the 'observational system'. The quantum theory of the hydrogen atom determines its transition energies in terms of μ; and (27·1) is then derived by introducing the relation $E=h\nu$ giving the frequency of the light emitted. The relation is often used conversely to define a quantum energy; energies so defined may be said to be *optically controlled*. The system used in § 27 is such that the masses of neutral particles are molarly controlled, and quantum masses or energies optically controlled. *We therefore define the observational system to be that in which these two conditions are satisfied simultaneously.* By its most fundamental and widely accepted formulae, quantum theory is committed to the observational system.

The proof that $E=h\nu$ (i.e. that molar e.m. waves of frequency ν are quantizable into photons of energy $h\nu$) belongs to the theory of radiation ...no difficulty in verifying that..the quantum theory of radiation postulates system A; but the scrutiny would be out of place here. (System B is the primary reference system for Fermi-Dirac, A for Einstein-Bose particles. Since photons are E.-B. particles, the natural development of radiation theory follows system A.) Meanwhile we base the identification of the observational with system A on the experimental comparison (27·7).

There are other methods of determining constants observationally ...how do these stand in relation to the observational system? For example, the deflexion determination of e/m_e and the oil-drop...of e do not introduce optical control. But the 'observational system' refers to a combination of methods of experiment and of reduction; for it depends entirely on the computer whether he extracts from the data the constants of system A or of B or of no intelligible system at all....

The argument now follows **F** § 31.

29. **Calculated values of the microscopic constants:** close to **F** § 32. It is pointed out that $\Re = \frac{1}{2}(137)^{-2}\mu c/h$ (**F** (29·1)) is the usual expression simplified by using $hc/2\pi e^2 = 137$. The probable error of m_e is given as 32 (22 in **F**).

30. **The two-particle transformation:** the substance of **F** §§ 26 and 33 (The Coulomb energy). The early development follows **F** § 26 to (26·43), the one change being that the additive relation (26·22) is derived from (26·14) for a newly defined *energy tensor operator* (cf. **7·2** (5·141))

$$(\mathbf{T}_{\alpha\beta})_m = -\frac{\hbar^2}{m}\frac{\partial^2}{\partial x_\alpha \partial x_\beta}, \tag{30·21}$$

and not for $T_{\alpha\beta} = p_\alpha p_\beta / m V_0$ (**F** (23·1)).

As the concepts of 'bound' and 'free' particles (**F** § 27) have already appeared in § 26 of this MS., Eddington now uses them and proceeds to develop 'hamiltonian' theory for later applications:

The hamiltonian η_0 [**F** (26·42)] represents a bound intracule; if this is replaced by a free intracule with hamiltonian $\eta = \mu + \varpi^2/2\mu$,

$$H - \mu + \eta = h + h', \tag{30·44}$$

and the corresponding form of (30·22) [**F** (26·22)] is

$$(T_{\alpha\beta})_{\text{ext.}} - (T_{\alpha\beta})_f + (T_{\alpha\beta})_{\text{int.}} = (T_{\alpha\beta})_{\text{proton}} + (T_{\alpha\beta})_{\text{electron}}, \tag{30·45}$$

where

$$(T_{\alpha\beta})_f = 0, \quad \text{except} \quad (T_{44})_f = \mu. \tag{30·46}$$

Up to here the equations have been tensor equations invariant for Lorentz transformations, but (30·46) is not invariant and introduces a fixed time direction. By taking the time axis in the direction of the external momentum vector, we are able to amalgamate $(T_{\alpha\beta})_{\text{ext.}}$ and $(T_{\alpha\beta})_f$ into a new external energy tensor; so that

$$(T_{\alpha\beta})_{\text{ext.}} + (T_{\alpha\beta})_{\text{int.}} = (T_{\alpha\beta})_{\text{proton}} + (T_{\alpha\beta})_{\text{electron}}, \tag{30·47}$$

and

$$(T_{\alpha\beta})_{\text{ext.}} = 0 \quad \text{except} \quad (T_{44})_{\text{ext.}} = M - \mu. \tag{30·48}$$

For a free particle $p_4^2 - p_1^2 - p_2^2 - p_3^2 = m^2$, so that $p_4 = (m^2 + p^2)^{\frac{1}{2}}$. We call p_4 the vector (or first-order) hamiltonian, as contrasted with $h = m + p^2/2m$, which is the tensor (or second-order) hamiltonian. For small p the two agree; but each plays its part in the theory, and it must not be supposed that h is merely an approximation to p_4.

Since $p_\alpha = -i\hbar\,\partial/\partial x_\alpha$, the relation $p_4^2 - p_1^2 - p_2^2 - p_3^2 = m^2$ gives a wave equation

$$(-\hbar^2\Box - m^2)\psi = 0, \tag{30·51}$$

where $\square$ is the wave operator $\partial^2/\partial t^2 - \nabla^2$. The quantal theory of the free intracule, leading to..the levels of hydrogen and..atomic structure, begins with this; the development...in VIII [**F** X]. Although (30·51) has Lorentz-invariant form, the wave functions ψ are not L.-invariant; for it is only when the time axis is chosen to have the direction of the external momentum vector that (30·51) applies to the intracule.

Setting $\alpha, \beta = 4$ (in 30·47),

$$M - \mu + \varpi_4^2/\mu = p_4^2/m + p_4'^2/m'. \tag{30·52}$$

There is no linear relation between the first-order hamiltonians p_4, p_4', ϖ_4, corresponding to that (30·44) between the second-order h, h', η. In particular, $P_4 + \varpi_4$ would be a meaningless combination. The non-additivity of the quantum energies p_4 makes it impossible to apply molar control, even if we could in practice separate them from the energies of the extracules and of the radiation field. They are connected with molar measurements by optical control; this does not involve addition of energies, the frequency of the molar waves being independent of the number of radiating intracules.

When the spatial momentum is large enough to make it necessary to distinguish the hamiltonians h and p_4, we come to a parting of the ways. If we are interested in mechanical effects, e.g. the additional energy-density or pressure contributed by microscopic reactions, we must keep to the tensor hamiltonian h, which is additive to other tensor hamiltonians. The principal investigation of this kind is the calculation of the pressure in white dwarf matter. If we are interested in optical effects (spectroscopy) we must use the vector hamiltonian p_4. As this energy is not additive to other particle energies, the intracule is then treated in isolation; in fact, the whole purpose of freeing the intracule is to isolate an oscillator which intersects only with the radiation field. When only the tensor hamiltonian is to be used, it makes little difference whether we work with the free form η or the bound form η_0.

The distinction between h and p_4 is closely associated with the separation, customary in radiation theory, of the e.m. field into a longitudinal and transverse field, of which the former is representable by an energy tensor and the latter is not.

The argument of **F** § 33 now follows, to explain how the interchange energy is introduced into the wave equation. The only idea of **F** § 33 missing is the use of 'recoil' ideas to explain why the interchange a.m. is $\hbar$ and not $\frac{1}{2}\hbar$. This is regarded here mainly as a doubling of a purely internal energy to conform with the conventional double reckoning of mutual energies of particles and environment. The final remarks are:

After deriving the interchange term by unified theory, we pass over to the separated mechanical and electrical theory, in which the hydrocule is treated as a V_{136} particle with electrical corrections. These include this term (now regarded as Coulomb energy), and also the various adjustments, including freeing the intracule, treated in §§ 26–29, which result directly from the change of the number of d.f.'s.

8·4. Draft H *8* IV. Gravitation, Exclusion and Interchange

This covers the material of **F** IV (except §§ 44, 45) and V; Eddington expanded and split this chapter in the final text.

31. **Time:** the substance of **F** § 35, with an insert representing the first half of **F** § 36. The subject is time, or time analogues, with reference to p.d.'s. The section ends by linking p.d. functions with wave functions; this link bears on the later development:

The necessity for treating x and p as a joint-observable, owing to their non-commutation, is the principal reason for employing wave rather than distribution functions...Associated with the usual wave function $\mathfrak{f}(x)$.. there is a reciprocal function $\mathfrak{F}(p)$, the probabilities in the ranges dx, dp being respectively

$$2\pi\hbar\,|\,\mathfrak{f}(x)\,|^2\,dx, \quad 2\pi\hbar\,|\,\mathfrak{F}(p)\,|^2\,dp. \qquad (31\cdot1)\ [\mathbf{F}\ (37\cdot6)]$$

They are reciprocal Fourier integrals

$$\mathfrak{f}(x) = (2\pi\hbar)^{-\frac{1}{2}} \int_{-\infty}^{\infty} \exp\,(ipx/\hbar)\,\mathfrak{F}(p)\,dp, \qquad (31\cdot21)\ [\mathbf{F}\ (37\cdot71)]$$

$$\mathfrak{F}(p) = (2\pi\hbar)^{-\frac{1}{2}} \int_{-\infty}^{\infty} \exp\,(-ipx/\hbar)\,\mathfrak{f}(x)\,dx. \qquad (31\cdot22)\ [\mathbf{F}\ (37\cdot72)]$$

These formulae have a somewhat limited application because, even when $\mathfrak{F}(p)$ is a regular function of p, its complex argument (phase) may be altogether erratic and the integral (31·21) may not exist. We explain erratic (incoherent) phase as due to the inclusion of several independent particles in the distribution....When, for some origin, the phase of $\mathfrak{F}(p)$ is constant for all p, the distribution forms a symmetrical wave packet about this origin, and corresponds to a single fully observed particle.

32. **The weight function:** like **F** §§ 37, 38—the relation of physical to geometrical momentum distributions. There is an obiter dictum of great interest linking this with the later 'origin of mass':

...It will be noticed that, although we can derive $h(\xi)$ from $g(x)$ when the uncertainty of the physical origin is given, we cannot derive $H(\varpi)$ from $G(p)$ without the additional condition (32·62) [$K(P) = \delta(P)$, **F** (38·6)]. The reason is that the extension from distribution functions of position to wave functions giving a combined distribution of position and momentum introduces phases of the Fourier components, and we should need additional information as to the phases to be associated with the distribution of the physical origin. For this reason we do not employ the physical origin directly as a dynamical reference body, but use comparison particles instead. The result is that we have developed separate physical reference systems for positional theory and for dynamical theory. It will be our task in §§33, 34 to connect the two systems by showing how the mass m_0 of a comparison particle is related to the uncertainty constant σ of the physical origin.

33. **The genesis of proper mass:** close to **F** §39. One amplification is:

...In a field of gravitational potential ϕ, the rest energy m' is $m' = m - m\phi$. The energy m' corresponds to the whole inertial-gravitational field of the environment, which is separated into an inertial field due to the standard uranoid and a gravitational field due to actual deviations from the standard uranoid. Thus the change of temperature of the actual environment affects the actual rest mass m', but not the proper (or inertial) mass m, which comes from the uranoid alone. In the foregoing we do not employ..an actual change of the mean temperature of the uranoid...We only consider the effect of adopting a zero- instead of an infinite-temperature environment as our standard.

34. **Determination of m_0:** close to **F** §40.

35. **Exclusion,** 36. **The negative energy levels:** these two sections cover the ground of **F** §§41–43, but the arrangement is more like **D** §§25, 26. Section 35 begins with the substance of **F** §41 ('Exclusion'). In this, a footnote asserts that the exclusion principle for *extra- and intracules* is fundamental, and the customary transformation into principles for protons and for electrons serves no particular purpose.

The section continues as **F** §43: one passing remark checks a misconception often held of Eddington's dictum 'Exclusion is a wave-mechanical substitute for gravitation':

Exclusion is the wave-mechanical substitute for the particular gravitational field concerned in wave-mechanical problems, namely, the inertial field of the uranoid which is responsible for the masses of the quantum particles.

The calculation ends with the formula **F** (43·6).

Section 36 corresponds to **F** §42 ('The negative energy levels'), and ends by correcting **F** (43·6) by the factor found in this argument, so as to obtain the final formula **F** (43·7) for m_0.

Note. **F** §§44, 45 are not represented, although (in §35 here) a detailed theory of superdense matter is promised for a later section.

37. **The planoid:** like **F** (Chapter V) §46, introducing the planoid, and the basis of **F** §48, introducing the special planoid. The latter part is quoted.

In mechanical theory it is unnecessary to fix N_1 and R_1 separately, since only the combination $\sqrt{N_1}/R_1$ is required. But it is well known that the relative scale of electrical and gravitational force between elementary particles depends on $\sqrt{N}$. For combined electrical and mechanical theory we adopt a *special planoid* defined in the following way:

In the special planoid $N_1, R_1 = N, R_0$. This involves a change of units of length and mass; so that after working out results we have to apply factors to transform to standard units. We can include in these factors the selection factor, already treated in § 36 [**F** 42], which replaces $\hbar'$ [$\hbar_1$ in **F** § 42] by $\hbar$. The factors transforming planoid lengths (or times) L_1 and masses (or energies) M_1 into standard L, M are then..(37·4) [**F** (48·2)]. The transformation of L is determined by $\sigma^2 = R_0^2/4N$, $\sigma_1^2 = R_0^2/5N$, and is due to the modification of the environment; the transformation of M is then fixed by the condition that the planoidal constant $\hbar_1$ must be equal to $\hbar'$ [i.e. the $\hbar_1$ of **F** § 42], since otherwise we should have to make a further transformation on account of the selection factor which is introduced in the standard units. For $\hbar$ and e^2, which are of dimensions LM, we have by (37·4)

$$\hbar = (\tfrac{5}{3})^{\frac{1}{2}} \hbar_1, \quad e = (\tfrac{5}{3})^{\frac{1}{4}} e_1, \tag{37·5}$$

so that by (36·1) [**F** (42·2)], $\hbar_1 = \hbar'$ [i.e. = the $\hbar_1$ of **F** § 42] as was required.

The reason for keeping the same number of particles in the uranoid and the special planoid is that we are concerned with both quadratic and linear characteristics of particles; and it would greatly complicate their mutual relations if we varied the number of carriers. Taking $R_1 = R_0$..gives the appropriate volume relation (24·1) [$V_3 = \frac{3}{4}V_4$] between three- and four-dimensional theory.

The special planoid will be used in § 39,..in electrical theory. The next section involves mechanical characteristics only; and it is then easier to use values of N_1, R_1 adjusted to the standard value of σ, so as to avoid the transformation of lengths in (37·4). In this case we have still to reckon with the transformation of $\hbar$, and this is considered at an appropriate stage.

38. Interchange of extracules: close to F § 47. An argument is added on a planoid of V_{10} particles (in F the planoid is of V_3's):

The straightforward application of the half-quantum rule gives a planoid of V_3 particles. When other kinds of particles are considered the representation is complicated by multiplicity factors. We can, however, exhibit the rest energy of a planoid of V_{10} particles in a physically significant way. Since $M = \frac{3}{10}m_3$, we obtain from $p_s = \frac{1}{2}\hbar/r_s$, $m_3 = \Sigma p_s^2/2\mu$;

$$M = \tfrac{1}{2}\Sigma p_s'^2/2\mu, \quad p_s' = \tfrac{1}{2}\hbar'/r_s \quad [\hbar' = \sqrt{\tfrac{3}{5}}(\hbar)]. \tag{38·8}$$

Since $\hbar'$ [the $\hbar_1$ of **F** (42·2)] is the primitive constant, this is really the purest form of the result. The factor $\frac{1}{2}$ arises from the double reckoning of energy and momentum, which results from the definition of quantum particles as superpositions on a rigid environment. We do not attempt to eliminate the double reckoning...but a double reckoning of the momentum p'_x and linear energy p'_0 leads in (38·7) [$E_3 = .. = p'^2_0/2\mu$: **F** (47·63)] to a quadruple reckoning of E, and the factor $\frac{1}{2}$ is required to reduce this to the standard double reckoning.

We conclude then that primitively the half-quantum rule gives a planoid of particles which by their masses are identified as actual hydrogen extracules. Selection and other factors introduced in our common super-threshold outlook have an effect equivalent to stabilizing them as V_3 particles.

39. **Non-Coulombian energy:** like **F** §49 and most of §50. The calculation stops at the results **F** (50·3, 50·4) for k, A, without proceeding to use $\mathfrak{R}$. These observations on β-factors follow:

The experimental values of A and k will probably differ from those here calculated by β-factors. To determine these definitively it would be necessary to examine critically the whole procedure of observation and computation by which the so-called experimental values are determined. It is outside our purpose to enter on such specialized investigations. We may expect, however, that the β-factors will be those which correspond to the transformation from system A to B in §27. If so, R_0 will be replaced by $R_0\beta^{-\frac{1}{6}}$ as in (34·8) [**F** (40·8)] and σ consequently by $\sigma\beta^{-\frac{1}{6}}$. Provided the quantum constant e (not the molarly controlled constant e') is used, no other change is introduced. Provisionally therefore the experimental constants are identified as

$$k' = \beta^{-\frac{1}{6}} R_0/\surd N, \quad A' = \left(\frac{16}{3\pi}\right)^{\frac{1}{2}} \frac{m_p}{m_0} \frac{\beta^{\frac{1}{6}} e^2}{\sigma}. \tag{39·93}$$

The section then ends with estimates of the ratios of Coulomb to non-Coulomb energy for protons and for electrons.

40. **Calculated values of the molar and nuclear constants:** this contains the matter of **F** §51 and of the table **F** §52. I quote (i) an account of the table, (ii) the table and (iii) the final paragraphs on N and the mass of the universe. In (i) and (ii) the reference numbers 17–28 have been changed to 16–27 to conform with **F**. (The numbering continues that of a table, lines 1–15, in **F** §32; the corresponding earlier table in **8·3** §29 contained an additional line, later deleted.)

(i) In nos. 16–23 the value [$\frac{3}{2}.136.2^{256}$] of N is employed. Nos. 24–27 do not involve the separation of N and R_0 and are found directly from (40·2) [namely, **F** (51·2), which is **F** (40·8) put in the form

$$\frac{R_0}{\surd N} = \frac{136}{10}\left(\frac{9}{20}\right)^{\frac{1}{2}} \frac{\beta^{\frac{1}{6}} h}{2\pi c M}\Bigg].$$

The authority for the observational values of k and A is Breit, Thaxton and Eisenbud (*Phys. Rev.* **55**, 1018 (1939)). The scattering experiments determine Ak^2 with much greater accuracy than A or k. We therefore give first (in no. 24) a comparison of the theoretical k with the rather rough observational value and then, *adopting the theoretical value of k*, we compare (in no. 26) the theoretical A with the rather accurate observational value of A which results.

Theoretical investigators of the range and intensity of non-Coulombian forces have generally assumed the distribution $e^{-\lambda\rho}$, following the Yukawa theory; but fortunately the principal practical investigations have been reduced with the correct Gaussian distribution function, and furnish a direct comparison. The calculated values of A and k do not include the β-factors provisionally given in (39·93).

(ii) **Molar and nuclear constants**

Ref. no.	Symbol	Description	Calculated	Observed
16	κ	Constant of gravitation	$6{\cdot}6665 \times 10^{-8}$	6·670
17	N	Particles in the universe	$2{\cdot}36216 \times 10^{79}$	—
18	R_0	Einstein radius of space	$9{\cdot}33544 \times 10^{26}$	—
19	R_0	(In megaparsecs)	302·38	—
20	M_0	Mass of the universe	$1{\cdot}97675 \times 10^{55}$	—
21	ρ_e	Density of Einstein universe		
22	V_0	Nebular recession (km.sec.$^{-1}$mp.$^{-1}$)	572·36	560
23	$e^2/\kappa m_p m_e$	Force constant		—
24	k	Nuclear range constant	$1{\cdot}9208 \times 10^{-13}$	1·936
25	A	Nuclear energy constant	$4{\cdot}2572 \times 10^{-5}$	—
26	$A/m_e c^2$	—	52·01	52·26
27	σ	The uncertainty constant	$9{\cdot}604 \times 10^{-14}$	—

[For nos. 21, 23, **F** supplies the calculated values $1{\cdot}23088 \times 10^{-27}$ and $2{\cdot}2714 \times 10^{39}$.]

(iii) The value $\frac{3}{2}.136.2^{256}$ of N is derived in § from a more fundamental constant $N_4 = \frac{4}{3}N$, called the *cosmical number*, which represents the total number of independent quadruple vector wave functions at any point of space. These are non-integrable in spherical space, so N_4 is essentially a constant introduced in the analysis of a local system. Its local character is emphasized, because the breakdown of the representation through non-integrability means the local wave functions are never completed and 'full occupation' is meaningless; consequently the usual rule that each wave function contains one excluding particle does not apply. By substituting $\frac{3}{4}N_4$ for N in (40·3) [**F** (51·4)], the equation becomes explicitly a relation between local constants; and the remote environment is seen to be irrelevant.

In cosmological theory it has been usual to distinguish three possibilities according as the mass $\boldsymbol{M}$ of the universe is greater than, equal to or less than the Einstein mass M_0. The present discussion rules out the cases $\boldsymbol{M} \neq M_0$, which give a universe with no possible equilibrium state. In microscopic theory this would mean that the universe is a system which has no ground state. The provision of a standard of mass (quantum-specified, so as to be available everywhere) is outside the scope of molar relativity; so that it has been overlooked that the cases $\boldsymbol{M} \neq M_0$ conflict with the definition of mass.

CHAPTERS **9–13**

E-NUMBER THEORY AND APPLICATIONS

CHAPTER 9

SUMMARY OF F CHAPTERS VI–VIII

9·0. Preliminary notes

Chapters VI–VIII of **F** are an exposition of Eddington's E-number theory and of its general relation to physics. In summarizing this, the most mathematical part of the book, I confine attention primarily to mathematical definitions and results and to the physical concepts; much variant material on the latter will be found in **11–13**. The reader intent on mastering the mathematics will be helped by the text of **RTPE** and **F**, and some of the papers mentioned in the Bibliography.

The general reader will find this Summary more comprehensible if, once he has met the E-symbols in § 53, he bears these remarks in mind:

(i) Each E-symbol has square -1; one alone commutes in multiplication with all the others; this 'algebraic' symbol (later a diagonal matrix) is E_{16}, sometimes replaced by $+i$ or $-i$. The other fifteen symbols are written *either* with two suffixes as $E_{\mu\nu}$ ($0 \leqslant \mu < \nu \leqslant 5$; when the mathematics leads to a symbol $E_{\alpha\beta}$ with $\alpha > \beta$ this is to be interpreted as $-E_{\beta\alpha}$) *or* with one suffix as E_{μ} ($\mu = 1, 2, \ldots, 15$). Single suffix notation is not used in manipulative work, and so is not correlated with the primary double-suffix notation.

(ii) As explained in § 60, E_{12}, E_{23}, E_{31}, E_{04}, E_{45}, E_{50} are to be regarded as *real*, and the other ten as *imaginary*; their behaviour in multiplication is determined by this.

(iii) In the double-suffix notation, suffixes 1, 2, 3 are linked with three spatial dimensions and 4 with time and analogous quantities. The suffix 5 occurring with these generally denotes mechanical characteristics, e.g. (p_{15}, p_{25}, p_{35}, p_{45}) is the linear momentum-energy 4-vector. The suffix 0 is associated with *electrical* characteristics; see § 55 below.

(iv) The following *conventions* are used in the Summary and in quotations from MSS.: *suffixes* 1, 2, 3 *occurring in round brackets are to be permuted*. In particular (systematizing Eddington's occasional usage) $E_{(12)}$ denotes the set E_{12}, E_{23}, E_{31}; and $(E_{14}p_{14})$ the *sum* $\sum_{1}^{3} E_{\mu 4}p_{\mu 4}$. The *summation convention* applies only to differently placed suffixes; i.e. $T_{\mu}{}^{\mu}$ denotes $\Sigma T_{\mu}{}^{\mu}$, but $T_{\mu\mu}$ is not summed.

9·1. F VI. The Complete Momentum Vector

For relevant drafts, see **11·1, 12·1, 13·1**.

53. **The symbolic frame.** Chapters I–V and VI–VIII describe the 'statistical' and 'spin' theories; some results of the latter have been used in Chapters I–V without proof.

A V_{10} particle has a linear momentum-energy 4-vector with components which will be designated $p_{(1)5}, p_{45}$ and a 'spin' a.m. 6-vector, written $p_{(23)}, p_{(1)4}$. These 10 mechanical characteristics are brought together as a *complete momentum vector*; and this is symbolized by an *E-number*

$$P = \Sigma E_{\mu\nu}p_{\mu\nu} \quad (1 \leqslant \mu < \nu \leqslant 5), \tag{53·1}$$

where the $E_{\mu\nu}$ are symbolic coefficients (*E-symbols*) akin to the **i**, **j**, **k** units of a space vector $\mathbf{r} = \mathbf{i}x + \mathbf{j}y + \mathbf{k}z$.

To allow for additional e.m. characteristics and for other purposes, we generalize (53·1) to an *extended momentum vector* of 16 components, which is symbolized by a *general E-number*

$$P = E_{16}p_{16} + \Sigma E_{\mu\nu}p_{\mu\nu} \quad (0 \leqslant \mu < \nu \leqslant 5), \tag{53·2}$$

where 16 means sixteen, and not a pair 1, 6 of suffixes. The $E_{\mu\nu}$ have square -1; one, E_{16}, commutes (i.e. $E_{\mu\nu}E_{16} = E_{16}E_{\mu\nu}$) with the others and may be denoted by $\pm i$ as an 'algebraic' root of -1. The other symbols anti-commute *or* commute if they have *or* have not just one suffix in common. In full

$$\left.\begin{aligned} (E_{\nu\mu} \equiv -E_{\mu\nu}), \quad E_{\mu\nu}{}^{2} = -1, \quad E_{\mu\sigma}E_{\nu\sigma} = -E_{\nu\sigma}E_{\mu\sigma} = E_{\mu\nu}, \\ E_{\mu\nu}E_{\sigma\tau} = E_{\sigma\tau}E_{\mu\nu} = \pm E_{16}E_{\lambda\rho}, \end{aligned}\right\} \tag{53·4}$$

with + or − as $\mu\nu\sigma\tau\lambda\rho$ is an odd or even permutation of 0, 1, .., 5. To conform with a later matrix representation, the 'algebraic' term $E_{16}p_{16}$ in (53·2) is called the *quarterspur*:

$$\mathrm{qs}\, P = E_{16}p_{16} \equiv \pm ip_{16}. \tag{53·3}$$

The E-numbers P for complex $p_{\mu\nu}$ have a closed algebra; the 'group structure' of the $E_{\mu\nu}$ is that of rotations in a 6-space.

Five symbols with a common suffix, e.g. $E_{0(1)}, E_{04}, E_{05}$, form a *pentad* of anti-commuting symbols; four of these, a *tetrad*. Three anti-commuting with three suffixes, e.g. $E_{(12)}$, form a cyclic *triad*. Pairs of cyclic triads like $E_{(12)}$ and $E_{45\ 50,04}$ are called *conjugate*.

General references on E-numbers are listed in the Bibliography. See a note on the general exposition in **11·1** § 2·1, and a remark on the 'spin 6-vector' at the end of **12·1** § 53.

54. **Miscellaneous properties of E-symbols.** We sometimes use a one-suffix notation

$$P=\sum_1^{16} E_\mu p_\mu \quad (E_{16}=\pm i \text{ as before}). \tag{54·1}$$

If $P=0$, all the $p_\mu=0$. Each of $E_1, \ldots, E_{15}$ anti-commutes with 8 symbols, and commutes with the rest. If P commutes with a tetrad, it is algebraic (i.e. ip_{16}).

$$p_\mu=-\operatorname{qs}(E_\mu P), \tag{54·2}$$

$$\operatorname{qs}(PQ)=-\Sigma p_\mu q_\mu, \tag{54·3}$$

$$\Sigma E_\mu P E_\mu=-16 \operatorname{qs} P. \tag{54·4}$$

55. **Equivalence and chirality.** An *equivalent* set E'_μ satisfying (53·4) is given by

$$E'_\mu=qE_\mu q^{-1}, \tag{55·1}$$

where q is any non-singular E-number (i.e. one having a reciprocal q^{-1}). If $P=\Sigma E_\mu p_\mu$ becomes $\Sigma E'_\mu p'_\mu$, in terms of the E'_μ, then by (54·2),

$$p'_\mu=-\operatorname{qs}(qE_\mu q^{-1}P), \tag{55·2}$$

and we regard p_μ, p'_μ as components of an entity in two equivalent *frames* E, E'. If physical systems S and S' are described by E-numbers $P=\Sigma E_\mu p_\mu$, $P'=\Sigma E'_\mu p_\mu$, they are equivalent systems. In this case, by (55·1),

$$P'=qPq^{-1}. \tag{55·3}$$

We describe this as a *rotation* (of P in the frame E, or of P with the frame to a frame E').

No transformation $q(\,.\,.\,)q^{-1}$ gives $E'_{16}=-E_{16}$. Thus E-frames form two distinct or 'anti-chiral' systems, right- and left-handed, with $E_{16}=+i$ and $-i$. *A right-hand frame E_μ and a left-hand F_μ with $E_{(1)5}$, E_{45} in common have*

$$E_{0(1)},\quad E_{04},\quad E_{05},\quad E_{16} \tag{55·7}$$

of opposite sign. As chirality corresponds to *sign of electric charge,* we expect $P=\Sigma E_\mu p_\mu$ and $P^\dagger=\Sigma F_\mu p_\mu$ to represent electrically or magnetically opposite particles; the 6 components $p_{0(1),04,05,16}$ to be e.m., and the other 10 mechanical. In a neutral environment e.m. characteristics are dormant, leaving a V_{10} with 10 mechanical components and the 'complete' momentum vector (53·1); this is contrasted with the 'extended' vector (53·2).

56. **Rotations**. Defining $e^x = \sum_0^\infty x^n/n!$,

$$e^{\pm E_\mu \theta} = \cos\theta \pm E_\mu \sin\theta. \tag{56·21}$$

The rotation $q = \exp(\frac{1}{2}E_{12}\theta)$ in (55·1) converts

$$P = \Sigma E_\mu p_\mu \quad \text{into} \quad P' = \Sigma E'_\mu p_\mu = \Sigma E_\mu p'_\mu,$$

where $$\left.\begin{aligned} p_{01} = p'_{01}\cos\theta + p'_{02}\sin\theta, \quad p_{02} = p'_{02}\cos\theta - p'_{01}\sin\theta, \\ p_{03} = p'_{03}, \quad p_{04} = p'_{04}, \end{aligned}\right\} \tag{56·5}$$

i.e. it rotates p_{01}, p_{02} in their plane, and similarly the pairs p_{31}, p_{32}; p_{41}, p_{42}; p_{51}, p_{52}; and leaves the other 8 components unchanged. There are 15 simple rotations $\exp(\frac{1}{2}E_{\mu\nu}\theta)$ (excluding E_{16}) like this. Components $p_{\mu\sigma}, p_{\nu\sigma}$ with a common suffix can be rotated together by $E_{\mu\nu}$ and are called *perpendicular*; pairs $p_{\mu\nu}, p_{\sigma\tau}$ are *antiperpendicular*.

If we represent four mutually perpendicular components $p_{0(1)}, p_{04}$ as a vector in 4-space, the rotations for

$$E_{(12)}, \quad E_{(1)4} \tag{56·6}$$

correspond to rotations in the coordinate planes. For these, the components of P behave like

(*a*) a 4-vector $p_{0(1)}, p_{04}$,
(*b*) a 4-vector $p_{(1)5}, p_{45}$,
(*c*) a 6-vector (antisymmetrical second rank tensor) $p_{(23)}, p_{(1)4}$, and
(*d*) two invariants p_{05}, p_{16}. (56·7)

Of these, by (55·7), (*a*), (*d*) are electrical and (*b*), (*c*) mechanical.

Identifying this 4-space with space-time associates E-numbers with ordinary vectors and tensors; and of the transformations (56·6), those in $E_{(12)}$ become ordinary space rotations and those in $E_{(1)4}$ Lorentz transformations.

The operation $q(\ldots)q$ is a 'pseudo-rotation'; it gives symbols E'_μ which do not satisfy (53·4), and formally it rotates pairs of *anti-perpendicular* components. Thus it represents a *strain* or deformation of the system.

See **13·1** § 44 on the significance of 'rotations'.

57. **Five-dimensional theory**. For the 10 rotations

$$q = \exp(\tfrac{1}{2}E_{\mu\nu}\theta)$$

involving $$E_{(23)}, \quad E_{(1)4}, \quad E_{(1)5}, \quad E_{45}, \tag{57·2}$$

the components of P behave like tensors in a 5-space, namely,

(*a*) an (electrical) 5-vector $p_{0(1), 04, 05}$,
(*b*) a 10-vector of mechanical components, and
(*c*) an invariant p_{16}. (57·1)

In a six-dimensional representation with all 15 rotations, $p_1, \ldots, p_{15}$ form an antisymmetrical tensor; the $E_{\mu\nu}$ are associated with the coordinate planes. When we descend to five dimensions, the five $E_{0\mu}$ become associated with coordinate axes, the corresponding vector $p_{0\mu}$ ((57·1) (*a*)) being 'dormant' (electrical). When we descend further to four dimensions, the axis symbols are changed from

$$E_{0(1)}, \quad E_{04}, \quad E_{05} \quad \text{to} \quad E_{(1)5}, \quad E_{45}. \qquad (57\cdot3)$$

As a pentad (representing five mutually perpendicular axes) contains at least one electrical suffix 0, we cannot extend mechanical vectors beyond four dimensions.

58. **Ineffective relativity transformations.** We call relativity transformations *effective* if they correspond to those of current theory. The common view that there are six effective rotations (three spatial, three Lorentz) is based on flat space-time. Normally our microscopic theory employs only six, analogous to these but not identical because of the role of phase and interchange (cf. **3·4** §35). We shall find, however, that *strains* are as prominent as rotations.

59. **Strain vectors.** The strain vector S for a given vector P is defined to be

$$S = -PE_{45}. \qquad (59\cdot1)$$

S corresponds, in tensor character but not in physical dimensions, to the density of P in the three-space, x_1, x_2, x_3; for the reciprocal V^{-1} of a three-volume V is a space-time vector in the 4-direction which has the symbol E_{45} [cf. note **9·0** (iii)]; or, by **3·1** §8, V^{-1} is a mass and so of character p_4, again with symbol E_{45}. Multiplying (59·1) by E_{45},

$$P = SE_{45}. \qquad (59\cdot2)$$

Setting $P = \Sigma E_\mu p_\mu$, $S = \Sigma E_\mu s_\mu$ and $E_{16} = +i$, for

$$\left.\begin{array}{ll} \text{momentum:} & p_{(1)5}, p_{45} = -s_{(1)4}, is_{16}, \\ \text{spin:} & p_{(23)}, p_{(14)} = is_{0(1)}, s_{(1)5}, \\ \text{electrical:} & p_{0(1)} = is_{(23)}, \quad p_{04,05,16} = s_{05}, -s_{04}, is_{45}. \end{array}\right\} \qquad (59\cdot3)$$

Strain vectors form a step from isolated particles towards observable systems. A system introduces the idea of 'simultaneity'—a time axis along the resultant momentum. Associating the P of a particle with the time-symbol E_{45} of a system gives a relation S. With now a fixed time axis, we cannot eliminate an x_1-momentum $E_{15}p_{15}$ by a Lorentz (E_{14}) transformation; thus the system of particle-plus-time-direction is *strained* in comparison with a system with $p_{15} = 0$. The strain component s_{14}, having the symbol E_{14} of the inhibited rotation, now replaces the momentum p_{15}. Thus the change from vector to strain vector corresponds to the change from an unobservable geometrical frame to an observable deformation.

S is also approached via 'ordered volumes' in **F**: see **F** and **13·1** §49 below. An early approach is given in **11·1** §§2·3, 2·4.

60. **Real and imaginary E-symbols.** From (53·4) we can choose either (*a*) 6 E_μ real, 10 imaginary or (*b*) vice versa. We choose (*a*), which also allows a corresponding matrix representation. If conjugate triads (§53) are real, namely,

$$\zeta_{1,2,3}=E_{\mu\nu,\nu\sigma,\sigma\mu}, \quad \theta_{1,2,3}=E_{\tau\lambda,\lambda\rho,\rho\tau}, \tag{60·1}$$

the 16 E's are

$$\zeta_\alpha, \quad \theta_\alpha, \quad i\zeta_\alpha\theta_\beta, \quad i. \tag{60·2}$$

Real 'ζ-numbers' $\left(\sum_1^4 a_\nu\zeta_\nu,\ \zeta_4\equiv 1,\ a_\nu \text{ real}\right)$ form a closed real (quaternion) algebra. Its direct square is a closed algebra of real '$\zeta\theta$-numbers' or E-numbers

$$\Sigma E_\mu p_\mu=\Sigma a_{\alpha\beta}\zeta_\alpha\theta_\beta \quad (\alpha,\beta=1,\dots,4). \tag{60·3}$$

With the E_μ of (60·2), p_μ is real for the 6 symbols (60·1), and imaginary for the others. *We now choose suffixes in* (60·1) *such that*

$$E_{(23),04,05,45} \text{ (only) are real,} \tag{60·4}$$

and

$$\zeta_{(1)}=E_{(23)}, \quad \theta_{1,2,3}=E_{45,50,04}. \tag{60·5}$$

For the rotation $q(..)q^{-1}$ to give an E'-frame '*fully equivalent*' (i.e. with the same real or imaginary character for each suffix) to the E-frame,

$$q \textit{ must be real.} \tag{60·6}$$

This 'full equivalence' will be implied in relativistic equivalence.

See **11·1** §2·5(*a*), **12·1** §45, **13·1** §46 for amplifications. The quaternion basis has been developed by Kilmister and by Milner, and a useful introduction provided by McCrea; see the Bibliography.

61. **Reality conditions.** Imaginary or complex numbers representing physical characteristics are restricted by *reality conditions*, limiting the domain of p.d.'s in the representation space of the characteristics. These conditions follow from invariance under relativity transformations. The combination of physically real simple rotations shows that they must be of the form $q=\exp(\frac{1}{2}E_{\mu\nu}\alpha_{\mu\nu}\theta)$, with θ real and $\alpha_{\mu\nu}=1$ or i. Now by (60·6), rotations giving *full* equivalence are of the form $q=\exp(\frac{1}{2}E_{\mu\nu}\beta_{\mu\nu}\theta)$, where $\beta_{\mu\nu}=1$ or i according as $E_{\mu\nu}$ is real or imaginary. We identify these with physically real rotations by taking $\alpha_{\mu\nu}=\beta_{\mu\nu}$ (and not $i\beta_{\mu\nu}$). Accordingly,

$$\textit{A rotation } q(\dots)q^{-1} \textit{ is physically real if } q \textit{ is real.} \tag{61·3}$$

See **11·1** §2·5(*a*), **12·1** §45.

62. **Distinction between space and time.** If all the terms of a sum are real, or all imaginary, the sum is called *monothetic*; and two such expressions are *homothetic* or *antithetic* according as they have the same or opposite character.

The real rotations are $\exp(\frac{1}{2}E_\mu\theta)$ or $\exp(\frac{1}{2}E_\mu iu)$ according as E_μ is real or imaginary. If $E_{\mu\sigma}$, $E_{\nu\sigma}$ are homothetic, their product $E_{\mu\nu}$ is real and $\exp(\frac{1}{2}E_{\mu\nu}\theta)$ gives a circular rotation of $p_{\mu\sigma}$ with $p_{\nu\sigma}$. If $E_{\mu\sigma}$, $E_{\nu\sigma}$ are antithetic, $E_{\mu\nu}$ is imaginary and $\exp(\frac{1}{2}E_{\mu\nu}iu)$ gives a hyperbolic rotation of $p_{\mu\sigma}$ with $p_{\nu\sigma}$. Thus real and imaginary E_μ give circular and hyperbolic relativity rotations. We infer that for 'reality' the terms $E_\mu p_\mu$ (if connected by effective rotations) are *homothetic.*

Since a pentad (representing 5 perpendicular axes) contains 3 imaginary symbols and so allows only 3 homothetic components, 'space' defined by circular rotations is limited to 3 dimensions; a fourth dimension, being linked by hyperbolic (Lorentz) rotations, must be time-like. Thus (as in (53·1)) we associate 3 imaginary symbols $E_{(1)5}$ with space axes and the real E_{45} with time. The position and momentum vectors are then

$$X=(E_{15}ix_1)+E_{45}t, \quad P=(E_{15}ip_1)+E_{45}\epsilon. \tag{62·2}$$

The space-time axes are now 'realizations' of $E_{(1)5}i$, E_{45}. There is no separate realization of i; physical meaning is given only to real magnitudes associated with realized symbols, e.g. x_1 with $E_{15}i$, θ (rotation angle) with E_{23}. Thus P_m (the ten mechanical components (53·1)) *must be real to have physical meaning.*

See **11·1** § 2·5 (*a*), **12·1** § 46.

63. **Neutral space-time.** In (62·2) we have not used the fifth perpendicular axis E_{05}. In a *neutral* uranoid we could add to X a fifth term $E_{05}E_{16}x_0$ (*not* $E_{05}x_0$ which would be chiral) representing space curvature.

As electrical vectors are undetectable in a neutral environment, the electrical terms P_e of an extended vector P must be antithetic to P_m (63·3), and so imaginary. The *strain vector* $S=-PE_{45}$ will be physically real if P is; it follows (cf. (59·3)) that

$$\textit{all the components } s_\mu \textit{ are imaginary.} \tag{63·4}$$

See full discussions in **11·1** § 2·3, **12·1** § 47 and **13·1** § 48.

64. **Congruent spaces.** The above results apply to a standard neutral uranoid. The E-frame is completely 'realized' only in a monochiral uranoid (particles all of like charge); this has 5 dimensions (3-space + 2-time) $E_{0(1)}$, E_{04}, E_{05}. Neutral space-time corresponds to two chirally opposite and superposed frames of this sort; the resulting

achiral axes are: space $E_{(1)5}$ corresponding to $iE_{0(1)}$, time E_{45} to iE_{04}. This identification can also be seen as a correspondence of vector densities in space-time with vectors in the 5-space.

There is a long discussion in **11·1** § 2·5 (*a*), (*b*); see also **12·1** § 47 and **13·1** § 48.

65. **Determinants and eigenvalues.** Some results from **RTPE** are stated. The *determinant* of $P = \Sigma E_{\mu\nu} p_{\mu\nu}$ is defined to be

$$\det P = \Sigma p_\mu^4 + 2\Sigma' \pm p_\mu^2 p_\nu^2 + 8\Sigma p_{\mu\sigma} p_{\mu\tau} p_{\nu\sigma} p_{\nu\tau} + 8\Sigma'' \pm p_{\mu\nu} p_{\sigma\tau} p_{\lambda\rho} p_{16}, \quad (65·1)$$

with + or − in Σ' according as E_μ, E_ν anticommute or commute, and in Σ'' as $\mu\nu\sigma\tau\lambda\rho$ is an even or odd permutation. A *singular* P (one with no reciprocal) has $\det P = 0$.

The *characteristic equation* of P is $\det (P - \lambda) = 0$, which is a quartic

$$f(\lambda) \equiv \prod_1^4 (\lambda - \lambda_i) = 0. \quad (65·52)$$

If for symbols X, ϕ and a number α, $X\phi = \alpha\phi$, then ϕ is an eigensymbol and α an eigenvalue of X. The eigenvalues of P are the λ_i of (65·52); of E_μ are $\pm i$. If P has a zero eigenvalue, it is singular.

By analogy with ordinary vectors we define the *magnitude* of P as (*a*) qs P or (*b*) $(\det P)^{\frac{1}{4}}$, and the unit E-number as P/(magnitude).

See **12·1** § 49 on 'space- and time-like' quantities; also **5·1** § 18, **5·2** § 2·6 on 'magnitudes' of tensors.

9·2. F VII. Wave Vectors

Drafts are in **11·2**, **12·2**, **12·5**, **13·2**.

66. **Idempotency.** A quantity X is idempotent if

$$X^2 = X; \quad (66·1)$$

thus $X^n = X$; X has eigenvalues 0 and 1 and so is singular (§ 65). Carriers of idempotent variates may be treated as 'individuals', since their non-linear characteristics are determined by their linear characteristics. Thus when the analysis of matter reaches idempotent carriers, it has gone as far as is necessary.

If P is idempotent, so is its 'image' $1 - P$. The idempotents $1, 0, \frac{1}{2}(1 + iE_{\mu\nu})$ with quarterspurs $1, 0, \frac{1}{2}$ are called *trivial*; *non-trivial* idempotents have quarterspurs $\frac{1}{4}$ or $\frac{3}{4}$. Idempotency is invariant under rotations $q(\,..\,)\,q^{-1}$.

See **11·2** § 3·1, **13·2** § 51 for notes on significance. In later developments the *spur* of an idempotent is called its *rank*.

67. Standard form of idempotent vectors. It is proved that every non-trivial idempotent P can be reduced, by rotation of the frame, to

$$P = -\tfrac{1}{4}E_{16}(E_{\mu\nu} + E_{\sigma\tau} + E_{\lambda\rho} + E_{16}), \tag{67·1}$$

or its image, where $\mu\nu\sigma\tau\lambda\rho$ is an *even* permutation of 0, 1, ..., 5. The proof uses *pentadic parts* ϖ_α of P such as

$$\varpi_0 = \sum_1^5 p_{0\mu} E_{0\mu}. \tag{67·31}$$

The square

$$\varpi_0^2 = -\Sigma p_{0\mu}^2 \tag{67·33}$$

is algebraic. Cases of (67·1) are

$$P = -\tfrac{1}{4}i(\pm E_{01} \pm E_{23} \pm E_{45} + E_{16}), \tag{67·7}$$

where two or no negative signs are to be chosen.

For example, $-E_{01} - E_{23} + E_{45} = E_{10} + E_{32} + E_{45}$, giving an *even* permutation as required by (67·1).

68. Spectral sets. As a momentum vector satisfying the reality conditions has ip_{16} imaginary, it cannot be idempotent. We consider instead a vector P with 'imaginary idempotency', $P^2 = iP$ (or $(-iP)^2 = -iP$; this i suggests quantum theory with real momenta i times the classical) and qs $P = \tfrac{1}{4}E_{16}$; this is said to represent a *pure particle*.

The vector

$$P = \tfrac{1}{4}(E_{01} + E_{23} + E_{45} + E_{16}) \tag{68·1}$$

satisfies these requirements and the reality conditions (§§ 62, 63). The carrier of P is a particle at rest (since $p_{(1)5} = 0$) with mechanical components mass $\tfrac{1}{4}E_{45}$ and a.m. $\tfrac{1}{4}E_{23}$. The electrical invariant $\tfrac{1}{4}E_{16}$ evidently represents charge and $\tfrac{1}{4}E_{01}$ magnetic moment.

The strain vector

$$S = -PE_{45} = -\tfrac{1}{4}i(E_{23} + E_{01} + E_{16} + E_{45}) \tag{68·2}$$

is idempotent. Thus by (67·7) we can reverse any *pair* of signs of $E_{23,01,45}$ without losing idempotency, and so obtain 4 strain vectors. The corresponding momentum vectors are

$$\left.\begin{array}{lccccccc} & \text{mag. moment} & & \text{spin} & & \text{mass} & & \text{charge} \\ P_a = \tfrac{1}{4}(& +E_{01} & + & E_{23} & + & E_{45} & + & E_{16}), \\ P_b = (& - & - & & + & & + &), \\ P_c = (& - & + & & + & & - &), \\ P_d = (& + & - & & + & & - &). \end{array}\right\} \tag{68·4}$$

The sign of mass is constant; that of magnetic moment reverses with spin or charge, confirming our identification.

The corresponding strain vectors form a *spectral set*, i.e.

$$S_a^2 = S_a, \quad S_a S_b = 0, \quad \text{etc.}, \quad S_a + S_b + S_c + S_d = 1. \tag{68·5}$$

They show the *scalar particle* $S = 1$ (i.e. pure mass density, neutral and spinless) to be composed of four kinds of 'pure particle', which will later develop into protons and electrons when we supply an environment.

Eddington defines a 'pure particle' as having qs $P = \frac{1}{4}$, an obvious slip for $p_{16} = \frac{1}{4}$ (cf. (68·1)). See **11·2** § 3·2 on 'neutral' particles and **13·2** § 53 on 'spectral analysis'.

69. **Catalogue of symbolic coefficients.** The *electrical* part of a momentum vector is

$$P_e = (E_{01} p_{01}) + E_{04} i\varpi_{04} + E_{05} i\varpi_{05} + E_{16} p_{16} \tag{69·1}$$

(p, ϖ real). We identify $p_{0(1)}$ as magnetic moment (cf. (68·4)), ϖ_{04} as magnetic energy, ϖ_{05} as pole strength. We can now catalogue the E-coefficients of real classical physical quantities as those given by (68·4) and others derived by rotations and multiplications of symbols. For example

momentum $E_{(1)5} i$, a.m. $E_{(23)}$, volume E_{45}, magnetic moment $E_{0(1)}$.

In this *classical* description conjugate variables (e.g. momentum, position) have the same coefficient.

For a pure particle at rest, as $P = SE_{45}$, $S^2 = S$,

$$-S = P^2. \tag{69·3}$$

This is a case of an energy tensor (represented by S as a mass density) being the square of a momentum vector. This holds in both fixed-scale and 'natural' (cf. **3·1** § 8) units.

See **12·1** § 53 and **13·2** § 54 on details; also **13·5** § 92.

70. **The wave identities.** We seek idempotency conditions not restricted like (67·1) to a special frame. It is shown that if $P^2 = P$ and qs $P = \frac{1}{4}$, then

$$(\varpi_\alpha - ip_{16}) P = 0 \quad (\alpha = 0, 1, .., 5), \tag{70·1}$$

i.e. P is an eigensymbol of every pentadic part ϖ_α (cf. (67·31)) of P. Conversely, *sufficient* conditions for idempotency are

(*a*) Three equations of (70·1) with (*b*) qs $P = \frac{1}{4}$. (70·6)

The two sets

$$(\varpi_\alpha - ip_{16}) P = 0, \quad P(\varpi_\alpha - ip_{16}) = 0 \tag{70·7}$$

are called the *wave identities*: they are satisfied by any multiple of an idempotent with qs $P = \frac{1}{4}$, e.g. by the momentum vector. For the strain vector $S = -PE_{45}$, they become

$$\{E_{45}(\varpi_\alpha - ip_{16})\} S = 0, \quad S\{E_{45}(\varpi_\alpha - ip_{16})\} = 0. \tag{70·8}$$

See **11·2** § 3·3, **12·2** § 54.

71. Matrix representations of E-numbers. The 16 $E_{\mu\nu}$ can be represented by 4 by 4 matrices, in particular by 'four point' matrices having just four non-zero elements ± 1 or $\pm i$; for example

$$\begin{array}{llll} E_{15}=i(3\bar{4}1\bar{2}), & E_{25}=i(\overline{4321}), & E_{35}=i(12\overline{34}), & E_{45}=(\overline{34}12), \\ E_{01}=i(2143), & E_{02}=i(1\bar{2}3\bar{4}), & E_{03}=i(4\overline{32}1), & E_{04}=(\bar{2}14\bar{3}), \\ E_{14}=i(1\overline{23}4), & E_{24}=i(\overline{21}43), & E_{34}=i(\overline{3412}), & E_{05}=(\bar{4}3\bar{2}1), \\ E_{23}=(\overline{43}21), & E_{31}=(\bar{3}41\bar{2}), & E_{12}=(2\bar{1}4\bar{3}), & E_{16}=i(1234). \end{array}$$

The matrices are here in Kilmister's notation; numbers denote columns of the non-zero elements of rows 1, 2, 3, 4 in turn; bars denote elements -1. E.g. in E_{15} the elements $e_{\alpha\beta}$ of row α column β are $e_{13}=e_{31}=i$, $e_{24}=e_{42}=-i$. For verifying the multiplication table, the reader may treat the brackets as permutations.

This is a 'true representation', in that real/imaginary $E_{\mu\nu}$ (§ 60) have real/imaginary matrices; these are antisymmetrical/symmetrical respectively. A general E-number $T=\sum_{1}^{16} t_\mu E_\mu$ now becomes a general 4 by 4 matrix, say $T_{\alpha\beta}$ ($\alpha, \beta=1, .., 4$), whose spur (diagonal sum) is 4 times the quarterspur (i.e. the algebraic term in E_{16}) of the E-number:

$$\text{spur}\, T=\Sigma T_{\alpha\alpha}=4it_{16}=4\,\text{qs}\, T. \qquad (71{\cdot}2)$$

Moreover, the 'determinant' (65·1) of $\Sigma t_\mu E_\mu$ equals $\det\{T_{\alpha\beta}\}$. These results hold for any 'true' representation.

As by § 63 the s_μ of a momentum strain vector $S=\Sigma s_\mu E_\mu$ are all imaginary, the *real* part of the matrix form of S (corresponding to the *imaginary* E-symbols) is a symmetrical matrix, and the imaginary part is antisymmetrical. Thus

The momentum strain vector is a hermitian matrix. (71·6)

The reader is cautioned that in **RTPE** the matrix suffices are first, number of column, second, number of row—the reverse of the usual convention. There are notes on strain vectors in **11·2** § 3·4, **12·2** § 55.

72. Factorization of E-numbers. If for a matrix T, $T_{\alpha\beta}=\psi_\alpha \chi_\beta$, then $(T^2)_{\alpha\beta}=T_{\alpha\beta}\,\text{spur}\, T$. Hence

A factorizable matrix with unit spur is idempotent. (72·1)

Conversely,

An idempotent E-number with quarterspur $\frac{1}{4}$ has a factorizable matrix. (72·2)

Thus,

The momentum and strain vectors of pure particles are factorizable, (72·4)

Any factorizable momentum vector, unless spurless, is a multiple of the momentum vector of a pure particle. (72·5)

We call the four-valued quantities ψ, χ (such as $\psi_1, .., \psi_4$, a factor of T) *wave vectors*; ordinary physical vectors may be distinguished as *space vectors*. We 'star' *initial* wave vectors; thus

$$\left.\begin{aligned} \chi^*\psi &= \text{inner product} = \sum_\alpha \chi_\alpha \psi_\alpha = \text{spur}\,(\psi\chi^*), \\ \psi\chi^* &= \text{outer product} = \text{matrix}\,(\psi_\alpha \chi_\beta), \end{aligned}\right\} \quad (72\cdot 6)$$

$$\chi^* E_\mu = \text{wave vector} \sum_\alpha \chi_\alpha (E_\mu)_{\alpha\beta},$$

$$E_\mu \chi^* = \text{third rank wave tensor } (E_\mu)_{\alpha\beta}\chi_\gamma.$$

A hermitian factorizable matrix must be of the form $\psi\psi^\dagger$ (the dagger, denoting complex conjugate, serves also as a star for 'initial'). Thus by (71·6), for a pure particle

$$S = \psi\psi^\dagger, \quad P = \psi\chi^*, \quad \text{where} \quad \chi^* = \psi^\dagger E_{45}. \quad (72\cdot 7)$$

Writing $P = \psi\chi^*$, the wave identities (70·7, 70·8) (which are conditions for 'purity' and therefore for factorizability) become

$$(\varpi_\alpha - ip_{16})\,\psi = 0, \quad \chi^*(\varpi_\alpha - ip_{16}) = 0, \quad (72\cdot 81)$$

$$\{E_{45}(\varpi_\alpha - ip_{16})\}\,\psi = 0, \quad \psi^\dagger\{E_{45}(\varpi_\alpha - ip_{16})\} = 0, \quad (72\cdot 82)$$

i.e. equations *to determine the factors* of a factorizable P. In particular, for $\alpha = 5$, ϖ_α includes the momentum $p_{(1)5}$ and energy $p_{45} = \epsilon$; thus setting $p_{16} = m$, p_{05} (magnetic charge) $= 0$, (72·81) and (72·82) give

$$\{(E_{15}\,ip_1) + E_{45}\,\epsilon - im\}\,\psi = 0, \quad (72\cdot 83)$$

$$\{(E_{14}\,ip_1) + \epsilon + imE_{45}\}\,\psi = 0, \quad (72\cdot 84)$$

which corresponds to Dirac's linear wave equation.

73. Wave tensors of the second rank. By (54·2), the components of a factorizable $P = \Sigma E_\mu p_\mu = \psi\chi^*$ are

$$p_\mu = -\tfrac{1}{4}\chi^* E_\mu \psi. \quad (73\cdot 1)$$

Thus $$ip_\mu/p_{16} = \chi^* E_\mu \psi/(\chi^*\psi). \quad (73\cdot 21)$$

If $$\chi^* \mathbf{a} \psi/(\chi^*\psi) = a, \quad (73\cdot 22)$$

we call the number a the *expectation value* of the operator **a** for the system carrying ψ, χ. Thus by (73·21)

$$\mathbf{p}_\mu = -ip_{16}E_\mu, \tag{73·23}$$

so *the E-frame is the operational form of the (particle) momentum vector.*

If a space vector $P=\psi\chi^*$ is rotated to $P'=qPq^{-1}$, then $P'=\psi'\chi'^*$, where

$$\psi'=q\psi, \quad \chi'^*=\chi^*q^{-1}. \tag{73·3}$$

This is the law of transformation of wave vectors; the two types in (73·3) are called *co-* and *contravariant*. The distinction is independent of 'initial' or 'final' nature; for (73·3) can be written

$$\psi'^*=\psi^*\bar{q}, \quad \chi'=\bar{q}^{-1}\chi \quad (\bar{q}\equiv\text{transpose}). \tag{73·4}$$

For $q=\exp(\frac{1}{2}E_\mu\theta)$, the co- and contravariant transformations agree if $\bar{q}=q^{-1}$, i.e. if E_μ is antisymmetrical and therefore real (§71); then the rotation is *circular*. Thus as in Special Relativity,

Co- and contravariant wave vectors behave similarly/oppositely in circular/hyperbolic rotations. (73·5)

Let ψ, ϕ be co-, χ, ω contravariant. Second-rank wave tensors (i.e. outer products, or sums of outer products, of wave vectors) are of the types

(1) $\psi\phi^*\rightarrow q(\phi\psi^*)\bar{q}$ covariant wave tensor (=covariant strain vector),
(2) $\psi\chi^*\rightarrow q(\psi\chi^*)q^{-1}$ mixed wave tensor (=covariant space vector),
(3) $\chi\psi^*\rightarrow \bar{q}^{-1}(\chi\psi^*)\bar{q}$ mixed wave tensor (=contravariant space vector),
(4) $\chi\omega^*\rightarrow \bar{q}^{-1}(\chi\omega^*)q^{-1}$ contravariant wave tensor (=contravariant strain vector). (73·6)

The interpretation of (2), (3) as *space* vectors follows from the law $P'=qPq^{-1}$ for such vectors, with $\bar{q}^{-1}$ for q in (3); the identification of (2) and (3) as co- and contravariant implies that the *axes* are treated as *co*variant. On the other hand, the interpretation of (1), (4) as *strain* vectors $S=-PE_{45}$ is strictly valid only for the 10 mechanical rotations $E_{(12),(1)4,(1)5,45}$, under which $q(\ldots)\bar{q}$ leaves E_{45} invariant. To allow all rotations we should redefine strain vectors as E-numbers transforming by the law $q(\ldots)\bar{q}$.

See **11·2** §3·5 (clarifying the relation with special relativity co- and contravariance) and **12·2** §57.

74. Wave tensors of the fourth rank. These are equivalent to second-rank physical tensors. We denote contravariant *wave* suffixes (in the sense of (73·3)) by upper or underlined symbols, covariant by lower plain symbols. Contra- and covariant character of a *physical* (space) tensor will be denoted by an upper or lower 0. Thus in (73·6), $P_{\underline{\alpha}\beta} = P^{\alpha}{}_{\beta} = \chi^{\alpha}\psi_{\beta}$ is a contravariant space vector P^0. (The wave suffixes α, β do not correspond to the space-vector suffixes, which are not shown). There are 24 types of second rank space and strain tensors; examples are

$$\left.\begin{aligned} T_1 &= T_{\alpha\underline{\beta}\gamma\underline{\delta}} = \psi_\alpha \chi^\beta \psi_\gamma \omega^\delta = -T_{00} && \text{(covariant space tensor),} \\ T_2 &= T_{\alpha\underline{\beta\gamma}\delta} = \psi_\alpha \chi^\beta \omega^\gamma \phi_\delta = T_0{}^0 && \text{(mixed space tensor),} \\ T_5 &= T_{\alpha\beta\underline{\gamma\delta}} = \psi_\alpha \phi_\beta \chi^\gamma \omega^\delta = -Z_0{}^0 && \text{(mixed strain tensor),} \\ T_7 &= T'_{\alpha\beta\underline{\gamma\delta}} = \psi_\alpha \phi_\beta \omega^\gamma \chi^\delta = Z_0'^0 && \text{(mixed strain tensor).} \end{aligned}\right\} \tag{74·1}$$

The change $T_1 \to T_2$ is an interchange of ϕ, ω; this transposes the matrix $\phi\omega$ and so reverses the sign of the *antisymmetrical* E_μ terms in $\phi\omega$. But 'raising' a space-tensor suffix in Galilean coordinates reverses the space-like terms which have *symmetrical* matrices. Thus if T_1 and T_2 are to be forms T_{00}, $T_0{}^0$ of one space tensor, we require the negative sign shown in (74·1). The strain tensors $Z_0{}^0$ and $Z_0'^0$ are T_{00} with the second and third, and $T_0{}^0$ with the second and fourth, wave vectors interchanged: they are the 'cross' and 'straight' duals of $T_0{}^0$ (cf. §§ 82, 83).

Concerning 'cross duals', see **11·2** § 3·6, **12·2** § 58 and **13·2** § 60. The significance of this mixing of the pairs of wave vectors in (74·1) will be clear in Chapter VIII, where the two pairs are associated with two E-frames.

75. Phase space. The 'complete' strain vector S_m for a 'complete' (i.e. mechanical) momentum vector is by (59·3) of the form iX, where

$$X = \sum_{(10)}' E_\mu x_\mu, \tag{75·2}$$

the sum being over the 10 imaginary (symmetrical) E_μ, and the x_μ *real*. Values of S_m describe states of strain involving a V_{10} mechanical particle; so p.d.'s of these states are representable in a real Euclidean 10-space x_μ; this is *phase space*. The rotations in this space correspond to *all* E-number rotations (except E_{16}) of the form $qS\bar{q}$ of strain-vector transformations; for these keep distinct the mechanical strain vector terms.

Details in **F** are omitted; but see **11·1** §§ 2·7, 2·8, **12·2** § 57 and **13·2** § 59.

76. **Relative space.** The two-particle transformation (26·12) [**3·3**] with $\alpha=1,2,3,4$ is extended to a fifth dimension by the phases x_0, x_0' with conjugate scale momenta p_0, p_0'; after transformation these give an extracule scale momentum P_0 and an intracule interchange momentum ϖ_0 (conjugate to the scale of the relative ξ-space). (i) For molar extracule theory, 4-space-time results from stabilizing scale, so dropping X_0. (ii) For quantal intracule theory, interchange is indispensable, and the 'relative time' ξ_4 is dropped, thus giving 'stabilized simultaneity' of the particles. (This implies that in treating extra- and intracule together, we choose a frame in which the external momenta $P_{(1)}$ vanish). The resulting 4-spaces (i) of $X_{(1)}$ (space), X_4 (time) and (ii) $\xi_{(1)}$ ('space'), ξ_0 (phase) are structurally similar. This suggests the *quantum-classical analogy*, in which the intracule non-spatial component ϖ_0 is *called* energy like the P_4 of the classical extracule. This analogy, introduced by the two-particle problem, has a wider application when we change *molar space* (space-time) into *micro-space* by replacing time by phase, or equivalently by stabilizing the energy and de-stabilizing scale.

For general implications of 'micro-space', see **12·5** §79.

77. **Vectors in micro-space.** In the momentum vector

$$P=(E_{15}p_{15})+E_{45}p_{45}+E_{05}p_{05}+E_{16}p_{16} \qquad (77\cdot 11)$$

in molar space, the stabilized scale is the algebraic component p_{16}, and the energy p_{45}. To convert to micro-space, we stabilize the energy by multiplying by E_{45} (making p_{45} now the algebraic term); this also de-stabilizes the scale component. Thus the representation of P in micro-space is

$$P_q=PE_{45}=-S \qquad (77\cdot 3)$$

by (59·1); so

the vectors of micro-space are the strain vectors of molar space. (77·4)

In quantum theory, however, we give components of P_q the names their E-symbol coefficients would suggest for a *molar* momentum vector; this, again, is the *quantum-classical analogy*.

The method in **F** makes the transformation via the vector density $\mathfrak{P}$: for this, see **13·2** §62. For §§76–78 see also **12·2** §§60, 61.

78. **The quantum-classical analogy.** To show the analogy, we take in micro-space a frame E'_μ, formed from E_μ by changing suffixes

$$4, 5, 0 \quad \text{to} \quad 5, 0, 4. \qquad (78\cdot 1)$$

The 'quantum vector' $P_q=PE_{45}=(\Sigma E_\mu p_\mu)E_{45}$ becomes (when the notation $E'_{\mu\nu}$ is applied to the $E_\mu E_{45}$ *after* these products have been reduced by the rules in (53·4))

$$\left.\begin{array}{ll}(E'_{15}p_{15})+E'_{45}p_{05} & \text{(momentum),}\\ (E'_{23}ip_{01})-(E'_{14}ip_{23}) & \text{(spin),}\\ (E'_{01}p_{14})+E'_{04}p_{04}-E'_{05}ip_{16}+E'_{16}ip_{45} & \text{(dormant).}\end{array}\right\} \quad (78\cdot2)$$

If now we write $P_q=\Sigma E'_\mu p'_\mu$, identifying coefficients gives

$$\left.\begin{array}{lllllll}p'_{(1)5}, & p'_{45}, & p'_{(23)}, & p'_{(1)4}=p_{(1)5}, & p_{05}, & ip_{0(1)}, & -ip_{(23)},\\ p'_{0(1)}, & p'_{04}, & p'_{05}, & p'_{16}=p_{(1)4}, & p_{04}, & -ip_{16}, & ip_{45}.\end{array}\right\} \quad (78\cdot3)$$

The names 'momentum', 'spin' are given to P_q components by *analogy*; since, for example, $(E_{15}p_{15})+E_{45}p_{45}$ is the molar momentum-energy vector, p'_{45} is called 'energy' by analogy, although p_{05} is *not* molar energy. (In unifying molar and quantum physics, however, p_{05} corresponds to the 'quantum unit p'_{45}' of energy and so is taken zero (78·6) in the scale-free approximation).

It is found that

the relative reality conditions between analogous components (p'_μ) *are the same as in the molar vector*; (78·4)

the absolute conditions differ by a universal factor i. (78·5)

Thus in the strain $S=-P_qE'_{45}$, the coefficients s'_μ are all real (contrast (63·4)).

The following components are imaginary (the rest real) in A classical, B quantum designation for 1 molar particle or extracule, 2 intracule:

$$\left.\begin{array}{ll}\text{A1}\ p_{(1)5,\,(1)4,\,04,\,05}, & \text{A2}\ p_{(23),\,(1)4,\,16,\,05},\\ \text{B1 All}\ p'_{\mu\nu}, & \text{B2}\ p'_{(23),\,0(1),\,45,\,16}.\end{array}\right\} \quad (78\cdot7)$$

Conditions A1 are (63·3), with the mechanical vector $\Sigma p_\mu E_\mu$ chosen real; A2 is the 'analogous' set of reality conditions for P_q, applying to the $p'_{\mu\nu}$ and translated back to the $p_{\sigma\tau}$ by means of (78·3). B1 is as for a classical space vector regarded as a quantum *strain* vector $(P=-P_qE'_{45})$; B2 is for the 'native' intracule vector of micro-space corresponding to A1 with a universal factor i ((78·5)).

Persistent components satisfy the same reality condition in 1 and 2; these are $p_{0(1),\,(1)4,\,45,\,05}$ or $p'_{(23),\,0(1),\,16,\,45}$. (78·8)

The persistence of $p_{0(1)}$ gives the intracule a real classical magnetic moment.

An attempt has been made here to clarify (78·7). For the discussion of 'relative' and 'absolute' reality conditions, the reader is referred to **F** and to **13·2** § 63. The significance of these results will be apparent later in the theory.

9·3. F VIII Double Frames

For drafts, see **11·3, 12·3, 12·4, 13·3.**

Some *notes* added here after § 79 are important later in the chapter.

79. **The *EF*-frame.** For the second-rank space tensors (e.g. the energy tensor) we use a double or *EF*-frame, with 256 symbols $E_\mu F_\nu$ ($\mu, \nu = 1, \ldots, 16$), where the E_μ, F_ν form two equivalent E-frames with $E_\mu F_\nu = F_\nu E_\mu$; thus $(E_\mu F_\nu)^2 = +1$. An *EF-number* is

$$T = \sum_1^{16}\sum E_\mu F_\nu t_{\mu\nu} \quad \text{or} \quad \sum_1^{256} EF_\mu t_\mu. \tag{79·2}$$

An equivalent double frame is given by a rotation (with q an *EF*-number):

$$EF'_\mu = qEF_\mu q^{-1}. \tag{79·3}$$

The ordinary relativity rotations, in which the E and F-frames rotate together (or *cogrediently*), are given by

$$q = \exp\{\tfrac{1}{2}(E_\mu F_{16} + F_\mu E_{16})\, i\theta\} \quad \text{or} \quad \exp\{\tfrac{1}{2}(E_\nu F_{16} + F_\nu E_{16})\, u\}, \tag{79·4}$$

with $\mu = (23)$, $\nu = (1)\,4$ and θ, u real.

The outer product of two momentum 4-vectors $P = \sum_1^4 E_{\mu 5} p_{\mu 5}$ and $P' = \sum_1^4 F_{\mu 5} p'_{\mu 5}$ is the *EF*-number

$$P \times P' = \sum_1^4 \sum E_{\mu 5} F_{\nu 5} p_{\mu 5} p'_{\nu 5}. \tag{79·61}$$

Correspondingly the ordinary (16-component) energy tensor is

$$T_{00} = \Sigma\Sigma E_{\mu 5} F_{\nu 5} t_{\mu 5, \nu 5} = \Sigma\Sigma E_{\mu 5} F_{\nu 5} T_\mu{}^\nu, \tag{79·71}$$

where the terms $T_\mu{}^\nu$ with μ *or* $\nu = 4$ differ from normal (Special Relativity) notation by a factor $\pm i$. This T_{00} is regarded as *covariant* because only cogredient rotations of the E's and F's are allowed here. The associated mixed and contravariant tensors are

$$-T_0{}^0 = \Sigma E_{\mu 5} \bar{F}_{\nu 5} T_\mu{}^\nu, \quad T^{00} = \Sigma \bar{E}_{\mu 5} \bar{F}_{\nu 5} T_\mu{}^\nu; \tag{79·72}$$

for 'raising' a suffix in ordinary notation reverses the sign of space ($\nu = (1)$) but not time ($\nu = 4$) components; thus as the matrices $F_{\nu 5}$ are symmetrical for $\nu = (1)$ and antisymmetrical for $\nu = 4$, to 'raise' suffix ν we take the transpose $\bar{F}_{\nu 5}$ and reverse all signs.

The tensor (79·71) can be regarded as part of an *extended energy tensor*

$$T_{00} = \sum_1^{16}\sum E_\alpha F_\beta t_{\alpha\beta} = \sum_1^{16}\sum E_\alpha F_\beta T_\alpha{}^\beta. \tag{79·73}$$

See **12·3** § 61 on variants of these tensors, **13·3** § 64 on contragredient frames.

Notes on tensor nomenclature

(1) In mechanical vectors and tensors, space (1, 2, 3) and time (4) are associated respectively with the imaginary $E_{(1)5}$ and real E_{45} (and $F_{(1)5}, F_{45}$); compare **9·0** and §§ 60, 71 above. As a generalization, any of the 10 imaginary $E_{\alpha\beta}$ may be called *space-like*, and a real symbol, $E_{(12)}$, $E_{04,45,50}$, *time-like*. The matrix representation (§ 71) predominates from now on, with *symmetrical* matrices for imaginary $E_{\alpha\beta}$ and *antisymmetrical* for real $E_{\alpha\beta}$.

(2) The ordinary mechanical energy tensor of Special Relativity (with $g_{(11)}=1, g_{44}=-1$) for velocity $v^{(1)}$ and density ρ is

$$T^{\sigma\tau}=\rho v^{\sigma}v^{\tau}, \quad T^{\sigma 4}=\rho v^{\sigma}, \quad T^{44}=\rho, \tag{A}$$

where suffixes σ, τ take the values 1, 2, 3 *throughout these notes*. As the terms $T^{\sigma\tau}$ give rise to the pressure $p^{\sigma\tau}$ of internal motion, they are called *pressure terms*, and $T^{\sigma 4}$, T^{44} *momentum, density terms*. Associated with the space and time symbols (1) 5, 45, the energy tensor is

$$T=\Sigma\Sigma E_{\mu 5}F_{\nu 5}T^{\mu\nu} \quad (\mu, \nu=1, \ldots, 4). \tag{B}$$

Thus (i) pressure, (ii) momentum and (iii) density terms are associated with products $E_{\mu}F_{\nu}$ which have E_{μ}, F_{ν} respectively (i) both imaginary, (ii) antithetic and (iii) both real. By analogy, when a tensor is *extended* to the form (79·2), the names pressure, momentum, density are applied respectively to the 100, 120 and 36 terms with $E_{\mu}F_{\nu}$ of the types (i), (ii) and (iii). The corresponding symmetries of the double matrices EF_{μ} are easily inferred for these types.

(3) The combined EF-symbols in (B) are real except for the momentum terms; thus if T is to be a 'real' sum, the momentum components $T^{\sigma 4}$ in (B) differ by a factor $\pm i$ from the 'usual' $T^{\sigma 4}$ of (A). This was noted in § 79.

(4) The effects of raising and lowering tensor suffixes need not be discussed here; it is governed by the symmetry type of the E-matrices, as in (79·72).

80. **Chirality of a double frame.** We use only *homochiral* double frames, i.e. those with $E_{16}=F_{16}$. Of the 256 EF's, 120 are *chiral*, reversing sign when E_{16} and F_{16} both change from $+i$ to $-i$. These 120 'dormant' symbols do not appear in a neutral uranoid; so that as we do not go beyond double frames (save in the evaluation of the cosmical number), we drop them altogether.

See **13·3** § 65 for a pleasant remark on 'dormancy'.

The chiral EF-symbols are *not* those involving imaginary E_{μ}'s or F_{ν}'s, but those involving one of $E_{0\mu}$, E_{16} *or* $F_{0\mu}$, F_{16} (cf. (55·7)).

81. **The interchange operator:** this is

$$I=\frac{1}{4}\sum_{1}^{16}E_{\mu}F_{\mu}. \tag{81·1}$$

It is found $\quad I^2=1, \quad F_{\nu}=IE_{\mu}I, \quad E_{\mu}F_{\nu}=I(F_{\mu}E_{\nu})I. \quad$ (81·2)

Thus $I(..)I$ (or $I(..)I^{-1}$) interchanges the E- and F-frames. This is equivalent to the interchange of suffixes in **3·3** § 25. There is no *real* continuous interchange operation (with $q=\exp(\frac{1}{2}Ii\theta)$), just as there can be no steady Coulomb (interchange) configuration without a.m.

It is observed that $I=-I_1I_2$, where

$$I_1=\tfrac{1}{2}\{(E_{23}F_{23})-1\},\quad I_2=\tfrac{1}{2}\{E_{45}F_{45}+{}_{50}+{}_{04}-1\}, \tag{81·41}$$

or

$$I_1=-\tfrac{1}{2}\{1+(\boldsymbol{\sigma},\boldsymbol{\sigma}')\},\quad I_2=-\tfrac{1}{2}\{1+(\boldsymbol{\tau},\boldsymbol{\tau}')\}, \tag{81·6}$$

where (cf. (60·5))

$$\boldsymbol{\sigma}=i\zeta_{(1)}=iE_{(23)},\quad \boldsymbol{\tau}=i\theta_{(1)}=iE_{45,50,04}, \tag{81·5}$$

and $\boldsymbol{\sigma}'$, $\boldsymbol{\tau}'$ are similar in the F-frame. We call $\boldsymbol{\sigma}$ and $\boldsymbol{\tau}$ the *spin* and *co-spin*.

See **11·3** § 4·2 on 'other factorizations' of I; **12·3** § 63 on 'continuous interchange'; **12·4** § 62.

82. **Duals.** Representing both E_μ and F_μ by the same matrices, say $(\Gamma_\mu)_{\alpha\beta}$ $(\alpha,\beta=1,..,4)$ of § 71, we have for $(E_\mu F_\nu)$ a double matrix with general element $(\Gamma_\mu)_{\alpha\beta}(\Gamma_\nu)_{\gamma\delta}$. A *transpose* of this is the result of suffix interchange; thus $I(..)I$ interchanges $\alpha\beta$ with $\gamma\delta$. The *dual* of an EF-number or double matrix T is $\tilde{T}$, where

$$\tilde{T}_{\alpha\beta\gamma\delta}=T_{\alpha\delta\gamma\beta}. \tag{82·31}$$

The dual of $(E_\mu)_{\alpha\beta}(F_\nu)_{\gamma\delta}$ is written $(C_\mu)_{\alpha\beta}(D_\nu)_{\gamma\delta}$, as a symbol of a 'double CD-frame'. It is found that

$$C_\mu D_\nu=-E_\mu F_\nu I, \tag{82·5}$$

$$\tilde{T}=-TI. \tag{82·6}$$

The permutations of a double matrix $(\alpha\beta\gamma\delta)$ are thus

$$(\alpha\delta\gamma\beta)=-(..)I,\quad (\gamma\beta\alpha\delta)=-I(..),\quad (\gamma\delta\alpha\beta)=I(..)I. \tag{82·7}$$

An EF-number T is symmetrical or antisymmetrical with respect to the E- and F-frames according as $I=\pm ITI$ (or $IT=\pm TI$). Writing

$$\gamma_{\mu\nu}=\tfrac{1}{2}(E_\mu F_\nu+E_\nu F_\mu),\quad \zeta_{\mu\nu}=\tfrac{1}{2}(..-..) \tag{82·82}$$

(there are 136 independent $\gamma_{\mu\nu}=\gamma_{\nu\mu}$, and 120 $\zeta_{\mu\nu}=-\zeta_{\nu\mu}$), the symmetrical and antisymmetrical parts of a general T are

$$T_s=\sum^{(136)}\gamma_\mu(t_s)_\mu,\quad T_a=\sum^{(120)}\zeta_\mu(t_a)_\mu. \tag{82·83}$$

The duality relation $\tilde{T}=-TI$ is invariant only under *symmetrical* rotations $q=\exp(\gamma_\mu\theta)$.

Calling $E_{16}F_{16}t_{16,16}$ the quarterspur of T, if

$$T=\Sigma E_{\mu}F_{\nu}t_{\mu\nu}=T_{\alpha\beta\gamma\delta},$$

then $$t_{\mu\nu}=\text{qs}\,(E_{\mu}F_{\nu}T)=\tfrac{1}{16}(\Gamma_{\mu})_{\beta\alpha}(\Gamma_{\nu})_{\delta\gamma}T_{\alpha\beta\gamma\delta} \quad (82\cdot91, 92)$$

(summed for $\alpha\beta\gamma\delta$).

For these symmetry results, cf. **RTPE** § 10·3. See also **11·3** § 4·3. For matrix manipulation, it is to be noted that

$$I_{\alpha\beta\gamma\epsilon}=-\delta_{\alpha}{}^{\epsilon}\delta_{\beta}{}^{\gamma}.$$

83. **The *CD*-frame.** We do not use separate C_{μ} and D_{ν} 'frames', as this would involve new multiplication rules, but merely use $C_{\mu}D_{\nu}$ as the dual of $E_{\mu}F_{\nu}$.

The *outer product* $P\times P'$ of two vectors in an E-frame is formed by transferring one to an F-frame: thus

$$P\times P'=P.IP'I=\tilde{P}\tilde{P}', \quad (83\cdot3)$$

i.e. it is the straight product of their duals.

With matrix suffixes, an E-number P as an EF-number is $P=-iPF_{16}$, i.e. $P_{\alpha\beta(\gamma\epsilon)}=P_{\alpha\beta}\delta_{\gamma\epsilon}$, where δ is the Kronecker symbol. Thus, $\tilde{P}_{\alpha\beta\gamma\epsilon}=P_{\alpha\epsilon}\delta_{\gamma\beta}$ and $\tilde{P}'_{\alpha\beta\gamma\epsilon}=P'_{\alpha\epsilon}\delta_{\gamma\beta}$. Thus using chain multiplication as in (84·1), the Kronecker delta properties, and summing repeated suffixes,

$$(\tilde{P}\tilde{P}')_{\alpha\beta\gamma\epsilon}=\tilde{P}_{\alpha\mu\gamma\nu}\tilde{P}'_{\mu\beta\nu\epsilon}=P_{\alpha\nu}\delta_{\gamma\mu}P'_{\mu\epsilon}\delta_{\nu\beta}=P_{\alpha\beta}P'_{\gamma\epsilon}$$

or $$\tilde{P}\tilde{P}'=P\times P' \quad \text{as required.}$$

The *cross-dual* $T^{\times}$ is defined by

$$T^{\times}{}_{\alpha\beta\gamma\delta}=T_{\alpha\gamma\beta\delta}; \quad \text{so} \quad (C_{\mu}\bar{D}_{\nu})=(E_{\mu}\bar{F}_{\nu})^{\times}, \quad (83\cdot4)$$

the bar denoting the *transpose* as before. In the tensors of (74·1),

$$Z'_{0}{}^{0}=\tilde{T}_{0}{}^{0}, \quad Z_{0}{}^{0}=T^{\times}{}_{00}. \quad (83\cdot6)$$

Thus the dual and cross-dual permutations change space to strain tensors.

The matrix proof of (83·3) has been added here; compare **11·3** § 4·3. See also **12·3** § 64 and **12·4** § 63 on strain tensors.

84. **Double-wave vectors.** In $T_{\alpha\beta\gamma\delta}$, $\alpha\beta$ refers to E_{μ} and $\gamma\delta$ to F_{ν} matrices, and so form separate chains in matrix multiplication; thus the rotation $T'=qTq^{-1}$ means (*summed* on the right)

$$T'_{abcd}=q_{a\alpha c\gamma}T_{\alpha\beta\gamma\delta}q^{-1}_{\beta b\delta d}. \quad (84\cdot1)$$

If $T_{abcd}=\Psi_{ac}X_{bd}$, the *double-wave vectors* Ψ, X transform under (84·1) as

$$\Psi'_{ac}=q_{a\alpha c\gamma}\Psi_{\alpha\gamma}, \quad X'_{bd}=X_{\beta\delta}q^{-1}_{\beta b\delta d}. \quad (84\cdot2)$$

By analogy with (73·3) we call Ψ and X co- and contravariant, and T here a mixed double-wave tensor. The outer product $Z=\Psi\Phi$ of two covariant double-wave vectors is called a covariant double-wave tensor; its rotation law is

$$Z'=qZ\bar{q} \quad (84\cdot42), \quad \text{where} \quad \bar{q}_{\alpha\beta\gamma\delta}\equiv q_{\beta\alpha\delta\gamma}. \tag{84·32}$$

T and Z are to be identified with second-rank space and strain tensors. Their description as co- or contravariant or mixed implies that the rotations q are limited to

$$q_{\alpha\beta\gamma\delta}=q_{\alpha\beta}q_{\gamma\delta} \quad \text{or} \quad q_{\alpha\beta}\bar{q}_{\gamma\delta}{}^{-1}, \tag{84·5}$$

which we call *co-* and *contragredient*. We use mainly the former, in which the E- and F-frames rotate together.

A space tensor T_{00} as an outer product of four wave vectors has both space vector U, V and double-wave vector factors Ψ, Φ:

$$(T_{00})_{\alpha\beta\gamma\delta}=U_{\alpha\beta}V_{\gamma\delta}=\Psi_{\alpha\gamma}\Phi_{\beta\delta}. \tag{84·81}$$

Its cross-dual is $$(Z_0{}^0)_{\alpha\beta\gamma\delta}=\Psi_{\alpha\beta}\Phi_{\gamma\delta}=U_{\alpha\gamma}V_{\beta\delta}. \tag{84·82}$$

The strain-tensor interpretation, like the cross-dual itself, is not invariant for cogredient rotation.

On contragredient rotations, see **13·3** § 69.

85. **The 136-dimensional phase space.** Of the 256 $E_\mu F_\nu$'s, 136 are real (the 10 real E_μ with the 10 real F_ν, plus the 6 imaginary E_μ with the 6 imaginary F_ν); these are also the *achiral* symbols (those unaffected by reversing both E_{16} and F_{16} in the manner of § 55) multiplied by $E_{45}F_{45}$. Thus if strain and space tensors are related by

$$Z=TE_{45}F_{45}, \tag{85·1}$$

the achiral part of T corresponds to the part of Z with real product coefficients $E_\mu F_\nu$.

A 136-phase space is defined by the *real* products $E_\mu F_\nu$ as axes, the position vector being

$$X=\Sigma' EF_\mu x_\mu \quad (\Sigma' \text{ over } \textit{real } EF_\mu). \tag{85·2}$$

The transformations of phase space into itself are found to be of the type $qX\bar{q}$, with $\bar{q}$ as in (84·32). Thus the position vector in phase space is essentially a *strain tensor limited to the real EF_μ terms.* As real EF_μ are made of both symmetrical or both antisymmetrical E and F matrices, this tensor has the symmetry

$$(Z_0{}^0)_{\alpha\beta\gamma\delta}=(Z_0{}^0)_{\beta\alpha\delta\gamma}. \tag{85·51}$$

Now the classical energy tensor is symmetrical between the E and F frames, and so has the symmetry

$$(T_{00})_{\alpha\gamma\beta\delta} = (T_{00})_{\beta\delta\alpha\gamma}. \qquad (85\cdot52)$$

Thus its cross-dual $(T_{00})^{\times}$ (with second and third suffixes interchanged) has the symmetry (85·51) of $Z_0{}^0$. Moreover, the $E-F$ symmetry of T_{00} means that it has 136 independent EF components (cf. (82·83)). As $(T_{00})^{\times}$ and $Z_0{}^0$ have the same symmetry and number of independent components we make the fundamental identification

$$Z_0{}^0 = (T_{00})^{\times} \qquad (85\cdot6)$$

between the phase tensor (the X of (85·2)) and the classical energy tensor.

The above argument may be regarded as clarifying or supplementing **F**, which also uses wave-vector factors. It is to be observed that (i) the original definition (85·1) of 'strain' character is *not* equivalent to (85·6), which is based on the comparative behaviour of strain and space vectors or tensors in rotation; (ii) the co- and contravariant characters in (85·6) are determined for contragredient EF transformations, and so are invalid for the cogredient rotations used elsewhere.

For the opening part, compare **11·3** § 4·4; see also the end of **11·3**, and **12·3** § 66 and **13·3** § 70.

86. **Uranoid and aether.** An energy tensor $T_0{}^0$ gives two strain tensors:

$$Z_u = TE_{45}F_{45}, \quad Z_a = \tilde{T} = -TI. \qquad (86\cdot1)$$

The former corresponds to (85·1) or the original single-frame definition of 'strain vector' (§ 59). The latter is recognized as a strain tensor by its transformation behaviour (akin to that of $T^{\times}$ in § 85). We shall say that Z_u and Z_a represent strains produced by the carrier of T in two systems U and A. U is characterized solely by a time axis $(E_{45}F_{45})$, or energy tensor $E_{45}F_{45}\rho$ of matter at rest, and so is the standard uranoid or Einstein universe. Regarding $-I$ (in Z_a) by analogy as proportional to the energy tensor of system A, its recognized part is, by (81·1),

$$-\frac{1}{4}\sum_{1}^{4} E_{\mu 5}F_{\mu 5}, \qquad (86\cdot2)$$

or, by (79·71),

$$T_{A\mu}{}^{\nu} = -\tfrac{1}{4}\delta_{\mu}{}^{\nu}. \qquad (86\cdot3)$$

This is invariant for Lorentz transformations, so that A represents entirely uncertain momentum (or no criterion for determining it). Thus A is identified with emptiness or 'aether'—or a de Sitter universe.

Z_u is appropriate for a quantum particle (as a superposition on the uranoid), Z_a for a relativity particle (which has to create its own field).

Useful amplifications are in **12·3** § 67 and **12·4** § 65.

87. **The Riemann-Christoffel tensor.** The R.C. tensor $B^{\mu\epsilon}{}_{\nu\sigma}$, usually regarded as describing the curvature of space-time, is here interpreted in terms of the mechanical recoil of the physical reference frame for the motion of an object-particle. If a vector A^{μ} is carried by an object-particle (which has fixed characteristics in the local *physical* frame at each point) round a circuit $dS^{\nu\sigma}$, the total change of A^{μ} in the geometrical frame is

$$dA^{\mu} = \tfrac{1}{2}A_{\epsilon}dS^{\nu\sigma}B^{\mu\epsilon}{}_{\nu\sigma} \tag{87·1}$$

—the *definition* of B. If the circuit is a circle of small radius r in the $\gamma\delta$ plane (in rectangular coordinates), described by the object-particle with angular velocity $\omega_{\gamma\delta}$, the physical frame has angular velocity ω' (determined by the behaviour of dA^{μ}), where

$$\omega'^{\alpha\beta}/\omega^{\gamma\delta} = -\tfrac{1}{2}r^2B^{\alpha\beta}{}_{\gamma\delta}. \tag{87·3}$$

Thus for an object-particle in the x_2x_3 plane with a.m. $\Omega^{23} = mr^2\omega^{23}$, the physical frame recoils with angular velocity $\omega' = -(2m)^{-1}B^{23}{}_{23}\Omega^{23}$, which in an Einstein universe of radius R_0 is $-\Omega^{23}/(2mR_0^2)$. This is as if a.m. were *conserved*, the moment of inertia of the physical frame being

$$C = 2mR_0^2. \tag{87·42}$$

A linear (space or time) displacement is interpreted as rotation about the centre of curvature of space. For time displacement

$$\omega'^{45}/\omega^{45} = -\tfrac{1}{2}R_0^2B^{45}{}_{45}. \tag{87·51}$$

For a coherent time system of object and frame $\omega'^{45} = \omega^{45}$; so that

$$B^{45}{}_{45} = -2/R_0^2. \tag{87·52}$$

This determines a component of B beyond the usual 4-dimensional R.C. tensor.

See **12·3** §70, or the full discussion in **RTPE XI**. An objection to 'rotation about the centre of curvature' in (87·51) is that (87·3) was designed for *small* r.

88. **The de Sitter universe.** The R.C. tensor in the EF-frame is

$$B_{00} = \Sigma E_{\mu\epsilon}F_{\nu\sigma}B_{\mu\epsilon}{}^{\nu\sigma}, \tag{88·1}$$

where $B_{\mu\epsilon}{}^{\nu\sigma}$ agrees with the ordinary relativity notation in the 'pressure' and 'density' terms, but differs by i in the 'momentum' terms [see Notes after §79]. The recoil ω' is coplanar with and proportional to ω if B_{00} contains merely equal diagonal terms, i.e. if

$$B_{00} = k\Sigma E_{\mu}F_{\mu} = \alpha I. \tag{88·2}$$

This corresponds to a de Sitter universe which has

$$B_{\mu\epsilon}{}^{\nu\sigma} = \pm R^{-2} \quad \text{if} \quad E_{\mu\epsilon} = \pm E_{\nu\sigma}, \text{ and } = 0 \text{ otherwise.} \tag{88·3}$$

This universe, being empty, thus shows a non-existent object recoiling in step with a non-existent background.

For contragredient frames (§ 84), the de Sitter R.C. tensor is

$$B_0{}^0 = \alpha I. \tag{88·4}$$

The dual is the strain tensor

$$Z_0{}^0 = \tilde{B}_0{}^0 = \alpha \tilde{I} = \alpha E_{16} F_{16}, \tag{88·5}$$

which is the energy tensor (identifying the energy and R.C. tensors as in § 90) of particles at rest in the standard *planoid* (the Einstein (curved) uranoid has terms $B_{(2323)} = R_0^{-2}$ which are absent from (88·5)).

See **13·3** § 72.

89. **The tensor identities.** For any symmetrical double-wave vector, $\Psi_{ab} = \Psi_{ba}$, we can show

$$\left\{ \sum_{\sigma=0}^{5} E_{\mu\sigma ac} E_{\mu\sigma bd} - E_{16ac} E_{16bd} \right\} \Psi_{cd} = 0 \tag{89·31}$$

(a, b, c, d, *matrix* suffixes, summed when repeated). If T_{cadb} is symmetrical in $a \leftrightarrow b$, and in $c \leftrightarrow d$, then

$$\{ \sum_{\sigma} E_{\mu\sigma ac} E_{\nu\sigma bd} - \delta_\mu{}^\nu E_{16ac} E_{16bd} \} T_{cadb} = 0, \tag{89·35}$$

or by (82·92)

$$\sum_{\sigma} t_{\mu\sigma, \nu\sigma} = \delta_\mu{}^\nu t_{16,16}. \tag{89·4}$$

Since $T_{dacb} = T^{\times}_{dcab} = T^{\times}_{dcba} = T^{\times}_{cdab}$ (§ 83), $T^{\times}$ contains only 'pressure' terms (i.e. $T^{\times} = \Sigma t^{\times}_{\mu\nu} E_\mu F_\nu$ with E_μ, F_ν symmetrical and so imaginary; see notes after § 79); thus (89·4) holds if $T^{\times}$ is so restricted. It is found similarly that if $T^{\times}$ consists only of density terms ($= \Sigma t^{\times}_{\mu\nu} E_\mu F_\nu$ with E_μ, F_ν antisymmetrical and so real), so that T_{abcd} is antisymmetrical for $a \leftrightarrow c$, and $b \leftrightarrow d$, then

$$\sum_{\sigma} t_{\mu\sigma, \mu\sigma} + 3 t_{16,16} = -2 t^{\times}_{mn, mn}, \tag{89·7}$$

where for $\mu = (1)$, $m, n = (2, 3)$ and for $\mu = 4, 5, 0$, $mn = 05, 04, 45$.

A correction has been made in this last statement: see **13·3** § 73. See also note in **12·3** § 68. The tensor identities have been extended by C. W. Kilmister (see Bibliography).

90. **The contracted R.C. tensor.** We apply (89·4) to the R.C. tensor

$$T = B_{00} = \Sigma E_{\mu\epsilon} F_{\nu\sigma} B_{\mu\epsilon}{}^{\nu\sigma}. \tag{90·1}$$

The Ricci tensor $G_\mu{}^\nu$ and invariant G in space-time are

$$G_\mu{}^\nu = \sum_1^4 B_{\mu\sigma}{}^{\nu\sigma}, \quad G = \sum_1^4 G_\sigma{}^\sigma, \tag{90·2}$$

for $\mu, \nu = 1, .., 4$: we regard this as *defining* $G_\mu{}^\nu$ for $\mu, \nu = 0, 5$.

If $T^\times$ consists of pressure components, by (89·4)

$$\sum_\sigma B_{\mu\sigma}{}^{\nu\sigma} = \delta_\mu{}^\nu B_{16}{}^{16} \quad (\sigma = 0, ..., 5), \tag{90·23}$$

so that
$$G_\mu{}^\mu = B_{16}{}^{16} - B_{5\mu}{}^{5\mu} - B_{0\mu}{}^{0\mu} \text{ (no sum)}. \tag{90·24}$$

Hence
$$G_0{}^0 = G_5{}^5 = B_{16}{}^{16} - B_{05}{}^{05},$$
$$G = 2B_{16}{}^{16} + 2B_{05}{}^{05},$$

and
$$B_{\mu5}{}^{\nu5} + B_{\mu0}{}^{\nu0} = -G_\mu{}^\nu + \delta_\mu{}^\nu B_{16}{}^{16}$$
$$= -G_\mu{}^\nu + \tfrac{1}{2}\delta_\mu{}^\nu (G - 2B_{05}{}^{05}).$$

If we identify $B_{05}{}^{05}$ with the *cosmical constant* λ, this last expression is equal to $8\pi\kappa T_\mu{}^\nu$ $(\mu, \nu = 1, ..., 4)$, where $T_\mu{}^\nu$ is the relativity energy tensor, because of the law

$$8\pi\kappa T_\mu{}^\nu = -G_\mu{}^\nu + \tfrac{1}{2}\delta_\mu{}^\nu (G - 2\lambda).$$

We therefore write

$$B_{\mu5}{}^{\nu5} + B_{\mu0}{}^{\nu0} = 8\pi\kappa T_\mu{}^\nu \quad (\mu, \nu = 1, ..., 4), \tag{90·4}$$

and regard $T_\mu{}^\nu$ as the sum of mechanical and magnetic parts

$$(8\pi\kappa T_\mu{}^\nu)_{\text{mech.}} = B_{\mu5}{}^{\nu5}, \quad (8\pi\kappa T_\mu{}^\nu)_{\text{mag.}} = B_{\mu0}{}^{\nu0}, \tag{90·5}$$

since $E_{\mu5}$ and $E_{\mu0}$ are symbolic directions for mechanical and magnetic momenta.

We have thus identified the extended energy and R.C. tensors. (90·6)

The identification of $B_{05}{}^{05}$ with λ agrees with the fact that E_{05} is the scale dimension, and that λ correspondingly is a 'gauge standard' in relativity. This identification is useful only for theoretical molar physics; in observational and quantal physics a quantal, mass, standard p_{16} is necessary.

See **12·3** § 69, **12·4** § 66.

91. **States and interstates.** The classical energy tensor (formerly $T_\mu{}^\nu$ with $\mu, \nu = 1, .., 4$) and R.C. tensor (formerly $B_{\mu\epsilon}{}^{\nu\sigma}$) are included in one extended tensor T_{00} (which has also been called $8\pi\kappa B_{00}$) which we now call the energy tensor. The result (90·4),

$$T_{\mu5}{}^{\nu5} + T_{\mu0}{}^{\nu0} = T_\mu{}^\nu \quad (\mu, \nu = 1, .., 4) \tag{91·1}$$

requires $Z_0{}^0 = T^{\times}$ to contain only pressure components. Excluding the *momentum* components makes T symmetrical. The *density* components of $Z_{\mu}{}^{\nu}$ have $\mu = (23), 04, 45, 50$ and ν likewise; their exclusion reduces the number of independent components of T from 136 (the number for a symmetrical $T_{\mu\nu}$) to 100.

A steady distribution of probability—usually a circulatory flow—constitutes a *state*, occupied by a 'particle'. A steady circulation of probability between two states is an *interstate*, occupied by an 'oscillator'. A 'simple pure oscillator' carries a factorizable momentum vector, say $P = \psi\chi^{\dagger}$; the strain vectors of the states it connects are $U = \psi\psi^{\dagger}$, $V = \chi\chi^{\dagger}$. The internal circulation (e.g. from ψ to $\psi^{\dagger}$) in these states separately is represented by the outer products $UU^{\dagger}$, $VV^{\dagger}$, which as strain tensors correspond to the energy tensors of the states. The interstate circulation has the corresponding energy tensor $T = PP^{\dagger}$ $(= \psi\chi^{\dagger}\psi^{\dagger}\chi)$ whose cross-dual (interchanging the second and third vectors) is Z $(= \psi\psi^{\dagger}\chi^{\dagger}\chi) = UV^{\dagger}$. It is thus the *interstate* flow that is directly represented by a space vector or tensor, P or T, in the EF-frame.

The quantum momentum is classically a strain vector (§77), so that its energy is classically a strain tensor, and is to be identified with $Z_0{}^0$, the *cross-dual* of T_{00}. Thus (as a new view of the 'quantum-classical analogy')

The quantum energy tensor is the cross-dual of the classical energy tensor. (91·2)

Important preliminary views are provided by **12·3** §69 and **13·3** §§75, 76.

92. **The recalcitrant terms.** The quantum energy tensor (considered as the full 'position vector' or 'phase tensor' of the 136-space of §85) has 100 pressure terms Z_p and 36 density terms Z_d—the 'recalcitrant' terms which were excluded in §91. The corresponding parts of the classical energy tensor are defined as $T_p = Z^{\times}{}_p$, $T_d = Z^{\times}{}_d$. Only T_p is related to the classical conserved energy tensor and R.C. tensor.

From §89, T_p and T_d correspond to symmetrical and antisymmetrical double-wave vectors, or (to generalize) *wave functions* ((92·1)). The corresponding density terms Z_d are built of products of real antisymmetrical $E_{\mu\nu}$, and so of strain vectors

$$U_d = (E_{23}s_{23}) + E_{04}s_{04} + E_{05}s_{05} + E_{45}s_{45}. \qquad (92·2)$$

Interstate energy tensors connecting strain vectors of the two types would involve *momentum* terms of Z and so are banned. Thus

There are no transitions between states with symmetrical and antisymmetrical wave-functions (92·3)

—except in complex conditions beyond the scope of the EF-frame.

Two types of antisymmetry occur in wave functions or vectors, the *spin* type we have been discussing, and the *functional* antisymmetry involved when we use relative coordinates $\xi_{(1)}$ [cf. **3·3** §26] for the intracule. These antisymmetry effects are typical of quantum theory. Classical mechanics and relativity suffice for scale-free aggregates of extracules and in general for systems with energy tensors T_p. Only when we superpose a microscopic system of intracules, or a 'recalcitrant' energy Z_d, does quantum theory necessarily enter.

See **12·3** §69 and **13·3** §76.

CHAPTER 10

SUMMARY OF F CHAPTERS IX–XI

10·0. Preliminary notes

Chapters IX and XI of **F** have as their main 'practical' aim the calculation of the masses and magnetic moments of some elementary particles, whereas X is concerned with the solution of the hydrogen wave equation. A note is added in **10·4** on XII. Some attention is paid in this Summary to recent determinations of the constants discussed.

The draft material for **F** IX is contained (apart from one section, §74(*a*), of **12·6**) wholly in **13·4**, and this should be read with this Summary.

10·1. F IX. Simple Applications

93. **The metastable states of hydrogen.** When the momentum-energy vector $\sum_1^4 E_{\mu 5} p_\mu$ (45 = energy, conjugate to time) and scale momentum $E_{16} p_5$ (conjugate to phase) are post-multiplied by iE_{05}, we obtain a vector $\sum_1^4 E_{0\mu} i p_\mu + E_{05} p_5$ with 5 perpendicular components, of which $p_{(1)}$ belong to space, p_4 to 'time' and p_5 to the 'phase' axis. A *time-tilt* is a rotation in this 5-space, with symbol

$$q = \exp\left(\tfrac{1}{2} E_{45} iu\right), \qquad (93·1)$$

involving time and phase, leaving space components unaltered. The tilt adds to a pure-energy component ($E_{04} p_4$) a phase or *interchange* component ($E_{05} p_5$), without which an intracule cannot exist in a stable—or metastable—state.

Consider an intracule with momentum vector (in 4-dimensional classical representation) $P = E_{45} \epsilon + E_{01} g\hbar$, and so having energy ϵ and magnetic moment $g\hbar$. Applying a time tilt $q(..)q^{-1}$ to the 5-dimensional form iPE_{05}, and then converting back to 4 dimensions, we obtain as the tilted vector

$$P' = q(iPE_{05})\, q^{-1} iE_{05} = qPq = q^2 P, \qquad (93·20)$$

where the commutation properties have been used. As

$$q^2 = \exp(E_{45} iu) = \cosh u + iE_{45} \sinh u,$$

then
$$P' = E_{45} \mu - E_{16} f + E_{01} j\hbar - E_{23} \alpha\hbar, \qquad (93·23')$$

where
$$\mu, f = \epsilon(\cosh u, \sinh u), \quad j, \alpha = g(\cosh u, \sinh u). \qquad (93·40)$$

The appropriate *quantum* vector $P_q = P'E_{45}$ (with the suffixes permuted, after multiplication, as in **9·2** (78·1)) is

$$P_q = E_{16}i\mu + E_{05}if + E_{23}ij\hbar + E_{14}i\alpha\hbar. \tag{93·3}$$

The new terms introduced by the tilt are those in E_{14} (interchange circulation a.m.) and E_{05} (magnetic energy). The E_{23} term is spatial a.m. (on the quantum-classical analogy) and is assumed to be quantized as a multiple of $\hbar$, so that $j = 1, 2, 3, \ldots$. The interchange a.m. $\alpha\hbar$ is assumed to be quantized at one unit $\hbar$, but as the quantization must be done in the *rigid field* (in which alone the free intracules can be added to extracules to form hydrogen), and also the phase is the quantum analogue of time, the term $E_{14}i\alpha\hbar$ appears divided (**3·2** §19) by the multiplicity $k = 137$ (as for System B in **3·3** §27, with 'free intracules'). Thus

$$j = 1, 2, 3, \ldots, \quad \alpha = \tfrac{1}{137}. \tag{93·7}$$

From (93·40) $\epsilon/f = g/\alpha$ and the energy ϵ for state j is given by

$$\epsilon/(\mu^2 - \epsilon^2)^{\frac{1}{2}} = (j^2 - \alpha^2)^{\frac{1}{2}}/\alpha. \tag{93·5}$$

This is Sommerfeld's formula for the metastable states ϵ of hydrogen; $j = 1$ is the ground state. The formula for general states,

$$\epsilon/(\mu^2 - \epsilon^2)^{\frac{1}{2}} = \{n + (j^2 - \alpha^2)^{\frac{1}{2}}\}/\alpha \tag{93·81}$$

or

$$\epsilon/f = (n + g)/\alpha \tag{93·82}$$

requires the use of the wave equation (Chapter X).

The usual arrangement of (93·81) (e.g. Condon and Shortley, *Theory of Atomic Spectra*, p. 128, or Sommerfeld, *Atomic Structure and Spectral Lines*, 3rd ed. Ch. VIII, §2 (24)) is

$$\epsilon/\mu c^2 = \{1 + \alpha^2[n + \sqrt{(j^2 - \alpha^2)}]^{-2}\}^{-\frac{1}{2}}.$$

The name 'metastable states' is not a common usage for the states with $n = 0$.

An attempt has been made to clarify the argument and algebra of **F**. The manipulation inserted in (93·20) shows that instead of E_{05} we could use any $E_{\mu\nu}$ which anticommutes with E_{45}, as the symbol transforming P before and after 'tilting'. The use here of a fifth momentum p_5 with p_4, instead of the coordinate pair x_0, x_4 of **F**, was suggested by the MSS. See **12·6** §74 (a) and the later **13·4** §77.

94. **Neutrium and deuterium.** The essential part of the intracule is the quantum a.m. $E_{23}ij\hbar$ (or classical magnetic moment E_{01}, as in §93); the momentum vector is thus

$$P_1 = [E_{16}p_{16}] + (E_{23}p_{23}). \tag{94·1}$$

(The bracket [ ] denotes the arbitrary or additive nature of this energy term, classically E_{45}; the ( ) denotes a cyclic sum as usual.)

This is part of the 'recalcitrant' momentum vector U_d of (92·2). For the rest of U_d we require another type of intracule, with vector

$$P_2=[E_{16}p_{16}]+(E_{45}p_{45}+{}_{50}+{}_{04}). \tag{94·3}$$

We attribute P_1, P_2 to *spin* and *co-spin intracules* (cf. §81). Observe that P_1, P_2 are also ζ- and θ-numbers (§60):

$$P_1=[a_0 i]+(a_1\zeta_1),\quad P_2=[b_0 i]+(b_1\theta_1). \tag{94·4}$$

Combined with an extracule, the spin and co-spin intracule give a hydrogen atom and a neutron. Having a.m. (E_{23}) zero, the neutron is 'a hydrogen atom in the 0-quantum state'; but the binding of the electron is different in its nature from that in hydrogen proper. Nuclear (or neutron) and atomic binding [of electrons] correspond to co-spin and spin intracules.

The dissociation of a neutron involves energy emission. Emission energy tensors may involve 'recalcitrant' or Z_d terms (§92); this occurs in the conversion of a co-spin to a spin intracule, where the emission is a 'neutrino'. Systems differing only in the Z_p part of the energy tensor are called *modularly equivalent*; their interconversion can be reckoned as transitions involving only energy and momenta.

The E- and F-frames are outer products of ζ- and θ-frames and (say) ζ'- and θ'-frames. Thus the combination (normally multiplicative) of two spin (or two co-spin) intracules has by (94·4) an EF-number; but the combination of a spin and a co-spin intracule, called a *double intracule* or D.I., may have a pure E-number. If in (94·4) we take $a_0=b_0=1$,

$$P_1\times P_2=i(P_1+P_2)+[Q], \tag{94·5}$$

where Q has only symmetrical $E_{\mu\nu}$; thus the multiplicative combination of P_1 and P_2 (strictly, of P_1/i and P_2/i) is *modularly equivalent* to addition.

If to a D.I. we add a double extracule (or two simple extracules, since these are scale-free), we have a system of two protons and two electrons, one with atomic and one with nuclear binding; that is, a *deuterium atom* (94·7).

See **13·4** §78.

95. **Mass of the neutron.** A free intracule is a particle of rest mass μ and multiplicity 10 (cf. **3·2** §18). The co-spin (or neutron) intracule is, however, equivalent in multiplicity to an intracule restricted to a single state and so has multiplicity 4 (as that of (94·1)). According to the scale-free theory of **3·2**, this makes its energy

$$\mu_4=\tfrac{10}{4}\mu. \tag{95·1}$$

Thus the difference of n', H$'$, the neutron and hydrogen masses, is

$$n'-\mathrm{H}'=(\tfrac{10}{4}-1)\mu=1{\cdot}5\mu. \tag{95·2}$$

This becomes, with the correcting factor $\beta=\frac{137}{136}$ to be found in §98,

$$n'-\mathrm{H}'=1{\cdot}5\beta\mu=0{\cdot}0008236 \text{ atomic weight units,} \tag{95·5}$$

where $\mu=10M/136^2=0{\cdot}00054505$ a.w.u. The present observational value $0{\cdot}00082\pm0{\cdot}00003$ is rough.

See **13·4** §79. A recent value (based on data in J. W. M. DuMond and E. R. Cohen, *Rev. Mod. Phys.* **25**, 691 (1953)) is $n'-\mathrm{H}'=0{\cdot}000840$.

96. **Double intracules.** If two pure particles have wave vectors ψ, χ, the double-wave vectors

$$\Psi^s=\tfrac{1}{2}(\psi_\alpha\chi_\beta+\chi_\alpha\psi_\beta) \quad \text{and} \quad \Psi^a=\tfrac{1}{2}(\psi_\alpha\chi_\beta-\chi_\alpha\psi_\beta)$$

represent double extra- and intracules, corresponding to the Z_p and Z_d parts of an energy tensor with 100 and 36 independent EF components. Thus an element of extended energy tensor may be analysed into *either* (a) a standard particle—a simple extra- and intracule (proton and electron) *or* (b) a double extra- and intracule (two protons and two electrons). The extracules correspond in (a) and (b), but the intracules differ. Those in (a) are V_{10}'s as in **3·2** §18. The D.I.'s in (b) are V_{36}'s, corresponding to a physically inseparable combination of two V_6's, or the *electrical* parts of two momentum vectors. If the V_{10}'s, of mass μ, of (a) were stabilized to correspond as V_6's, their mass (by **3·2** §16) would become

$$\mu_6=\tfrac{10}{6}\mu=\tfrac{5}{3}\mu. \tag{96·1}$$

This corresponds to the selection factor $\frac{5}{3}$ in the 'sub-threshold analysis' of **3·4** §42. In fact the analysis (b) gives the 'sub-threshold' distribution of *excluding particles* (cf. the exclusion unit cell of four particles in §41). Analysis (a) corresponds to the gravitational theory of **3·2**, (b) to electrical theory.

In the gravitational-mechanical analysis (a) we require an extra, or 137th, d.f. for electrical properties (cf. **3·3**); this is not required in the electrical analysis (b). Thus the mass unit in (b) is $\beta=\frac{137}{136}$ times that in (a) for simple particles, and β^2 as great for double particles; it is the unit in (a) that is connected with observation.

See **13·4** §81.

97. **Comparison with field theory.** The mechanical-symmetrical and electrical-antisymmetrical tensors $G_{\mu\nu}$ and $F_{\mu\nu}$ (with 10 and 6 independent components) of field theory are to be compared with the 10 mechanical-symmetrical and 6 electrical-antisymmetrical components of the quantum momentum vector (**9·2** §78). The combination

$$H_{\mu\nu}=G_{\mu\nu}+iF_{\mu\nu} \tag{97·1}$$

corresponds to the extended momentum vector. The *linear* and *quadratic* analyses (*a*) and (*b*) of § 96 depend on the action invariants

$$K_1 = H = H_{\mu\nu} g^{\mu\nu} = G, \tag{97·21}$$

$$K_2 = H_{\mu\nu} H^{\mu\nu} = G_{\mu\nu} G^{\mu\nu} - F_{\mu\nu} F^{\mu\nu}. \tag{97·22}$$

For the latter, the equation

$$G_{\mu\nu} G^{\mu\nu} = H_{\mu\nu} H^{\mu\nu} + F_{\mu\nu} F^{\mu\nu} \tag{97·23}$$

corresponds to the sum $Z = Z_p + Z_d$ in particle theory (*b*). The absence of *cross-terms* in $H_{\mu\nu}$, $F_{\mu\nu}$ corresponds to the absence of momentum terms in Z.

98. **Mass of the deuterium atom.** The mass defect of deuterium is the energy difference between the 'upper' (two protons and two electrons at rest) and 'lower' (a deuteron and electron at rest) state. For a pseudo-discrete distribution the former is indistinguishable from hydrogen. For the lower state we stabilize the intracules to obtain a D.I., thereby changing from 136 to 36 d.f.'s, so that the intracule rest mass μ becomes μ_2, where

$$\mu_2/\mu = \tfrac{136}{36}. \tag{98·1}$$

The mass defect is the loss

$$2\mathrm{H} - \mathrm{D} = \mu_2 - \mu = \tfrac{100}{36}\mu = 0{\cdot}001514 \text{ a.w.u.} \tag{98·2}$$

But the D.I. energy is multiplied by β^2 in the observational system, and the intracule energy by β (cf. § 96); so that more accurately

$$2\mathrm{H}' - \mathrm{D}' = \beta^2 \mu_2 - \beta\mu = 0{\cdot}0015404 \text{ a.w.u.} \tag{98·3}$$

The observational value (Mattauch) is $0{\cdot}001539 \pm 0{\cdot}000002$ a.w.u.

The β-factor in (95·5) arose as a similar consequence of the argument of § 96.

See **13·4** § 80 for a fairly clear approach. Recent observational values are near 0·001549 a.w.u.

99. **Mass of the helium atom.** Two deuterium atoms, *rigidly coupled* to have anti-parallel spins and co-spins, form a 'balanced atom'. The coupling energy will equal the interaction energy of any natural state of eight particles with such spins. We think the helium atom is close to this system.

Consider an assemblage of balanced particles in a pseudo-discrete state; the constrained balance is obtained by omitting the Z_d energy terms. The assemblage has an 'upper' state (four standard bi-particles) and a 'lower' with the electrical or intracule part Z_d eliminated. Changing from this 'quadratic' to 'linear' analysis (or by §§96, 97 from (*b*) to (*a*) or from sub- to super-threshold analysis), the difference of the states is still the elimination of intracule energy. We view this elimination then as a change of the upper-state standard particles to *hydrocules* (**3·3** §27); thus the mass defect of the lower ('balanced') state is

$$\tfrac{4}{137}M. \tag{99·1}$$

Now the two satellite electrons in helium provide a 'two-legged intracule' to carry transition energy. The corresponding 'electric' energy is $\mu\sqrt{2}$ (since as in **3·5** §47 interchange momenta add by summing squares, and there are here two sets of interchanging particles for the two legs). Thus the total mass defect of helium is

$$d = 4\text{H} - \text{He} = \tfrac{4}{137}M - \mu\sqrt{2} = 0{\cdot}02867 \text{ a.w.u.} \tag{99·2}$$

$$= 52{\cdot}589\mu. \tag{99·3}$$

The nuclear energy constant A of **3·5** (50·4, 50·6), corrected by a factor $\beta^{\frac{1}{6}}$ (i.e. β as in §95, divided by the $\beta^{\frac{5}{6}}$ of **3·3** (29·5) which is here irrelevant) is

$$A = 52{\cdot}416\mu. \tag{99·4}$$

A is regarded here as the energy of two *coincident* protons (**3·5** (50·2) with $r=0$); and (99·3, 99·4) may be taken as limits to the mass defect. Thus

$$4\text{H} - \text{He} = (52{\cdot}502 \pm 0{\cdot}086)\,\mu = 0{\cdot}02862 \pm 0{\cdot}00004 \text{ a.w.u.} \tag{99·5}$$

The observational value is 0·02866 a.w.u.

See **13·4** §82 for variants and discussion. The observational value has since changed very little.

100. **The separation constant of isobaric doublets.** A nucleus of mass number n and atomic number N_p is commonly said to contain N_p protons and $N_n = n - N_p$ neutrons. My atom has n protons and n electrons, N_n of the latter with co-spin and N_p with spin binding. If the particles are grouped, there is continuous interchange between the groups. Now the D (deuterium) group forms a unit cell of 4 particles in exclusion theory; thus with D groups we can calculate exclusion energy as the equivalent (cf. **3·4**, **3·5** §47) of this interchange energy.

Consider an atom of $\frac{1}{2}n$ D groups, so that $N_n = N_p = \frac{1}{2}n$. Applying the exclusion energy of **3·4** §43 to the $\frac{1}{2}n$ D.I.'s, the top quantum number is

$$\mathfrak{k} \approx (3n/4)^{\frac{1}{3}}, \tag{100·1}$$

and the top energy for varying n is

$$\mathfrak{E} = A\mathfrak{k}^2. \tag{100·2}$$

The case $\mathfrak{k}=1$ is a D atom, its D.I. having energy μ_2 (§ 98). Now (§ 43) at $\mathfrak{k}=1$ the energy is half rest mass and half exclusion; so in (100·2) $A=\frac{1}{2}\mu_2$, and

$$\mathfrak{E}=\tfrac{1}{2}\mu_2\mathfrak{k}^2\approx\tfrac{1}{2}\mu_2(\tfrac{3}{4}n)^{\frac{2}{3}}. \qquad (100\cdot3)$$

We consider $\mathfrak{E}_0=\mathfrak{E}-\frac{1}{2}\mu_2$, namely, the exclusion energy of the (top) D.I. reckoned from the ground 'state' $\mathfrak{k}=1$;

$$\mathfrak{E}_0\approx\tfrac{1}{2}\mu_2\{(\tfrac{3}{4}n)^{\frac{2}{3}}-1\}. \qquad (100\cdot4)$$

We use this in comparing atomic masses, regarding the top particle as comparatively 'loose'. In particular, we consider changing a spin and co-spin electron into two spin (i.e. satellite) electrons. Writing

$$n=N_n+N_p, \quad T_w=\tfrac{1}{2}(N_n-N_p), \qquad (100\cdot5)$$

the atom is called (n, T_w), and the change to be treated is $(n,0)\rightarrow(n,-1)$. In § 98 we had $D\rightarrow H_2$, or $(2,0)\rightarrow(2,-1)$. For the present change in a *top* particle, the change in $\mathfrak{E}_0$ is proportional to the change d in μ_2 in the D transformation; thus

$$\Delta\mathfrak{E}_0=\mathfrak{E}_0 d/\mu_2. \qquad (100\cdot61)$$

If this is the *only* change, the mass change is

$$\Delta\mathfrak{E}_0=M(n,-1)-M(n,0)=\tfrac{1}{2}d\{(\tfrac{3}{4}n)^{\frac{2}{3}}-1\}. \qquad (100\cdot62)$$

By (98·3) $d=1{\cdot}540$ mMu. (i.e. 10^{-3} a.w.u.); so that

$$M(n,-1)-M(n,0)=\tfrac{1}{2}\,1{\cdot}272n^{\frac{2}{3}}-0{\cdot}770. \qquad (100\cdot63)$$

To improve this, (*a*) we multiply $n^{\frac{2}{3}}$ by $(n-1)/n$, to replace the final exclusion field by the mean field in the change of top particle; (*b*) we consider only n odd, and the change from $\frac{1}{2}(n-1)$ to $\frac{1}{2}(n+1)$ satellite electrons. These pairs, $(n,\frac{1}{2})$, $(n,-\frac{1}{2})$, are called *isobaric doublets.* From (100·63), with (*a*), (*b*),

$$M(n,-\tfrac{1}{2})-M(n,\tfrac{1}{2})=\tfrac{1}{2}\,.\,1{\cdot}272(n-1)\,n^{-\frac{1}{3}}-0{\cdot}770, \qquad (100\cdot7)$$

which is to be compared with Wigner's formula:[a]

$$M(n,-\tfrac{1}{2})-M(n,\tfrac{1}{2})=\tfrac{1}{2}\,.\,1{\cdot}27(n-1)\,n^{-\frac{1}{3}}-0{\cdot}79, \qquad (100\cdot8)$$

with the factor 1·27 chosen to fit results for n up to 41.

See **13·4** § 83. Wigner's formula is still considered a good representation.

101. **Isotopic spin.** Nuclear theory has extended the concept of atomic spin a.m. to (nuclear) co-spin, by quantizing the resultant co-spin and a component in a control plane. As this must be a plane of the 3-space $E_{45,50,04}$, the effect corresponds to the 'time-tilt' of § 93.

[a] E. P. Wigner and E. Feenberg, *Reports on Progress in Physics*, 8, 274 (1941).

Converting the plane E_{45} of the tilt in the 5-dimensional representation by the symbol E_{05} (as in §93), we have

the control plane for co-spin momentum is E_{04};
the quantized component is the magnetic energy p_{04}. (101·1)

By analogy with the atomic spin a.m. of an atom, we assume that the resultant (total) co-spin a.m. T is (in units of $\hbar$) integral or half-integral, and that the controlled component, or *isotopic spin*, has values $T_w = T, T-1, \ldots, -T$; this corresponds to $2T+1$ different isobaric atoms. Isobaric doublets (as in §100) correspond to $T=\frac{1}{2}$, $T_w = \frac{1}{2}, -\frac{1}{2}$.

This quantization will be found in Wigner and Feenberg, loc. cit.

102. **Radii of nuclei.** Observation shows the protons and co-spin electrons form a compact 'nucleus'. We follow Wigner in assuming each particle has a uniform p.d. over a sphere of radius r, adjusting r to suit the observed energies. The electrostatic energy of the net charge $N_p e$ is $\frac{3}{5}N_p^2 e^2/r$. We adopt this (uniform p.d.) result as consistent with the smoothed treatment which gave (100·63) for the exclusion energy; and we identify this electric energy with the term in n in (100·63). (The constant term in (100·63) represents the difference between spin and co-spin binding.)

Since $N_p = \frac{1}{2}n - T_w$, the change of electrostatic energy in the change $\frac{1}{2}$ to $-\frac{1}{2}$ of T_w is $\frac{3}{5}ne^2/r$. This corresponds to $\frac{1}{2}.1{\cdot}272n^{\frac{2}{3}}$ in (100·63). Hence

$$r = 1{\cdot}460 \times 10^{-13} n^{\frac{1}{3}}\,\text{cm.} \qquad (102{\cdot}1)$$

The number density of protons in the nucleus is then independent of n, namely,

$$\sigma_p = n/\tfrac{4}{3}\pi r^3 = 7{\cdot}674 \times 10^{37}\,\text{cm.}^{-3}, \qquad (102{\cdot}2)$$

although this calculation requires T_w to be small.

See the last part of **13·4** §83 for an earlier sketch.

103. **The nuclear planoid.** Wigner's 'peripheral' theory of the nucleus can be annexed to our theory. Here we touch on the theory of the core.

Quantum nuclear theory implies a nuclear planoid (cf. **3·5** §46) as a rigid background, such as the spherical distribution in §102. Thus the non-Coulombian energy $-E$ per proton in a distribution σ_p (102·2) of protons is given by

$$E = \frac{1}{4}\int A\,\mathrm{e}^{-r^2/k^2}\sigma_p dV = \tfrac{1}{4}\pi^{\frac{3}{2}}Ak^3\sigma_p = 21{\cdot}631\,\text{mMu.} \qquad (103{\cdot}1, 2)$$

[cf. **3·5** (50·2)]. The energy is halved to avoid double reckoning if later we summed E over the protons; and is halved again because half the charge is 'neutralized', as compared with the planoidal calculation in §50, by co-spin bound charges. A similar result,

$$23\cdot116\,\text{mMu.}, \qquad (103\cdot5)$$

is obtained by a direct exclusion calculation using a 'white-dwarf' formula of §45.

Regarding E as a 'top energy' (§§16, 41), the mean exclusion energy is $\frac{3}{5}E$, so that there is a deficit of $\frac{2}{5}E$ in the nucleus as compared with the standard uranoid. This suggests that a *nucleus of mass number n should have a mass defect*

$$\tfrac{2}{5}En = 8\cdot65n\,\text{mMu.}, \textit{ compared with hydrogen},$$
$$0\cdot52n \textit{ compared with oxygen} \text{ (a 'packing fraction' } 5\cdot2). \qquad (103\cdot6)$$

This corresponds as a linear representation to observations for $30 < n < 150$ approximately.

Fundamental theory has thus given for the nucleus: (1) the law and constants A, k of non-Coulombian energy (§50); (2) the mass defect of deuterium and (3) of helium; (4) the isobaric doublet formula and nuclear radii; (5) the binding energy of a co-spin electron (the mass excess for a neutron); and (6) the 'whole-number rule' (103·6). The theory also (7) covers isotopic spin theory, but (8) rejects meson-field theory and the constitution of the nucleus of protons and *neutrons*; the latter hypothesis may, however, be a useful approximation.

Concerning the last point, a MS. fragment adds: 'If employed as an analytical device, it normally does little harm, the binding energy of an electron in the nucleus being not much different from its binding energy in the neutron.'

104. Mass of the mesotron. Energy tensors Z_m of momentum components were previously barred by reality conditions (§91), but may arise in atoms more complex than helium. Regarding Z_p and Z_d tensors as mechanical and electrical, we have *forbidden oscillators* (carriers of tensors Z_m connecting symmetrical Z_p and antisymmetrical Z_d tensors) of the two types

$$Z_{m1} = \Sigma E_\mu{}^s F_\nu{}^a p_{\mu\nu}, \quad Z_{m2} = \Sigma E_\mu{}^a F_\nu{}^s p_{\mu\nu} \qquad (104\cdot1)$$

(s, a = symmetrical, antisymmetrical matrices); each has 60 components. We identify these with transient particles or *mesotrons*; they are charged, as Z_m reverses with the chirality of the frame. They decay into the charged bi-particles (V_{136}'s) of §22—an electron m_e (or proton) with a comparison particle m_0 which disappears into the uranoid. By the multiplicity law the mass of a Z_m carrier is

$$\mathfrak{m} = \tfrac{136}{60}(m_0 + m_e);$$

so the energy radiated in the decay is

$$\mathfrak{m}-(m_0+m_e)=\tfrac{76}{60}(m_0+m_e). \quad (104\cdot 3)$$

The observed mass $\mathfrak{m}_0$ of the mesotron is this plus the mass m_e of the resultant electron; but we correct $\mathfrak{m}$ relative to m_0+m_e by the factor $\frac{137}{136}$. Thus

$$\mathfrak{m}_0=\tfrac{77}{60}(m_0+m_e)+m_e=173\cdot 98m_e, \quad (104\cdot 5, 6)$$

where $m_0=\frac{10}{136}M$ (**3·2** (18·5)). There should also be heavy mesotrons, decaying into protons and negatrons; their mass (with m_p for m_e in (104·5)) would be $2\cdot 38m_p$.

See **13·4** §85. There was formerly believed to be a meson with mass near the value calculated in (104·6), but the nearest is now at $210m_e$. The mass $2\cdot 38m_p$ has not been observed.

10·2. F X. The Wave Equation

For relevant drafts see **11·2, 12·2–12·6, 13·5.**

105. **Field momentum.** A small rotation $q=\exp(\frac{1}{2}E_{12}d\theta_{12})$ changes a covariant wave vector (73·3) ψ into $\psi'=q\psi$ (relative to the original frame), and a position vector $X=(x_1, \ldots, x_4)$ (symbolically $X=\sum_1^4 E_{\mu 5}x_\mu$) to $X'=qXq^{-1}$. By **9·1** (56·5),

$$x_1'=x_1-x_2d\theta_{12}, \quad x_2'=x_2+x_1d\theta_{12}, \quad x_3'=x_3, \quad x_4'=x_4. \quad (105\cdot 12)$$

Suppose now the wave components of ψ are functions of $x_1, \ldots, x_4$; $\psi=f(X)$. The rotation changes f to a new functional form f': $\psi'=f'(X')$. By (105·12) and Taylor's Theorem,

$$f'(X')=(1+\boldsymbol{l}_{12}d\theta_{12})f'(X)=\exp(\boldsymbol{l}_{12}d\theta_{12})f'(X),$$

where

$$\boldsymbol{l}_{12}=x_1\partial/\partial x_2-x_2\partial/\partial x_1. \quad (105\cdot 21)$$

Thus the change in the functional form is given by

$$f'(X)=\exp\{-\boldsymbol{l}_{12}d\theta_{12}\}f'(X')=\exp\{[-\boldsymbol{l}_{12}+\tfrac{1}{2}E_{12}]d\theta_{12}\}f(X),$$

since $f'=\psi'=q\psi$. The change in form per unit rotation $d\theta_{12}$ is $\mathbf{L}_{12}f$, where

$$\mathbf{L}_{12}=-\boldsymbol{l}_{12}+\tfrac{1}{2}E_{12} \quad (\text{i.e.} \lim_{d\theta\to 0}\{[\exp(\mathbf{L}_{12}d\theta)-1]/d\theta\}).$$

We identify this operator with *momentum*, writing

$$\mathbf{p}_{\mu\nu}=-i\hbar\mathbf{L}_{\mu\nu}=-i\hbar\{x_\mu\partial/\partial x_\nu-x_\nu\partial/\partial x_\mu-\tfrac{1}{2}E_{\mu\nu}\}, \quad (105\cdot 5)$$

although the full identification is only for the a.m. components $\mu, \nu=1, .., 4$. We call the parts

$$\left.\begin{array}{l} -i\hbar(x_\mu\partial/\partial x_\nu-x_\nu\partial/\partial x_\mu) \quad \textit{field a.m.,} \\ -ih(-\tfrac{1}{2}E_{\mu\nu}) \quad \textit{particle a.m.} \end{array}\right\} \quad (105\cdot 6)$$

The latter (the only part when $\mathbf{L}_{\mu\nu}$ operates on a simple wave vector ψ) agrees with the operational form (73·23) of the momentum vector $\mathbf{p}_\mu = -ip_{16}E_\mu$, with

$$p_{16} = \tfrac{1}{2}\hbar. \tag{105·42}$$

The 'field' part arises here as a *correction* to the particle or spin momentum, because of the dependence of a wave-vector *function* on position.

Being associated with a rotation about the origin, the field a.m. implies a *field linear momentum*

$$\mathbf{p}_\mu = -i\hbar\,\partial/\partial x_\mu, \tag{105·7}$$

which is limited to $\mu = 1,..,4$. For operation on a *contravariant* wave-vector function, we write

$$\mathbf{p}_\mu = i\hbar\,\delta/\delta x_\mu, \tag{105·9}$$

where $\delta/\delta x_\mu$ is written *after* its operand.

Eddington also defines the momentum operator as

$$\mathbf{p}_{\mu\nu} = -2ip_{16}\,\partial\mathbf{q}_{\mu\nu}/\partial\theta, \tag{105·3}$$

where $\mathbf{q}_{\mu\nu}$ denotes rotation through θ in the $E_{\mu\nu}$ plane; but he only uses infinitesimal rotations, and the summarized argument seems clearer without this. He uses this notion to discuss rotation as 'strain'. Earlier, 'field' momentum was called *recoil*. See **11·2** §§ 3·7 (*a*), 3·6 (*b*), 3·7 (*b*); also notes in **12·2** § 59, **12·5** § 73 and **13·5** § 87.

106. **The gradient operator.** Let wave-vector functions ψ, χ satisfy

$$(\mathbf{grad} - M)\,\psi = 0, \quad \chi^*(-\mathbf{darg} - M) = 0, \tag{106·21}$$

where $\mathbf{grad} = \sum_1^4 E_{\mu5}\partial/\partial x_\mu$, **darg** is the same operating to the left, and M is unspecified. Then

$$\chi^*(\mathbf{grad} + \mathbf{darg})\,\psi = 0, \tag{106·22}$$

i.e. $\sum_1^4 \partial/\partial x_\mu(\chi^* E_{\mu5}\psi) = 0$. Thus for a *particle* momentum

$$P = \Sigma E_\mu p_\mu = \psi\chi^*,$$

by (73·1)

$$\Sigma\partial p_{\mu5}/\partial x_\mu = 0. \tag{106·23}$$

For strain operators we take

$$\mathbf{grad}_s = -E_{45}\,\mathbf{grad} = (E_{14}\partial/\partial x_1) + \partial/\partial x_4. \tag{106·24}$$

For a pure particle (cf. § 72) the strain vector $-PE_{45} = S = \psi\psi^\dagger$, with $\chi^* = \psi^\dagger$. Thus (106·21) becomes

$$(\mathbf{grad}_s + E_{45}M)\,\psi = 0, \quad \psi^\dagger(-\mathbf{darg}_s + E_{45}M) = 0, \tag{106·3}$$

whence, if $S = \Sigma E_\mu s_\mu$ and $ix_4 = t$,

$$(\partial s_{14}/\partial x_1) + \partial s_{16}/\partial t = 0. \tag{106·4}$$

This is the conservation equation for fluid of density s_{16} and momentum density $s_{(1)4}$. We shall view it as *conservation of probability*, adding later a term for interchange circulation. Postponing this, we observe that in terms of the *field* momentum $\mathbf{p}'_{\mu 5}=\mathbf{p}_{\mu}$ of (105·7),

$$-i\hbar\,\mathbf{grad}=i\hbar\,\mathbf{darg}=\Sigma E_{\mu 5}\mathbf{p}'_{\mu 5}. \qquad (106·6)$$

Thus if M is algebraic (involving at most E_{16}), (106·21) become

$$(\sum_{\mu} E_{\mu 5}\mathbf{p}'_{\mu 5}+i\hbar M)\,\psi=0, \quad \chi^*(\ldots)=0. \qquad (106·7)$$

These are formally the same as the identities (72·81) for the *particle* momentum ($\varpi_5=\Sigma E_{\mu 5}p_{\mu 5}$ is the relevant 'pentad'). Comparison shows that the expectation values (73·22) $p'_{\mu 5}$ and $p_{\mu 5}$ of field and particle momenta are related by

$$p'_{\mu 5}+\hbar M p_{\mu 5}/p_{16}=0. \qquad (106·82)$$

Hence the wave equation,

$$(-i\hbar\,\mathbf{grad}-m')\,\psi=0, \qquad (106·91)$$

gives wave functions ψ containing a field momentum which is n times the particle 4-vector, where

$$n=m'/ip_{16}. \qquad (106·92)$$

Since there is no field component p'_{05}, the particle component p_{05}, representing extra-spatial probability flow, vanishes; this agrees with (106·21) (**grad** having no E_{05} component) and (106·23).

See first **12·2** §61; then **11·2** §§3·8, 3·7 (*b*) for remarks linking this section to later work.

107. **Isostatic compensation.** Wave functions give both particle and field momentum; the *wave equation* is the condition that these related momenta should leave the field self-consistent (i.e. that the total field should still correspond to the particles creating it). By taking n in (106·92) equal to the ratio of field and particle momentum as determined by the multiplicity factor (**3·2** (15·7)), we obtain the initial rigid (scale-free) field of §15; and the wave equation for the wave function ψ of such particles is (106·91), namely,

$$(-i\hbar\Sigma E_{\mu 5}\partial/\partial x_{\mu}-m')\,\psi=0, \qquad (107·1)$$

with a 'field' mass m' which is n times the particle mass. The solution is

$$\psi=\psi_0 e^{i\theta}, \quad \chi=\chi_0 e^{-i\theta}, \quad \theta=\sum_1^4 p'_{\mu 5}x_{\mu}\hbar^{-1}. \qquad (107·2)$$

The addition of particles with these wave functions to the 'uranoid' is analogous to the addition of mountains, plus their isostatic compensation, to the 'geoid' in terrestrial gravitation. These particles, however, are not individually identifiable, but are scale-free in accordance with the infinite plane wave form of (107·2). Moreover, the isostatic compensation does not cover all the components of a particle momentum vector, but is limited because the field momentum (105·7) has only four components.

The full wave equation has to regulate the addition, to the uranoid, of microscopic systems which (*a*) are in equilibrium (the wave equation serving as an equation of continuity of probability flow, including interchange), and (*b*) leave the uranoid in equilibrium (for this the wave equation regulates the 'isostasy' of added field and particle momentum).

Before the geodetic analogy was developed, the wave equation was already described as balancing the particle and field additions; see **11·2** §§ 3·8, 3·6 (*b*), **12·2** § 61. Later versions **12·5** § 74 (*b*), **12·6** § 74 (*b*) and **13·5** § 88 resemble **F** and are not quoted.

108. **Wave equation of the hydrogen intracule.** The internal wave equation of the hydrogen atom employs the relative coordinates ξ_α of electron and proton (§ 76); $\alpha=1,2,3$ gives space and $\alpha=4$ represents phase (instead of time). But we replace the ξ_α by x_α (treated as classical space-time coordinates), using the 'analogy' of (78·3) to reinterpret the results in quantum language. We write the wave equation (106·91) as

$$(\mathbf{W}-\mu)\psi=0, \quad \mathbf{W}=-i\hbar\,\mathbf{grad}, \tag{108·1}$$

where μ is the intracule mass. We do not insert a multiplicity factor for 'isostatic compensation', since in practice μ is determined from (108·1) spectroscopically, and so as a field quantity. But we insert a term (corresponding to the Coulomb energy in **3·3** § 33, and explained in § 110 below) to represent *interchange circulation*; thus (taking $\hbar=1$),

$$\mathbf{W}=-i\left\{\sum_1^4 E_{\mu 5}\partial/\partial x_\mu+E_{45}\alpha/r\right\}, \quad \alpha=\tfrac{1}{137}. \tag{108·2}$$

To solve (108·1), we seek four operators with common eigensymbols ψ—this implies that the operators should commute. One is $\mathbf{W}$; another is

$$\mathbf{U}_3=\partial/\partial x_4=i\partial/\partial t, \tag{108·3}$$

representing the *scale* (conjugate to phase). Two others commuting with these and each other are found to be [a]

$$\mathbf{U}_1=-iE_{45}\{(E_{23}[x_2\partial/\partial x_3-x_3\partial/\partial x_2])+1\}, \tag{108·4}$$

$$\mathbf{U}_2=-i\{x_2\partial/\partial x_3-x_3\partial/\partial x_2-\tfrac{1}{2}E_{23}\}. \tag{108·5}$$

[a] The bracket ( ) implies a sum in (108·4) and after (109·13).

These are (total and z) angular momenta; linear momenta would be out of place in a localized system. The *eigenvalues* are written

$$\mathbf{W}=\mu,\quad \mathbf{U}_1=j,\quad \mathbf{U}_2=u,\quad \mathbf{U}_3=\epsilon, \tag{108·6}$$

where μ is a given constant, j, u and ϵ are to be found.

On $\mathbf{U}_3$, see **12·2** § 62. On the nature of the equation: **12·5** § 80 and **13·5** § 89.

109. **Solution of the wave equation.** By (108·2–4)

$$iE_r\mathbf{W}=-iE_{45}r^{-1}\mathbf{U}_1-r^{-1}-\partial/\partial r+E_rE_{45}\{\mathbf{U}_3+\alpha r^{-1}\}, \tag{109·13}$$

where $r=\sqrt{(x_1^2)}$, $E_r=r^{-1}(E_{15}x_1)$. Putting this in (108·1), and using $\mathbf{W}\psi=\mu\psi$, $\mathbf{U}_1\psi=j\psi$, $\mathbf{U}_3\psi=\epsilon\psi$ (cf. (108·6)), and finally writing $\psi=r^{-1}\phi$, the wave equation becomes

$$\partial\phi/\partial r+(F+Gr^{-1})\,\phi=0, \tag{109·3}$$

where $$F=-E_rE_{45}\epsilon+iE_r\mu,\quad G=iE_{45}j-E_rE_{45}\alpha, \tag{109·21}$$

so that $$FG+GF=-2\alpha\epsilon,\quad F^2=f^2,\quad G^2=g^2, \tag{109·22}$$

where $$f^2=\mu^2-\epsilon^2,\quad g^2=j^2-\alpha^2. \tag{109·23}$$

If the wave function is of the form $\psi=\psi_0 f(x_{(1)},x_4)$ (ψ_0 a constant vector), the allowed states are those of (93·5), the metastable and ground states.

For general wave functions we write

$$y=2fr,\quad \chi=e^{\frac{1}{2}y}\,\phi, \tag{109·51}$$

so that $$\partial\chi/\partial y+G\chi/y+\tfrac{1}{2}\{Ff^{-1}-1\}\,\chi=0. \tag{109·52}$$

Assuming $$\chi=\sum_{s=0}^{n}C_s y^{s+p}, \tag{109·53}$$

we find for convergence of $\int\psi^\dagger\psi r^2\,dr$ (to give finite probability) that

$$p>-\tfrac{1}{2}; \tag{109·54}$$

and $$(s+p+G)C_s=-\tfrac{1}{2}\{Ff^{-1}-1\}C_{s-1}. \tag{109·55}$$

Thus for $s=0$, $GC_0=-pC_0$, so that $-p$ is an eigenvalue of G, and so (by (109·22)) equals $\pm g$. Later we shall find that $|j|\geqslant 1$, so by (109·23) $|g|>\frac{1}{2}$. Hence by (109·54) $p=+g$ (reckoned positive).

For $s=n+1$ in (109·55), $C_{n+1}=0$, so

$$(Ff^{-1}-1)C_n=0\quad\text{or}\quad FC_n=fC_n.$$

Putting $s=n$,

$$(Ff^{-1}+1)\,(n+g+G)C_n=-\tfrac{1}{2}(F^2f^{-2}-1)C_{n-1}=0, \tag{109·56}$$

by (109·22). Since $FG = -2\alpha\epsilon - GF$ [*note* signs],

$$\{(n+g)(Ff^{-1}+1)+G(1-Ff^{-1})-2\alpha\epsilon f^{-1}\}C_n=0, \qquad (109\cdot57)$$

or, since $FC_n = fC_n$, $\qquad n+g-\alpha\epsilon f^{-1}=0. \qquad (109\cdot58)$

Finally, by (109·23), $$\frac{\epsilon}{\sqrt{(\mu^2-\epsilon^2)}}=\frac{n+\sqrt{(j^2-\alpha^2)}}{\alpha}. \qquad (109\cdot6)$$

This is Sommerfeld's formula for all hydrogen states, with n a positive integer or zero.

By transforming to polar coordinates and using the single-valuedness of ψ, it is then found that the eigenvalues of the a.m. $\mathbf{U}_1$ and $\mathbf{U}_2$ are: j a (non-zero) integer and $u = -|j|+\frac{1}{2}$ to $|j|-\frac{1}{2}$ by unit steps.

The reader is referred to **F** for the last part, and to **RTPE** for the whole calculation. Some minor slips in **F** have been corrected in this Summary.

110. **The interchange momentum.** We now justify the extra term α/r in the wave equation (108·2). Without this term, the equation is $(\mathbf{W}-\mu)\psi=0$, with

$$\mathbf{W} = -i\hbar\sum_1^4 E_{\mu5}\partial/\partial x_\mu. \qquad (110\cdot1)$$

The eigenscale requirement is that $\partial\psi/\partial x_4$ reduce to an eigenvalue. There is arbitrariness of orientation between space $x_{(1)}$ and phase x_4 analogous to that between space and time given by the Lorentz transformation. For simplicity consider the plane of x_1, x_4. A wave vector ψ_0 becomes $\psi = q\psi_0$ for a rotation $q=\exp\{-\frac{1}{2}E_{14}\theta\}$ in this plane. By (105·5) the a.m. operating on ψ gives

$$-i\hbar\{\partial/\partial\theta-\tfrac{1}{2}E_{14}\}\psi=-i\hbar\{\partial/\partial\theta-\tfrac{1}{2}E_{14}\}\{\cos\tfrac{1}{2}\theta-E_{14}\sin\tfrac{1}{2}\theta\}\psi_0$$
$$=i\hbar E_{14}\psi,$$

a result independent of θ. Thus the arbitrariness of phase-to-space orientation corresponds to an a.m. $i\hbar E_{14}$, where here the 1 direction (space direction E_{15}) is chosen as the direction to be rotated with phase (E_{45}). If the intracule coordinates are given more generally by $\sum_1^3 E_{\mu5}x_\mu$, we may for this calculation, by means of a space rotation not affecting E_{45}, make the position vector become $E'_{15}r$ (i.e. $x'_1=r$, $x'_2=x'_3=0$ with $r=\sqrt{(x_1^2+x_2^2+x_3^2)}$). The a.m. $i\hbar E_{14}$ is then ascribed to a linear momentum $p_{45}=i\hbar/r$ in the phase direction E_{45}; for this gives an a.m. $E'_{15}x'_1 . E_{45}p_{45}=E'_{14}i\hbar x'_1/r=E'_{14}i\hbar$ as is required. Thus when we remove this interchange momentum (of the position-phase rotation) from the linear-momentum operator $\mathbf{W}$ of (110·1), it becomes $\mathbf{W}-E_{45}i\hbar/r$; and this is the operator to be used in the wave equation, regarded as determining continuity of probability flow. The new term is (as in §33) multiplied by $\alpha = \frac{1}{137}$ for the observational system.

Eddington's subtle conversion of $i\hbar/x_1$ into $i\hbar/r$ by way of volume elements has here been replaced by the use of an equally plausible but simpler concept—the independence of interchange from space orientation. Some amplifications of Eddington's treatment appear in **13·5** §91; see also **12·2** §63.

111. **The two-frame transformation.** With I the interchange operator (81·1), let

$$G_{\mu 5} = \frac{I(E_{\mu 5} + iF_{\mu 5})}{1+i}, \quad H_{\mu 5} = \frac{(E_{\mu 5} + iF_{\mu 5})\, I}{1+i} \quad (\mu = 1, \ldots, 4, 0), \tag{111·11}$$

so that

$$E_{\mu 5} = \frac{I(G_{\mu 5} - iH_{\mu 5})}{1-i}, \quad F_{\mu 5} = \frac{(G_{\mu 5} - iH_{\mu 5})\, I}{1-i}. \tag{111·12}$$

The pentads $G_{\mu 5}$, $H_{\mu 5}$ generate complete sets $G_{\mu\nu}$, $H_{\mu\nu}$ with the same interchange operator I; but (111·11, 111·12) hold only for the pentad $\mu 5$. Let momentum vectors in the four frames be

$$\mathbf{p} = \Sigma E_{\mu} p_{\mu}, \quad \mathbf{p}' = \Sigma F_{\mu} p'_{\mu}, \quad \mathbf{P} = \Sigma G_{\mu} P_{\mu}, \quad \boldsymbol{\varpi} = \Sigma H_{\mu} \varpi_{\mu}, \tag{111·51}$$

where

$$P_{\mu} = (p_{\mu} + p'_{\mu})/\sqrt{2}, \quad \varpi_{\mu} = (p'_{\mu} - p_{\mu})/\sqrt{2}, \tag{111·52}$$

and $\mu = (\nu 5)$ $(\nu = 1, \ldots, 4, 0)$. It is found from (111·12) that

$$i^{\frac{1}{2}}(\mathbf{P} + i\boldsymbol{\varpi}) = I(\mathbf{p} + i\mathbf{p}'). \tag{111·54}$$

This couples with the transformation from EF to GH-frame a transformation of two ordinary particles *of equal mass* into an external and internal particle. The limitation of (111·51) to the pentad 5 makes this appropriate to field momentum and the differential wave equation. The strain vector form of this equation is $(H-\epsilon)\psi = 0$ ((72·84)), where (with $i' = \pm i$)

$$i'H = (E_{14} p_1) + E_{45} m \quad (p_{(1)} = -i\hbar \partial/\partial x_{(1)}). \tag{111·61}$$

For non-interacting particles, the wave equations $(H_x - \epsilon)\psi = 0$, $(H_{x'} - \epsilon')\psi' = 0$ in the E and F-frames combine as

$$(H_x \pm iH_{x'} - \epsilon_0)\Psi = 0 \quad (\Psi = \psi\psi'). \tag{111·62}$$

We find (by replacing $\mu 5$ by $\mu 4$ in (111·11) and all the derived results) that this is equivalent to

$$(H_X \pm iH_{\xi} - \eta_0)\Phi = 0, \tag{111·63}$$

where the extra- and intracule 'hamiltonians' H_X, H_ξ are given by

$$I(H_x+iH_{x'})=\tfrac{1}{2}(1+i)(H_X+iH_\xi), \tag{111·71}$$

$$i'H_X=(G_{14}P_1)+G_{45}2m, \quad i'H_\xi=(H_{14}\varpi_1). \tag{111·72}$$

Here we are using $\quad P_\mu=p_\mu+p'_\mu, \quad \varpi_\mu=p'_\mu-p_\mu, \qquad$ (111·64)

so that H_ξ is *twice* the normal hamiltonian; this accounts for other factors 2 that occur.

We now insert the Coulomb term in H_ξ, which for two *like* charges (cf. **3·5** §49) gives an addition

$$\delta H_\xi=2\alpha\hbar/r=2e^2/cr. \tag{111·81}$$

The corresponding changes in the particle hamiltonians are, by (111·71) and the corresponding equation with $-i$ for i,

$$\delta H_x=-Ie^2/cr, \quad \delta H_{x'}=Ie^2/cr. \tag{111·83}$$

As we regard $\xi=x'-x$ as measured from the unaccented particle, we therefore take the perturbing Coulomb energy as

$$Ie^2/cr. \tag{111·9}$$

See **12·3** §71, and **RTPE** X.

112. **Electromagnetic potentials.** The hamiltonian of an electron in the presence of other electrons is, by (111·9),

$$H=H_0+\sum_s \delta H_s, \tag{112·11}$$

where
$$iH_0=-i\hbar(E_{14}\partial/\partial x_1)+E_{45}m, \tag{112·12}$$

$$\delta H_s=I_s e^2/cr_s, \quad I_s=\tfrac{1}{4}\Sigma E_\mu(F_\mu)_s, \tag{112·13}$$

where $(F_\mu)_s$ is the frame associated with the sth perturbing electron; we must also include *positrons* in the analysis, to obtain a neutral environment but keeping the perturbing particles all of equal mass.

To average over a large number of perturbing charges we replace $(F_\mu)_s$ by its expectation value $i\sigma_\mu/\sigma_{16}$ (cf. §73) for the strain vector $\Sigma F_\mu\sigma_\mu$ of the sth charge; but we retain only the linear momentum part with components $\sigma_{16}=e_s$ (as 'charge density'), $\sigma_{(1)4}=e_s u_{(1)}$ (as charge flow with $u_{(1)}$ velocity; compare also (106·4)). Thus $(F_{(1)4})_s$, $(F_{16})_s$ become $iu_{(1)}$, i, and I_s becomes

$$\tfrac{1}{4}i\{(E_{14}u_1)+E_{16}\}. \tag{112·2}$$

Then by (112·13),

$$(\sum_s \delta H_s)_{\text{average}}=\tfrac{1}{4}ie_0\{(E_{14}\kappa_1)+i\kappa_4\}, \tag{112·3}$$

where $e_0 = -e$ is the charge of the object-electron, and

$$\kappa_{(1)} = \sum_s e_s u_{(1)s}/cr_s, \quad \kappa_4 = \sum e_s/cr_s, \tag{112·4}$$

which is seen to be the classical (unretarded) e.m. potential of the 'field' of the other charges.

We infer that the wave equation $(H-\epsilon)\psi = 0$ of an electron in a molar e.m. field is obtained by adding (112·3) to the H_0 of (112·12). This has the effect of replacing as coefficients of $E_{(1)4}\,..\,\psi$ in the wave equation

$$-i\hbar\partial/\partial x_\mu \quad \text{by} \quad -i\hbar\partial/\partial x_\mu - \tfrac{1}{4}e_0\kappa_\mu \quad (\mu = 1, 2, 3), \tag{112·6}$$

and in the term $-\epsilon\psi$, $-i\hbar\partial/\partial x_4$ by $-i\hbar\partial/\partial x_4 - \frac{1}{4}e_0\kappa_4$. This agrees with current theory apart from the factor $\frac{1}{4}$.

These results will be used later. The fragment **7·2** §5·7 (*b*), despite its title, is not relevant and is not quoted.

10·3. F XI. The Molar Electromagnetic Field

A late draft is outlined in **13·6**. An earlier fragment **13·7** introduces the main tool–affine field theory and gauge transformation; there is a full account of this in **MTR**. It is to be remembered that Eddington writes dx_μ for the contravariant element $d(x)^\mu$, so that expressions like $\sigma_\mu dx_\mu$ are to be *summed* over μ.

113. **Gauge transformations (molar theory).** Let ds be the interval, $ds^2 = g_{\mu\nu}dx_\mu dx_\nu$, and $g = \det(g_{\mu\nu})$. In Weyl's theory a *gauge transformation* is a change $ds' = \gamma ds$ of the measure of interval, keeping the coordinates unchanged, so that in the new system

$$g'_{\mu\nu} = \gamma^2 g_{\mu\nu}, \quad \sqrt{-g'} = \gamma^4\sqrt{-g}. \tag{113·1}$$

Parallel transport of a vector of length l at x_μ to a point $x_\mu + dx_\mu$ changes its length (in terms of the standard at $x_\mu + dx_\mu$) by dl, where

$$dl/l = \sigma_\mu dx_\mu. \tag{113·21}$$

If we apply a gauge transformation, l (at x_μ) becomes $l' = \gamma l$ (at x_μ); and (113·21) gives for the gauge-transformed coefficients σ'_μ

$$\sigma'_\mu dx_\mu = dl'/l' = dl/l + d\gamma/\gamma; \tag{113·23}$$

so

$$\sigma'_\mu = \sigma_\mu + \partial(\log\gamma)/\partial x_\mu, \tag{113·24}$$

$$\operatorname{curl}\sigma'_\mu = \operatorname{curl}\sigma_\mu. \tag{113·25}$$

We now identify σ_μ/i with the e.m. vector potential κ_μ, and take $\gamma = e^{i\phi}$ (ϕ real). Then gauge transformation gives

$$\kappa'_\mu = \kappa_\mu + \partial\phi/\partial x_\mu, \tag{113·31}$$

$$|ds'| = |ds|. \tag{113·32}$$

Thus the usual indeterminacy of κ_μ to the extent of an additive gradient is represented by a gauge transformation. The invariance merely of the *modulus* of ds indicates that in a non-neutral region our previous neutral reality conditions (e.g. time-like ds real, space-like ds pure imaginary) fail, and we have to consider the body of gauge-transformed ds's as all equivalent. This restricts us to tensors which are *gauge-invariant*, for example, the *e.m. force*

$$F_{\mu\nu} = \operatorname{curl} \kappa_\mu, \quad F'_{\mu\nu} = F_{\mu\nu}. \tag{113·41}$$

From the gauge-invariant 'Christoffel symbol' defined as

$$^*\{\mu\nu, \alpha\} = \{\mu\nu, \alpha\} + i(g_{\mu\nu}\kappa^\alpha - g_\mu{}^\alpha \kappa_\nu - g_\nu{}^\alpha \kappa_\mu), \tag{113·42}$$

we form the gauge-invariant tensor $^*G_{\mu\nu}$ analogous to $G_{\mu\nu}$; the scalar $^*G = g^{\mu\nu}\, {}^*G_{\mu\nu}$ then satisfies

$$^*G' = \gamma^{-2}(^*G). \tag{113·45}$$

Using (113·1, 113·4), we construct *in-invariant* (i.e. invariant for gauge and coordinate transformations) integrals; such are

$$\int (^*G)^2 \sqrt{(-g)}\, d\tau, \quad \int F_{\mu\nu} F^{\mu\nu} \sqrt{(-g)}\, d\tau, \quad \ldots. \tag{113·5}$$

These suggest themselves as generalizations of the action integral $\int G \sqrt{(-g)}\, d\tau$ of neutral matter.

See **13·6** § 97, and the introduction **13·7** to affine theory, which is basic to 'parallel transport'.

114. **Action invariants.** A possible action density is

$$\mathfrak{A} = (^*G^2 - \alpha F_{\mu\nu} F^{\mu\nu}) \sqrt{-g}, \tag{114·1}$$

which in the absence of e.m. field reduces to $G^2 \sqrt{-g}$. There is a deep distinction between this and the ordinary mechanical action density $G\sqrt{-g}$, corresponding to the distinction between the attribution of energy as a mutual and a self-property of particles. The action density $\mathfrak{A}_2 = \sqrt{\{-\,|\,G^*{}_{\mu\nu}|\}}$ is fundamental, as it can be defined from an affine connexion without introducing a metric.

115. **Gauge transformations (microscopic theory).** The usual mechanical and electrical energy tensors are

$$T_\mu{}^\nu = -(8\pi\kappa)^{-1}\{G_\mu{}^\nu - \tfrac{1}{2} g_\mu{}^\nu G\}, \quad E_\mu{}^\nu = -F_{\mu\alpha}F^{\nu\alpha} + \tfrac{1}{4} g_\mu{}^\nu F_{\alpha\beta}F^{\alpha\beta}. \tag{115·1}$$

In a *uniform* gauge transformation $\gamma = e^{i\phi}$ (ϕ *constant*), $G_{\mu\nu}$, $F_{\mu\nu}$ are unchanged and $g^{\mu\nu}$ varies as γ^{-2}, so that

$$T_\mu{}^\nu \propto \gamma^{-2}, \quad E_\mu{}^\nu \propto \gamma^{-4}. \tag{115·2}$$

This difference of behaviour forbids the direct association of $T_\mu{}^\nu$ and $E_\mu{}^\nu$. The forms unchanged by gauge transformation are $T_{\mu\nu}$ and the *tensor density* of $E_\mu{}^\nu$:

$$\mathfrak{E}_\mu{}^\nu = E_\mu{}^\nu \sqrt{-g}. \tag{115·3}$$

Their association is suited to the *microscopic* (not molar) unification of mechanical and electrical theory in the spirit of §§89–91. The fundamental relation (85·6) or (91·2) is in fact, when the density factor $\sqrt{-g}$ appropriate to a strain tensor is included,

$$T^\times{}_{00} = \mathfrak{Z}_0{}^0 \quad \text{[printed } Z_0{}^0 \text{ in } \mathbf{F}], \tag{115·41}$$

which corresponds to the similar (invariant) behaviour here of $T_{\mu\nu}$ and $\mathfrak{E}_\mu{}^\nu$. Thus the electrical energy tensor density $\mathfrak{E}$ corresponds to the interstate (exchange circulation) energy $T^\times{}_{00}$, and not to the mechanical state energy T.

See **13·6** §98.

116. **Indices of wave tensors.** The *index* of a wave or space vector or tensor is here the power of γ by which the entity is multiplied in a gauge-transformation $\gamma = e^{i\phi}$. A quantum momentum vector P is shown to have the form $\psi_0 \chi_0{}^*$, where the wave vectors ψ_0, $\chi_0{}^*$ have indices $+1$, -1, and so become under transformation

$$\psi, \chi = \psi_0 e^{i\phi}, \quad \chi_0 e^{-i\phi}. \tag{116·31}$$

The vector of a particle with opposite charge is represented in a frame with E_{16} reversed; this is equivalent to reversing i, so that the corresponding wave vectors have indices -1, $+1$.

With $\kappa_\mu = 0$ in (113·31), a 'field' of potential $\kappa'_\mu = \partial\phi/\partial x_\mu$ is created artificially by the gauge transformation $e^{i\phi}$. For a constant ψ_0, by (116·31),

$$\hbar\kappa'_\mu \psi = \hbar(\partial\phi/\partial x_\mu)\,\psi = -i\hbar\,\partial\psi/\partial x_\mu.$$

By (105·7) we thus recognize $\hbar\kappa'_\mu$ as the field momentum vector. Let

$$p_\mu{}^0 = -i\hbar\,\partial/\partial x_\mu - \hbar\kappa_\mu. \tag{116·4}$$

For an 'artificial' field $\kappa_\mu = \kappa'_\mu$, $p_\mu{}^0 = 0$. For a general (or genuine) e.m. field κ_μ, $p_\mu{}^0$ is gauge-invariant. *Thus the wave equation is made gauge-invariant by substituting* (116·4) *for* $-i\hbar\,\partial/\partial x_\mu$.

This agrees with the result (112·6) for an electron in an e.m. field if we choose the units of the new κ_μ suitably. Since $\hbar c/e^2 = 137$, the last term in (116·4) is $\hbar\kappa_\mu = 137 e^2 \kappa_\mu / c$; the corresponding term in (112·6) is $\frac{1}{4} e_0 \kappa''_\mu$, where κ''_μ is the e.m. potential in customary units. Thus

$$\kappa''_\mu = 4\,.\,137 e_0 \kappa_\mu \quad (c = 1) \tag{116·7}$$

($e_0 = \pm e$ for $+$ or $-$ charge), where $\kappa_\mu = \sigma_\mu / i$, as in §113, is the gauge-transformation form of the e.m. potential. The factor 137 represents the multiplicity of the particle considered; thus we generalize (112·6) by saying: the operator replacing $-i\hbar\partial/\partial x_\mu$ in the wave equation of a particle of multiplicity k in an e.m. field κ_μ is

$$p_\mu{}^0 = -i\hbar\partial/\partial x_\mu - \tfrac{1}{4}ke_0\kappa_\mu. \tag{116·8}$$

Comparison with (108·2), for example, shows that the wave equation of the *intracule* is the same as that of an *electron* in a molar field $\kappa_4 = e/r$, if the electron is assigned multiplicity 4.

The explanation of (116·7) has been amplified here; see also **13·6** § 99. These results are used in the following magnetic-moment calculations.

There is no draft of the following sections, but they are summarized for completeness.

117. **Magnetic moments.** The wave equation in a molar e.m. field is $(\mathbf{W}-m)\psi=0$, where, with $p_\mu{}^0$ as in (116·8),

$$\mathbf{W} = \sum_1^4 E_{\mu 5} p_\mu{}^0. \tag{117·1}$$

The second-order wave equation (obtained by multiplying by $\mathbf{W}+m$) is

$$(\mathbf{W}^2 - m^2)\psi = 0, \tag{117·2}$$

where
$$\mathbf{W}^2 - m^2 = -\left\{\sum_1^4 (p_\mu{}^0)^2 + m^2\right\} + L, \tag{117·31}$$

$$L = -\tfrac{1}{4}ike_0\hbar\Sigma E_{\mu\nu}F_{\mu\nu}, \tag{117·32}$$

and $F_{\mu\nu}$ (*not* an 'F-frame') = curl κ_μ is the e.m. 6-vector.

For an 'artificial' field (gauge-induced as early in § 116, and so with $F_{\mu\nu}=0$), the energy $\epsilon^0 = ip_4{}^0$ is for small $p_{(1)}{}^0$

$$\epsilon^0 = m + (2m)^{-1}\sum_1^3 (p_\mu{}^0)^2. \tag{117·34}$$

An irreducible ('genuine') field adds the gauge-invariant term $-L/2m$. Thus *an applied magnetic force* $F_{23}=H$ adds the energy

$$-L/2m = ke_0\hbar H E_{23} i/8m = \mathfrak{M} H E_{23} i, \tag{117·4}$$

corresponding to the energy, in a field H, of a *magnetic moment*

$$\mathfrak{M} = ke_0\hbar/8m. \tag{117·5}$$

The magnetic moment of a *hydrogen intracule* is thus (with $k=4$ as at the end of § 116, mass $m=\mu$, and inserting c)

$$\mathfrak{M} = -e\hbar/2\mu c. \tag{117·8}$$

This is currently attributed to the *electron*.

The 'current' electron moment in Eddington's time was (117·8) with the electron mass for μ. Eddington's attribution of the moment to the intracule is consistent, however, with the circumstance that the Uncertainty Principle forbids the measurement of *free* electron moments by simple magnetic field experiments such as those Eddington envisages. The later refined molecular magnetic resonance experiments of Kusch and others indicate that the electron spin moment should be increased by 0·115 %. A first quantum electrodynamic approximation to this had been calculated by Schwinger as $\alpha/2\pi = 1/(2\pi . 137)$, a factor which is discussed in Eddington's chapter on Radiation.

118. **Magnetic moment of the hydrogen atom.** The extracule (the hydrogen atom as a unit) has mass $m = M = m_p + m_e$, effective multiplicity $k = 10$ and charge e; thus by (117·5) its magnetic moment is

$$\mathfrak{M} = \frac{5}{2}\frac{e\hbar}{2Mc}. \tag{118·3}$$

Experimentally this is measured gyromagnetically by an alternating field. This adds one measured d.f., so making the mass $M' = 10M/11$ (as in **3·2** (16·5)). For the observational system applied here to the second-order wave equation we introduce a factor $\beta^2 = (\frac{137}{136})^2$ [cf. the single β in **3·3** (29·4)], so that

$$\mathfrak{M} = \tfrac{5}{2}.\tfrac{11}{10}\beta^2 \frac{e\hbar}{2Mc} = 2·7906 \frac{e\hbar}{2Mc} \tag{118·5}$$

$$= 2·7899 e'\hbar/(2m_p c), \tag{118·6}$$

in terms of the observational 'nuclear magneton' $e'\hbar/(2m_p c)$ [cf. **3·3** (30·5)]. The 'observed' factor for (118·6) is $2·7896 \pm 0·0008$ (Millman and Kusch, *Phys. Rev.* **60**, 91 (1941)).

The moment (118·6) is here in fact being compared with the proton moment. The current (1956) value of this is 2·7927.

119. **Magnetic moment of the neutron.** The multiplicity of the neutron is normally regarded as 14 (extracule d.f.'s 10, co-spin intracule 4 as in §95); this is effectively halved, since the dimensionality l in **3·2** (15·51) of the classifying characteristics is to be taken as 2 (not 1 as everywhere else), as the classification treats the neutron as a double particle, but the present work treats it as single. The mass-energy correction $\frac{10}{11}$ is, however, assumed as in §118; and a single factor β is inserted to correct the field H of the experiment to the observational system. The magnetic moment of the neutron so found is

$$\mathfrak{M}_n = \tfrac{7}{4}.\tfrac{11}{10}.\beta.\frac{e\hbar}{2m_n c}, \tag{119·1}$$

or
$$\mathfrak{M}_n = 1{\cdot}9371\,\frac{e'\hbar}{2m_p c}, \tag{119·2}$$

in nuclear magnetons. The observed factor (Alvarez and Bloch) is $1{\cdot}935 \pm 0{\cdot}020$.

The present value (N. R. Corngold, V. W. Cohen and N. F. Ramsey (1956)) is $-1{\cdot}913138 \pm 0{\cdot}000045$.

10·4. F XII. Radiation

This fragmentary chapter of **F** (see **1·3** above) lies outside the scope of the present review, since there are no drafts extant. The sections are: §121, Radiation by a moving electron; §122, Transition probabilities; §123, Compton scattering; §124, Transverse self-energy of a particle.

The early draft section **6·2** §3·9 on Radiant Energy is not connected with this chapter.

CHAPTER 11

DRAFTS OF F VI–XI: (i) DRAFT A

11·0. Note

A separate chapter is now given to each of the drafts **A**, **G** and **H** (the later part of **H**) covering *E*-number theory and applications. Of these drafts, **A** is the earliest. It is fairly clear that **A** *16* (**11·1** below) was originally Chapter I of the book, and that it was changed to II to make way for a preliminary chapter on statistical theory. This preliminary chapter may have been **B** *17* (**4·2** above); but after writing **A** II–IV Eddington presumably decided to develop the statistical theory more fully (as in draft **B**) before giving *E*-number theory.

Draft **A** does not go beyond *E*-number theory to the later applications, although it treats angular momentum (**11·2** § 3·7) in the manner of **F** X. The interest of **A** lies in the development of the physical ideas, and the formulation of 'Reality Conditions' which vexed Eddington throughout his researches.

11·1. Draft A *16* II. The Complete Momentum Vector

2·1 [formerly 1·1]. **Particles with spin:** like **F** §§ 53, 55. The V_{10} particles are specified by p.d.'s of the 10 variates $p_1, \ldots, p_{34}$, and then by *E*-numbers. Concerning the latter:

> We cannot afford space for so leisurely a treatment [as in **RTPE**] of ..a mathematical tool...A treatment which follows the natural development of the mathematical calculus employed confuses the order of the physical concepts;..it would not be a hopeful way of gaining insight into the physical universe to consider first the phenomena which involve only simple equations, then those requiring quadratic equations, then logarithms, etc. Therefore, having in my earlier book done what I could to ease the difficulty of the mathematics, I am now more concerned to simplify and extend the physics.

'Equivalent' and 'chiral' frames are introduced as in **F** § 55.

2·2 [formerly 1·2]. **Relativity rotations:** like **F** § 56 to (56·7), and (57·1) of **F** § 57. The behaviour of 'perpendicular' and 'antiperpendicular' components under rotation is contrasted; and identifying $p_{0(1)}, p_{04}$ as a 4-vector the behaviour **F** (56·7) of *P* is found. The 5- and 6-dimensional representations (**F** § 57) are noted.

2·3 [formerly 2·2]. **Neutral space-time:** related to **F** § 63, but with differences of interpretation.

The suffixes 1, . ., 5 are associated with 5 axes in or normal to space-time. The mysterious sixth axis, associated with suffix 0, may be described as electrical.... So far as geometry is concerned, nothing distinguishes it from the others, which are interchangeable with it by relativity rotation. Not until we introduce a neutral uranoid do we give it a distinctive character.

If a particle is contemplated in its *actual* surroundings, we can apply to it and its whole surroundings any of the transformations (2·16) [**F** (55·3)] and obtain an equivalent, though apparently quite different, description from another point of view—a description of a particle with different characteristics in surroundings with different characteristics. But if we contemplate a particle in *standard* surroundings, the transformations are limited to those which do not introduce into the surroundings a deviation from the standard specification. In particular, the transformations of a particle in association with a uranoid are limited to those which transform the uranoid into itself. This limitation. . of the standard surroundings. . rules out arbitrary rotations between the axis 0 and the other axes.

. . . it is possible to consider a uni-chiral uranoid. . . of particles of one sign. . . [but] We adopt for all ordinary purposes a neutral uranoid. This, being non-chiral, requires only 5 axes for the specification of its characteristics; and the axes designated 1, . ., 5 are *defined* to be the axes used in the specification of the uranoid. Thus the axis 0 is by definition the additional axis required to describe the chiral characteristics of the small irregularities which in the actual universe are superposed on the uranoid, as mountains and cavities are superposed on the geoid.

Except in cosmology, it is the superposed irregularities rather than the uranoid that the physicist chiefly studies. . as particles, fields, etc., occupying the neutral space-time which corresponds to the neutral uranoid.... Neutral space-time. . has Riemannian geometry; for in R. geometry there is no provision for chiral properties.

'Neutral space-time' does not preclude an e.m. field. . as an addition to space-time. . not incorporated in its geometry as in Weyl's non-R. geometry. . . In the same way an irregular gravitational field does not disturb the uniformity of space-time if it is treated in the Newtonian way as an addition to space-time. Space-time becomes irregular if the field is incorporated into its geometry as in general relativity.

The uniform neutral uranoid. . and space-time. . is a bridge which connects the various branches of physics. . . microscopic particles, e.m., radiation and gravitational fields are all additions to the neutral uranoid....

The E-symbols are primarily associated with planes, not axes; but when the sixth dimension is not recognized explicitly, 5 of the planes in 6 dimensions degenerate into axes in 5 dimensions, and the symbols $E_{01}, \ldots, E_{05}$ of the planes are transferred to these axes.

If, however, we try to represent the directions of the axes by $E_{01}, \ldots, E_{05}$, . . the vector $E_{01}x_1 + \ldots + E_{05}x_5$ representing position in the coordinate frame, with the derived vectors representing displacement and velocity, is reversed when a r.h. frame is changed to a corresponding l.h. frame, although the momentum vector $E_{15}p_{15}+\,..\,+E_{45}p_{45}$ is not reversed.. a particle moving always in the direction of its electric current instead of..its momentum (which for negative particles is opposed to the current). To preserve the relation between kinematics and dynamics, the directions of the axes must be represented by

$$E_{16}E_{01}, \quad \ldots, \quad E_{16}E_{05} \tag{2·31}$$

which are unaltered when a l.h. frame is substituted.

The factor E_{16} $(= \pm i)$ makes no difference to the geometry of the axes ..but eliminates the unwanted chirality....A neutral uranoid..must be unaffected by reversing the chirality of its reference frame; or equivalently, being composed of balancing pairs of particles, it must be unaffected by changing the signs of all the particles....If we want to introduce a chiral particle into the 5-space,..the components are not purely algebraic..we use an enlarged field of numbers involving the chiral symbols E_{16}.

Since $E_{16}E_{01}.E_{16}E_{02}=E_{21}$, the symbols of the coordinate planes of the 5-space (2·31) are the same as for the set of axes $E_{01}, \ldots, E_{05}$, except that the positive direction of rotation in the plane is reckoned oppositely.

In later developments (IV) we shall come across (2·31) in the form

$$E_{01}F_{16}, \quad \ldots, \quad E_{05}F_{16} \tag{2·32}$$

..as part of the theory of a 'double frame'.

2·4 [formerly 2·3]. **Strain vectors:** mainly like **F** §59. The 'axes' of (2·31) above are used to introduce volume, which leads to the strain vector:

Consider a rectangular block δx_1, δx_2, δx_3 with scalar volume $\tau=\delta x_1\,\delta x_2\,\delta x_3$. When symbolic coefficients [(2·31)] are included, the volume is

$$V=\prod_{\alpha=1}^{3} E_{16}E_{0\alpha}\,\delta x_{\alpha}=E_{45}\,\tau \tag{2·411}$$

..independent of the chirality of the frame..[but dependent] on the order of the edges.

A *strain vector* is a three-dimensional vector density..

$$S=PV^{-1}=-PE_{45}\tau^{-1}\ldots=-PE_{45}, \tag{2·412, 2·42}$$

when we take $\tau=1$, so that S and P differ only in their symbolic coefficients. They are then said to 'correspond'....

The results **F** (59·3) follow, and the discussion of strain as observable deformation much as in **F**.

2·5 (*a*) [formerly 2·4]. **Reality conditions:** like parts of **F** §§56, 60–62. This long section is summarized.

The e.m. part of a strain vector involves $E_{(23)}$ and $E_{50,04,45}$—two conjugate triads (here called 'antitriads') A_r, B_r $(r=1,2,3)$. The mechanical symbols $E_{\mu\nu}$ are then of the form iA_rB_s, i. We make the *convention* that the A_r, B_r count as *real*; the remaining $E_{\mu\nu}$ are then imaginary.

For a complete (in **F**: 'extended') momentum or strain vector to be physically real, the electrical part must be antithetic (cf. **F** §62) to the mechanical. For this result, we observe first that a rotation operator $q = \exp(E_\mu\theta)$ is physically real if $E_\mu\theta$ is real. (Applying this first to 4 dimensions shows one axis, 4, to be connected with the others, 1, 2, 3, by Lorentz transformations—thus the axes of space-time are not 'isomorphic'.) Applying a (real) relativity rotation to the momentum vector $E_{45}p'_{45}$ of a particle at rest shows that the resultant vector $E_{45}p_{45}+E_{15}p_{15}$ must be monothetic. Similarly, all components $E_\mu p_\mu$ connected by relativity rotations must be homothetic. In the 6-space, 15 $E_{\mu\nu}$'s are so connected, and the 15 components $E_{\mu\nu}p_{\mu\nu}$ (and for simplicity the quarterspur $E_{16}p_{16}$) are homothetic. In the projection into a neutral 5-space [cf. §2·3 above], the electrical 5-vector and mechanical 10-vector remain separate under relativity rotations, and so form two homothetic groups. To see that the two groups are *antithetic* to one another, we return to the 6-space where the complete vector is monothetic:

The mechanical part P_m is unaltered by the projection, the planes having the same symbols as in the 6-space. But the electrical part P_e is changed from the real to imaginary (or vice versa), a component p_{01} being referred to an *axis* $E_{16}E_{01}$ $(= \pm iE_{01})$ in the 5-space instead of to a *plane* E_{01} in the 6-space. So for these terms the reality condition in the 5-space is opposite to that in the 6-space and..P_e must be antithetic to P_m....

For a strain vector..the coefficients s_μ are homothetic.

2·5 (*b*). **Flat space-time:** related to **RTPE** §6·2 and **F** §64.

Thus far we have confined attention to the reality conditions for momentum. The extension to vectors defining position is not straightforward...the complication..is the root of the characteristic differences between wave mechanics and molar theory. Out of these arises the uncertainty principle...

...The axes in the neutral 5-space have symbolic directions $E_{01}E_{16}$, etc., which we now write as $E_{01}i$, etc. The coordinate or..'position vector'..is

$$X = \sum_{1}^{5} E_{0\alpha}ix_\alpha \tag{2·521}$$

..with a gradient operator

$$\nabla = \Sigma E_{0\alpha}i\,\partial/\partial x_\alpha. \tag{2·522}$$

Let X be a physically real point, i.e. capable of being occupied by real matter...Real relativity rotations of the 5-space..give a real locus of 'equivalent' points answering our conception of a uniform space or space-time. The locus is 4-dimensional. If we choose axes so that the initial X is $(0,0,0,0,x_5)$, rotations in the planes $E_{15},\ldots,E_{45}$ give displacements $dx_1,\ldots,dx_4$; but a displacement dx_5 cannot be produced by relativity rotation. Each point on the x_5 axis generates a different locus. The loci are similar except in scale; and instead of specifying x_5, or more generally the radial coordinate, we can equivalently specify the scale of the space-time...Thus displacement along the x_5 axis is interpreted as a gauge transformation...so x_5 is treated..as a stabilized characteristic (§ 1·9) [**4·2** above]....

To obtain flat space-time an orthogonal instead of a radial projection of the 5-space is necessary...In other words we simply ignore the fifth coordinate in the rectangular frame...The resulting flat manifold is not a locus of equivalent points; it is made up of concentric annuli with non-equivalent gauge...round the centre the non-uniformity is insensible.

This orthogonal reduction from 5 dimensions to 4 is mathematically similar to our previous..from 6 to 5. The fifth dimension being no longer recognized, 4 planes degenerate into axes, and their symbols $E_{15},\ldots,E_{45}$ are transferred to those axes. The axes now have a duplicate set of symbols, having already been furnished with symbols by the degeneration of the planes E_{01}, etc. Thus we have two kinds of 4-vectors.

Vectors of opposite kinds with the same coefficients will be called *congruent*, e.g.

$$X=(E_{01}ix_1)+E_{04}ix_4 \tag{2·531}$$

and

$$X'=(E_{51}ix_1)+E_{54}ix_4. \tag{2·532}$$

The first is the position vector (2·521) in the neutral 5-space reduced to 4 components by the orthogonal projection. The second describes position in a congruent space-time. The gradient operator for X' is ∇'..where

$$-\nabla'=(E_{15}i\,\partial/\partial x_1)+E_{45}i\,\partial/\partial x_4. \tag{2·533}$$

By the definition (1·51) [see **4·2** § 1·5] of momentum in wave mechanics,

$$p_{15}=-i\beta\,\partial/\partial x_1. \tag{2·534}$$

Hence the ordinary momentum 4-vector $P=(E_{15}p_{15})+E_{45}p_{45}$ can be expressed in the operational form

$$P=\beta\nabla'. \tag{2·54}$$

The connexion between position and momentum vectors is defined by (2·54). It gives a fully invariant relation between momentum and position in the space X', which is extended to position in the space X by congruence. Expressing congruence by the symbol $\stackrel{c}{=}$, we have

$$P\stackrel{c}{=}\beta\nabla. \tag{2·55}$$

Congruence is invariant for relativity rotations in 4 dimensions, but is destroyed by rotation into the fifth dimension which transforms the two kinds of 4-vector differently. There is therefore no direct connexion between the rotations (2·51) $[(E_{15}iu_{15})+E_{45}\theta_{45}]$ as applied to P and the same rotations as applied to X and ∇.

First,..the congruence $X \overset{c}{=} X'$ is homothetic,..E_{01} with E_{51}, E_{04} with E_{54}. We make no change in the relation between mathematical and physical reality when we substitute the congruent space-time. Secondly, the factor i appears automatically in (2·531), being required to counteract the chirality of E_{01}, etc.; but it is merely *copied* in (2·532). The i does not naturally appear in vectors of this system; it is a transformation which we apply to them to make them congruent with the vectors of the other system. A factor i is therefore forced into our formulae by the conception of congruence. It is this i which appears so unexpectedly in the momentum operator (2·534), with the result that real phenomena are described by complex functions of position—the source of the characteristic difference between wave and classical mechanics.

From (2·531, 2·532),

$$X = X'E_{05}, \quad X' = -XE_{05}. \tag{2·561}$$

A 4-volume element has symbol $E_{01}i.E_{02}i.E_{03}i.E_{04}i = iE_{05}$. As in defining the 3-dimensional density or strain vector (§ 2·4), we define $\mathfrak{X} = XiE_{05}$ to be the vector density (4-dimensional) corresponding to X. Hence

$$X' = i\mathfrak{X}, \quad X = -i\mathfrak{X}'. \tag{2·562}$$

Thus a vector in one space-time is ($\pm i$ times) a vector density in the congruent space-time...the relation is reciprocal....Our previous investigations, in which P [momentum] has been treated as a vector.. mean that..we must take X [position] to be an imaginary vector-density...[thus] $E_{01}x_1$ and $E_{05}x_5$ must satisfy the reality conditions for ..a vector density. The corresponding components of the vector $X' = -XE_{05}$ are

$$E_{15}x'_{15} = -E_{15}x_1, \quad E_{16}x'_{16} = x_5.$$

Since $E_{15}x'_{15}$ belongs to the mechanical and $E_{16}x'_{16}$ to the electrical part of X', x_5 has to be antithetic to $E_{15}x_1$. Hence x_5 is homothetic with x_1, i.e. *the radius of curvature is space-like*. The character of the rotations is therefore opposite to that $[i(E_{15}u_{15})+E_{45}\theta_{45}]$ indicated in (2·51)....

2·6. **Determinants and eigenvalues:** like **F** § 65.

2·7. **Phase space:** equivalent to **F** § 75. The notion of 'anti-perpendicular' ['pseudo-' in **F** § 56] rotation is introduced in this text at this point. The last part is:

The transformations which give displacement in phase space are $S' = qS\bar{q}$. In a systematic development of wave-tensor calculus the transformation $q(\ldots)\bar{q}$ of phase space into itself is introduced at the very beginning. It is the transformation of a covariant wave tensor, whereas $q(\ldots)q^{-1}$ is that of a mixed wave tensor. Thus complete vectors are introduced as mixed ..and strain vectors as covariant wave tensors. Following this order, we first meet the mathematical concepts, then seek something in the physical universe for them to represent.

Phase space is not closed. This can be seen by considering strain vectors close to a singular value, e.g. $S_0 = E_{01} + E_{16}$ (..with eigenvalues $2i$ and 0, and so singular). Since $\det S \to 0$ as $S \to S_0$, the coefficients s_{01}, s_{16} in the normalized $S/(\det S)^{\frac{1}{4}}$ tend to infinity. The non-closure also follows from the fact that many of the transformations $q(\ldots)\,\bar{q}$ are hyperbolic rotations.

2·8. **Probability distribution of strain vectors** [originally '*in phase space*']. The arguments given are superseded or developed in other directions in later versions, but the following excerpts are of interest.

For a..molar fluid it is necessary to specify..(*a*) the energy, linear and spin a.m. per unit 3-volume—the 10 components $s_{\mu\nu}$ of S, the strain vector, and (*b*) the pressure system and the other components of the energy tensor $T_{\mu\nu}$. The specifications overlap...but both are required because (*a*) gives no indication of the pressure and (*b*)..no vorticity....
When we introduce the concept of probability, as in the kinetic theory of gases,..(*b*) is no longer required....The full specification is now.. a distribution function

$$f(s_1, \ldots, s_{10}; x, y, z; t)\, d\tau \quad [d\tau \equiv ds_1 \ldots ds_{10}] \tag{2·81}$$

..then description (*a*) is provided by mean.., (*b*) by mean square values....*Correlation* characteristics are included by specifying..the distribution of a 'double strain vector'...

$$f(s_1, \ldots, s_{136}; x, y, z, x', y', z'; t)\, d\tau. \tag{2·82}$$

...In the case of a single particle..(2·81) is supplemented by the datum that the distribution consists of one particle of identified character in the defined volume under consideration. We are therefore not dependent on f for knowledge of the proper mass...the *scale* of S is withdrawn from the p.d. Writing $S = \mu S_n$, the constant μ, whether known or uncertain, is set apart; and the p.d. to be specified is that of the normalized S_n. Since S_n is represented by a point in phase space, the p.d. we have to consider in connexion with a single particle is a distribution over phase space....

11·2. Draft A *35* III. Elementary Particles

3·1. **Idempotent vectors:** like **F** § 66 and (closely) § 67. It is observed that for an idempotent ($X^2 = X$),

for a polynomial $\phi(X)$, the mean $\overline{\phi(X)} = \phi(0) + \{\phi(1) - \phi(0)\}\,\overline{X}$.

On the significance of idempotency:

The classical dynamics of a particle is based on the expression of a quadratic characteristic (energy or hamiltonian) as a definite function of linear characteristics (momenta). We now know matter cannot be analysed into elements having the individual distinctness that the classical particle was supposed to have; there is no getting away from the statistical concept. But in the idempotent elements...mean quadratic have a determinate relation to mean linear characteristics independent of the p.d., so that they behave dynamically as individuals and not as averages are generally expected to do.

3·2. **Spectral sets of particles:** equivalent to **F** § 68. The elementary particle is defined by $S^2 = S$, leading to $P^2 = iP$ (inverting the order of **F**). After the identifications of components with charge, momentum, etc., and the 'spectral analysis' as in **F**, a 'neutral particle' (not a neutron) is introduced:

We can now connect the scalar theory in Chapter I [**4·2**] with the vector theory in II [**11·1** above]. It is necessary to decide whether the unit particle in the scalar theory is to be taken as the sum or the average of four vector particles with balanced charges and spins; i.e. to decide whether its strain vector is 1 or $\frac{1}{4}$...a *scalar* or a *neutral* particle ($S = 1, \frac{1}{4}$). ...Regard S_a, S_b, S_c, S_d as the strain vectors of four pure elementary states of an entity; if..wholly in one of these, it is an elementary particle; if it has equal probability of being in each it is a neutral particle....It is appropriate to treat the uranoid as a distribution of neutral particles. But in an object-system the inclusion of a neutral particle would be injudicious....

The strain vector 1 of a scalar particle is idempotent, whereas that, $\frac{1}{4}$, of a neutral particle is not. For this reason there is a closer mathematical correspondence between scalar and vector theory if we take the unit particle in the former to be the scalar particle. The scalar particle is the *analogue*, the neutral the *representative*, of the elementary vector particle. Care must be taken not to overlook the factor $\frac{1}{4}$ arising from the fact that the strain vector of a neutral particle satisfies $S^2 = \frac{1}{4}S$.

3·3. **The linear wave equation:** like **F** § 70 and the latter part § 72. The results **F** (70·1, 70·6) are established; it is added:

The 'normalizing condition' (*b*) [$\text{qs}\, P = \frac{1}{4}$] is invariant..customary to leave it aside..and work with unnormalized vectors. Thus in current theory we determine the momentum vector from [**F** (70·6)] (*a*) alone. The vector so defined is on an arbitrary scale and satisfies $P^2 = imP$, where m is an arbitrary number.

The latter part of **F** § 72 is represented by:

The name 'wave equation' is commonly restricted to the equations... $\alpha = 5$:

$$\{(E_{15}p_{15}) + E_{45}p_{45} - m\}P = 0, \tag{3·361}$$

$$\{(E_{14}p_{15}) + p_{45} + E_{45}m\}S = 0, \tag{3·362}$$

where $m = ip_{16}$. We have omitted $E_{05}p_{05}$ in (3·361), $E_{04}p_{05}$ in (3·362) because in practice..the component p_{05} normal to space-time is zero. Using (2·331) [**F** (59·3)], we can write (3·362)

$$\{(E_{14}s_{14}) - is_{16} + E_{45}s_{45}\}S = 0. \tag{3·363}$$

With the usual convention that p_{15} is real, the reality conditions for P require ip_{16} antithetic to the imaginary quantity $E_{15}p_{15}$; hence m is real. Multiplying (3·361) initially by $(E_{15}p_{15})+E_{45}p_{45}+m\ldots$

$$-p_{45}^2=m^2+(p_{15}^2). \tag{3·364}$$

Since ip_{45} is the energy, this identifies $\pm m$ with the rest [originally 'proper'] mass..the sign is determined by the charge. It should be understood that m (substituted for $E_{16}p_{16}$) in the wave equation does not 'represent' the rest mass. The fact that (apart from ambiguity of sign) m has the same value as the rest mass expresses the relation between electrical and mechanical units..in the same vector.

...It is rather misleading to apply the name 'wave equation' to (3·361) alone..it needs to be supplemented by two more of (3·34) [**F** (70·1)]. However, the practice of quantum physics is better than its theory would suggest..two other equations are introduced in the guise of integrals of a.m....

These wave equations will be reduced to a more familiar form in § 3·8.

3·4. **Matrix representation of E-numbers:** like **F** §71 and the early part of §72. The results **F** §71 on matrices and **F** (72·1–72·4) on factorization are set out. From (71·6) it is concluded:

a factorizable strain vector can be expressed in the form $S=\psi\psi^\dagger$ ((3·48)) ...and it is uniquely defined if *one* factor ψ is specified. We can therefore specify the characteristics of an elementary particle (in a neutral uranoid) by a 'wave vector' ψ instead of by S or P.

3·5. **Wave vectors and tensors:** like **F** § 72 (definition of wave vectors, and the star notation for initial vectors) and §73 ((73·3) onwards). The four types **F** (73·6) of wave tensors are set out, and the space-vector types (2), (3) identified as in **F**. As, however, phase space has already been discussed in the present version (**11·1** §2·7), the strain-vector types (1), (4) are identified thus:

No. 1 is the transformation $q(\ldots)\bar{q}$ of a strain vector displaced in phase space (§2·7). We therefore identify a strain vector with a covariant wave tensor...as the space vector traverses all relativistically equivalent orientations, the strain vector traverses phase space. No. 4 has the same transformations as no. 1, except that $\bar{q}^{-1}$ is substituted...for q; its transformations are therefore locked to those of a contravariant space vector in the same way as..no. 1..to a covariant...We therefore distinguish nos. 1 and 4 as covariant and contravariant strain vectors....

The following paragraph emphasizes a point made in **F** §74. Note that 'raising' suffixes here applies to *pairs* of suffixes in the case of 6-vectors.

We can show further that $\psi\chi^*$ and $-\chi\psi^*$ are *associated* vectors. If $\psi\chi^*=\Sigma E_\mu A_\mu$, $-\chi\psi^*=\Sigma E_\mu A^\mu$, ... [then] $E_\mu A_\mu=-\bar{E}_\mu A^\mu$. Thus the space-like components are of opposite sign and the time-like are the same in the two vectors; e.g. $A_{01} = -A^{01}$, $A_{04} = A^{04}$, $A_{12} = A^{12}$, $A_{14} = -A^{14}$, $A_{15} = -A^{15}$. These are the same relations as those obtained by lowering the suffix (or suffixes) of the 4-vectors (or the 6-vector) contained in A^{μ}. Thus reversing the order of the wave-vector factors is equivalent to lowering or raising the suffixes of the corresponding space vector in a Galilean frame.

3·6 (*a*). **Space tensors and strain tensors of the second rank:** equivalent to **F** §74. Four *space* tensors are introduced as in **F**, but only one, T_5, of the strain tensors. This leads to a discussion of *oscillators* (cf. **F** §91):

Consider..

$$T^{(5)}{}_{\alpha\beta\underline{\gamma\delta}} = \psi_{\alpha}\phi_{\beta}\omega^{\gamma}\chi^{\delta} \tag{3·63}$$

..the product of a covariant and a contravariant strain vector..a *mixed strain tensor*..$\Sigma_{\mu}{}^{\nu}$. We have [with $T^{(2)}$ a mixed space tensor]

$$T^{(5)}{}_{\alpha\beta\underline{\gamma\delta}} = T^{(2)}{}_{\alpha\underline{\delta\gamma}\beta} \tag{3·64}$$

..or

$$\Sigma_{\alpha\beta\gamma\delta} = T_{\alpha\delta\gamma\beta}. \tag{3·65}$$

...If T is factorizable [as here]..the space tensor is the product of two space vectors $P = \psi\chi^*$, $P' = \omega\phi^*$, ..., the strain tensor the product of two strain vectors $S^1 = \psi\phi^*$, $S^2 = \omega\chi^*$. We may describe the transformation $P, P' \rightarrow S^1, S^2$ as *crossing* the vectors of the systems which they represent.

If the entities represented by P and P' are called particles, the crossed entities represented by S^1 and S^2 have a nature in which two particles are..involved. We call these *oscillators*. But equally we can take S^1 and S^2 to be the strain vectors of particles; P and P' will then be the momentum vectors of oscillators...a particle corresponds to the mutual relations of oscillators in the same way as an oscillator...to particles. A less abstract entity is..represented by a physical tensor of the second rank which can be dissected as we choose into a pair of particles or of oscillators.

...the momentum or strain vector of an oscillator does not in general fulfil the reality conditions determined for particles.

3·7 (*a*). **Angular momentum:** like the beginning of **F** § 73, and the substance of **F** §105. Eddington is now moving on (in this version) to the formulation of the differential wave equation (§3·8 below), and in this section introduces a.m. operators. What is called 'field' a.m. in **F** §105 was originally called *recoil* a.m. in this section; the word 'recoil' has been crossed out and replaced by 'field' throughout. Planck's constant $\hbar$ is represented by β (as in MS. **B**, **4·2**, **5·2**, ... above).

For a factorizable E-number P, the operational form of the momentum is found (as in **F** §73) to be

$$\mathbf{p}_{\mu} = -ip_{16}E_{\mu}, \tag{3·722}$$

and hence in terms of the rotation operator $\mathbf{q}_{\mu}$ (as in **F** §105)

$$\mathbf{p}_{\mu} = -2ip_{16}d\mathbf{q}_{\mu}/d\theta \tag{3·73}$$

—the 'fundamental definition of momentum'. Applied to a vector wave function $\psi(x_1, ..., x_4)$ (cf. **F** § 105)

$$\mathbf{p}_{12} = 2ip_{16}\{x_1\,\partial/\partial x_2 - x_2\,\partial/\partial x_1\} - ip_{16}E_{12}. \tag{3·745}$$

The last term...the spin momentum contained in the complete momentum vector. (3·745) gives an additional a.m. contained in the wave function but not in the wave vector or momentum vector of the particle. ...We shall call the additional momentum incorporated in the wave function *field* [originally 'recoil'] momentum. Accordingly, the field a.m. is the 6-vector $\mathbf{p}'_{\mu\nu}$ corresponding to a linear momentum

$$\mathbf{p}'_\mu = 2ip_{16}\,\partial/\partial x_\mu \quad (\mu = 1, ..., 4). \tag{3·747}$$

...also found directly by treating displacement in space-time as rotation about the centre of curvature....

In current theory...

$$\mathbf{p}'_\mu = -i\beta\,\partial/\partial x_\mu \tag{3·748}$$

...therefore $\beta = -2p_{16}$....If the wave functions are normalized to represent a single elementary particle, $p_{16} = \pm\frac{1}{4}$...then $\beta = \frac{1}{2}$...in terms of the natural unit determined by..the idempotent strain vector. Except in special investigations we retain β or $-2p_{16}$ as a scale constant. The current form of (3·745) is therefore

$$\mathbf{p}_{12} = x_1(-i\beta\,\partial/\partial x_2) - x_2(-i\beta\,\partial/\partial x_1) + \tfrac{1}{2}i\beta E_{12}. \tag{3·76}$$

The usual proof that (3·76) is the total a.m...depends on the dynamical equations of quantum theory...a *conserved* quantity must commute with the hamiltonian...Since the a.m. $x_1\mathbf{p}'_2 - x_2\mathbf{p}'_1$ derived directly from (3·748) does not.., a supplementary a.m. $\frac{1}{2}i\beta E_{12}$, not associated with any linear momentum and having therefore the character of vorticity, is inserted....

...our derivation:...$x_1\mathbf{p}'_2 - x_2\mathbf{p}'_1$ appears as a supplement to $\frac{1}{2}i\beta E_{12}$ arising out of the substitution of wave functions for wave vectors... The total momentum 4-vector is accordingly

$$\mathbf{p}_\mu = -i\beta\,\partial/\partial x_\mu + \tfrac{1}{2}i\beta E_{5\mu}. \tag{3·77}$$

The interpretation of momentum as 'strain' follows in the vein of **F** § 105.

3·8. **The differential wave equation:** a short section touching on **F** §§ 72, 107. It is observed that for an elementary particle, having factorizable P and S, the wave equations **F** (70·7, 70·8) become of the form **F** (72·81, 72·82) in wave vectors. These equations are related to Dirac's.

...the wave equations ($\alpha = 5$) for a strain vector $S = \psi\psi^\dagger$, satisfying the reality conditions, are

$$\{(E_{14}p_{15}) + p_{45} + E_{45}m\}\psi = 0, \quad \psi^\dagger\{...\} = 0. \tag{3·82}$$

...the factors of S are indeterminate to an algebraic multiple a, a^{-1}. We partially remove the ambiguity by restricting them to complex conjugates ψ, $\psi^\dagger$; but that leaves them still indeterminate to a factor $e^{i\theta}$, $e^{-i\theta}$. By taking θ to be a function of coordinates in space and time the wave vectors ψ, $\psi^\dagger$ are turned into wave functions.

For a particle in neutral space-time the wave functions currently adopted are

$$\psi = \psi_0 \exp\{i(\sum_1^4 p_{\mu 5} x_\mu)/\beta\}, \quad \psi^\dagger = \psi_0^\dagger \exp\{-(\ldots)/\beta\}. \tag{3·83}$$

We can then replace $p_{\mu 5}$ in (3·82) by

$$p_{\mu 5} = -i\beta\,\partial/\partial x_\mu = i\beta\,\delta/\delta x_\mu, \tag{3·84}$$

(δ..acting retrospectively on $\psi^\dagger$). With this substitution (3·82) becomes Dirac's linear wave equation in its recognized form.

By (3·748) $-i\beta\,\partial/\partial x_\mu$ is the operational form of the field momentum of the wave function ψ. *Thus the solutions of Dirac's differential wave equation couple with the momentum* 4*-vector of the particle an equal field momentum* 4*-vector*. It must be presumed...the purpose of the differential wave equation is to provide wave functions representing such [momentum] combinations.

We have to accept (3·83) as the beginning of a general method of proved value in practical problems....It is necessary to understand the general purpose of duplicating the momentum vector in the particle and the field. The repetition is a consequence of the *rigid-field convention*, accepted throughout quantum theory. We have referred to this in § 1·6 [**4·2**], and noticed it is the point at which the methods of quantum and relativity theory diverge. [The section—and chapter—end here.]

3·6 (*b*). **The differential wave equation,** 3·7 (*b*). **Angular momentum:** these two sections originally followed § 3·5 in this draft; later the section § 3·6 (*a*) above was inserted and these two were written as § 3·7 (*a*), § 3·8 above, with considerable rearrangement of the argument. § 3·6 (*b*) begins as § 3·8 above, but after the opening sentence of the last paragraph quoted, continues thus:

But even at this stage it is necessary to inquire how our outlook is affected by this substitution of wave functions for wave vectors. What is the purpose of introducing over again, as a functional factor attached to ψ, the momentum vector already contained in $\psi\psi^\dagger$? If it is really the same momentum that is being represented, it seems a pointless reiteration. But the repetition acquires a physical significance if we regard the momentum shown in the functional factor, not as equal to, but as balancing the momentum shown in $\psi\psi^\dagger$.

As we have continually emphasized, there are a vast number of particles concerned in any experiment on atomic systems which are not referred to individually in the equations professing to embody the theory of the experiment. A few particles are *specified*, i.e. assigned distinctive wave vectors or functions; the remaining *unspecified particles* constitute the uranoid or the uranoid plus a disturbing field. It is a condition of the validity of this treatment that the particles must be freely transferable from one group to the other. In particular, 'specification' of a particle must not involve any change of the total energy and momentum of the system. We have therefore to distinguish between the addition to the system of a particle with energy and momentum $\psi\psi^{\dagger}$, and the assignment of this energy and momentum to a particle, previously unspecified, at the expense of the rest of the system. The wave function couples the momentum 4-vector of the particle with a balancing momentum 4-vector, so that the total energy and momentum of the wave function is zero. Wave functions can therefore be freely inserted to describe features of the system on which we concentrate our study, and freely dropped if so much particularity is found unnecessary, without affecting the total energy and momentum. The wave functions of an elementary particle in neutral space-time afford a rather trivial example of this principle; but it extends to all manner of wave functions determined by more complicated wave equations. The wave functions most commonly considered in practice specify, not particles, but correlations. The requirement is that these shall *distribute* the energy and momentum without *creating* fresh energy and momentum; and the differential wave equation is a formulation of this requirement.

This 'balancing momentum' $p_{\mu 5}$ as in (3·84) above is called 'recoil momentum'; the total momentum is now stated to be the sum of this and of the particle momentum $\mathbf{p}_{\mu 5} = -ip_{16}E_{\mu 5}$. The formal discussion of a.m. in § 3·7 (*b*) follows much as in § 3·7 (*a*) above. One remark:

...a.m. is not treated in the same way as linear momentum. The recoil momentum coupled to the linear momentum $p_{\mu 5}$ exactly balances it; but the recoil a.m. added to p_{12} is not introduced to balance it, and, as likely as not, is in the same direction...there may be disappearance or creation of a.m. in transitions between states with a different number of d.f.'s, involving a change in the number or character of the wave functions employed. This is an aspect of 'the principle of systematic observation' explained in § 1·3 [**4·2**]....

11·3. Draft A *21* IV. The Energy Tensor

4·1. **Double frames:** the basis of **F** § 79.

4·2. **Interchange:** like parts of **F** §§ 81–83. The operator $I(..)I$ interchanges the E-, F-frames; the outer product of E-numbers P, P' is the ordinary product of PI, $P'I$. I commutes with the symmetrical part of an EF-number and anticommutes with the antisymmetrical [**F** § 82]. The complete energy tensor, being symmetrical, is of the form T_s of **F** (82·83). I is factorizable, as **F** (81·41):

other ways of factorizing I are obtained by suitable interchanges of suffixes.

4·3. **The dual frame:** like parts of **F** §§ 82, 83. The topics are: double (EF) matrices; the dual $\tilde{T}$; the transformation from the EF to the CD-frame ('not a relativity rotation; a tensor $t_{\mu\nu}$ is not relativistically equivalent to its dual'); $\tilde{T} = -TI$. Finally:

...to form the dual of a simple vector $P = \Sigma E_\mu p_\mu$, we write it $-i\Sigma E_\mu F_{16} p_\mu$. Then $\tilde{P} = \Sigma C_\mu p_\mu \ldots$; the outer product of two vectors is the ordinary product of their duals in the same frame.

4·4. **Double phase space:** related to **F** § 85. The 136 real $E_\mu F_\nu$ are 'space-like' (i.e. homothetic with their eigenvalues), the 120 imaginary are 'time-like'. The electrical part of a double momentum vector $P \times P'$ reverses with the chirality of the double frame, and so consists of the 120 terms containing suffix 0 or 16 once only. The electrical part of a double strain vector $S \times S'$ consists of products of electrical (real E_μ or F_μ) and mechanical (imaginary) factors, and so has imaginary $E_\mu F_\nu$. Thus, as for simple strain vectors,

The mechanical part of a double momentum strain vector is space-like and the electrical part time-like.... (4·41)

An entity..specified by a double momentum space or strain vector will be called a *bi-particle*. In neutral environment the electrical components are dormant, and the bi-particle is adequately specified by the 136 mechanical components..a V_{136}..(certain 'elementary states'.. correspond to a pair of particles..)....

When the strain tensor is limited to 136 space-like components,.. the corresponding space tensor is symmetrical....The symmetry condition is $(T_{\mu\nu})_{abcd} = (T_{\mu\nu})_{cdab}$. Now $(T_{\mu\nu})_{abcd} = (T_\mu{}^\nu)_{abdc} = (\Sigma_\mu{}^\nu)_{acdb}$ [Σ the strain tensor as in (3·65) above]. Hence $(\Sigma_\mu{}^\nu)_{acdb} = (\Sigma_\mu{}^\nu)_{cabd}$; so that if the strain tensor is $\Sigma E_\mu F_\nu \sigma_{\mu\nu}$, $\Sigma E_\mu F_\nu \sigma_{\mu\nu} = \Sigma \bar{E}_\mu \bar{F}_\nu \sigma_{\mu\nu}$. Thus $\sigma_{\mu\nu} = 0$ unless

$$E_\mu F_\nu = \bar{E}_\mu \bar{F}_\nu. \tag{4·43}$$

Since $E_\mu = +\bar{E}_\mu$ for the imaginary and $-\bar{E}_\mu$ for the real symbols, (4·43) is satisfied only by the real $E_\mu F_\nu$, and the time-like (electrical) components of $\Sigma_\mu{}^\nu$ vanish. Thus the condition that $T_{\mu\nu}$ is symmetrical is equivalent to the condition that $\Sigma_\mu{}^\nu$ consists of mechanical components only. Hence

The momentum space tensor of a V_{136} bi-particle is symmetrical. (4·44)

...The reality conditions for a double momentum vector $P \times P'$: if P and P' are characteristics of a real system, so is $P \times P'$...From the reality conditions for P and P'..we deduce that in $P \times P'$ 120 terms (with suffix 0 or 16 once only) must be antithetic to the rest....For a double strain vector $S \times S'$...: since the coefficients s_μ of a simple strain vector have to be homothetic, the $\sigma_{\mu\nu} = s_\mu s'_\nu$ are all real. Thus

The reality condition for a momentum strain tensor in neutral space-time is that all the coefficients are real (4·45)

..or, the electrical part..is antithetic to the mechanical part....The reality conditions are not affected by the electrical interaction between the two particles...this has a *field* momentum, so that it appears only in the wave functions and not in the wave vectors and does not affect the reality conditions of the latter.

The theory of the phase space of a V_{136} bi-particle follows precisely the theory for a V_{10} particle in §§ 2·7, 2·8. It is the extension indicated in (2·82)....Denoting the strain tensor by $\Sigma = \Sigma \overline{EF}_\mu \sigma_\mu$, the σ_μ define a point in a 136-space....But, as usual, we reserve the *scale* of Σ..to be fixed by stabilizing; so that the p.d. to be considered is that of the normalized Σ_n,..a 135-locus,..the phase space of the V_{136}....

4·5. **The relation between mass and density:** like **F** § 8, ending with the remarks (cf. § 4·4 above):

the energy tensor has the same dimensions, as well as tensor form, as the outer product of two momentum vectors; it can therefore be identified with a double momentum vector. It is specialized by symmetry, so that its typical form is $T_{\mu\nu} = P \times P' + P' \times P$.

The section appears to end here; but three (earlier draft) pages follow, introducing some ideas used in **F** § 22, on the partition of mutual energy between object and uranoid or comparison particle. The last paragraph is:

As explained in § 3·6 we couple with the momentum vector of the object-particle a recoil vector; so that the whole addition is represented by a wave function instead of a wave vector. With a simple frame the best we could do was to balance the linear momentum and energy; the whole addition is then described by a Dirac wave function. With a double frame we can represent a complete balance, the particle vector being $\Sigma E_\mu p_\mu$ and the recoil vector $-\Sigma F_\mu p_\mu$. The self-energy $T = A_s(P \times P)$ has the form of a mutual energy of the particle and recoil vector.

4·6. [No title]: a section dealing with particle-plus-comparison-particle in terms of double frames. This begins:

We shall now amplify the investigations in II, III which are better expressed in terms of a double frame. They refer mainly to a particle with momentum vector $P = \Sigma E_\mu p_\mu$ in a neutral environment....Consider a double system (bi-particle) of an object-particle and one uranoid particle ...—the *comparison* particle...in..the E-frame and..the F-frame. If we adopt a zero-temperature uranoid, so that the comparison particle is at rest, its momentum vector will reduce to...$F_{45} p'_{45}$. The bi-particle has the double momentum vector

$$B = P \times P' = \Sigma_\mu E_\mu F_{45} p_\mu p'_{45}. \tag{4·611}$$

The formula (2·31) [**11·1**] for the symbolic axes in 5-space was obtained by substituting a neutral for a unichiral uranoid. In single-frame representation the chiral symbol F_{16} of the uranoid could not be distinguished from the chiral symbol E_{16} of the particle, so that the effect of neutralizing the uranoid was expressed by $E_{01} \to E_{01}E_{16}$. Now that the uranoid has its own symbolic frame the corresponding transformation is $E_{01} \to E_{01}F_{16}$; and the symbolic directions of the axes in the 5-space are the double symbols $E_{01}F_{16}, \ldots, E_{05}F_{16}$...Since $F_{16} = E_{16}$, the change is only formal, but clarifies the conception....The complete coordinate vector $\Sigma E_\mu F_{16} x_\mu$ can be written more symmetrically as $\Sigma E_\mu F_{16} x_\mu x'_{16}$ $(x'_{16} = 1)$. This is the outer product of two simple position vectors $X = \Sigma E_\mu x_\mu$, $X' = F_{16}x'_{16}$ associated with object and comparison particle...the coordinate vector $Y = X \times X'$...

We can therefore start with a general bi-particle characterized by a double momentum vector B and position vector Y, and obtain the system composed of an object and comparison particle by giving P' and X' the special values

$$P' = F_{45}p'_{45}, \quad X' = F_{16}x'_{16}. \tag{4·62}$$

For the general bi-particle we can know only the p.d. of the components of B and Y. Ignoring electrical components, which have no meaning when the environment is neutral, these p.d.'s have 136 d.f.'s. For the specialized bi-particle P' and X' are definitely prescribed, and the d.f.'s are reduced to 10, the variates being the non-electrical components of P and X. This reduction..is an example of stabilization. It occurs because we allow ourselves to prescribe the environment as part of the enunciation of the problem..instead of...as dependent on observational investigation.

The position vector $F_{16}x'_{16}$ is a pure scalar, invariant for all relativity rotations..indicating entirely uncertain position. This was to be expected since we assigned the comparison particle an exact momentum vector $F_{45}p'_{45}$. It is to be noticed that stabilization, eliminating.. probability, introduces differential treatment of object and comparison particles; an object with completely uncertain position vector *may* occupy any position, whereas a comparison particle with completely uncertain vector *does* occupy every position. The assigned distribution of the comparison particle is a uniform distribution (not a uniform p.d.) over all orientations; and the non-scalar components of X' drop out because they have no significance in such a distribution. [*End.*]

Fragmentary sections follow: (i) on mutual energy, and (ii) an earlier §4·5 with the result [cf. **F** §85]:

The dual of a mixed space tensor is a mixed strain tensor.

CHAPTER 12

DRAFTS OF F VI–XI: (ii) DRAFT G

12·0. Note

A general account of Draft **G** is given in **2·3** above. A plan made at this stage for the continuation of the book is quoted in **12·4**. Draft **G** was difficult to handle, having lost its 'statistical' chapters I–IV, and being less straightforward than other drafts; but many details of interest required quotation, and some of these are referred to in **9** above.

12·1. Draft G *28* V. The Complete Momentum Vector

41, 42, 43: like **F** §§53, 54, 55.

44. **Rotations:** like **F** § 56, the first results of § 57, and the essence of §58. 'Pseudo-rotations' are here called 'anti-perpendicular rotations'.

45. **Real frames:** the basis of **F** §§60, 61. The order and detail of the argument on the choice between systems (*a*) and (*b*) [**F** §60] is different, but the conclusions agree. To summarize:

We define a *real E-number* $\sum_1^{16} R_\mu p_\mu, \ldots, p_\mu$ real, $R_\mu = E_\mu$ or $E_\mu E_{16}$, the choice being such as to give a closed algebra of real E-numbers. .i.e. the products of R's are to be R's. We regard R_μ as 'real', so that E_μ is real or imaginary according as R_μ is E_μ or $E_{16}E_\mu$.

The multiplication table **F** (53·4) shows that the E must be either (*a*) or (*b*) of **F** §60. We specialize the E_μ to satisfy (*a*). This may be justified *either* (1) by invoking the representation by fourfold matrices, *or* (2) by considering triads

$$\zeta_{(1)} = E_{(23)}, \quad \theta_{(1)} = E_{45,50,04}. \tag{45·1}$$

Then (with $E_{16} = i$) the complete set of E-symbols is ζ_α, θ_α, $i\zeta_\alpha\theta_\beta$, i $(\alpha, \beta = 1, 2, 3)$, and *in assignment* (*a*) the R_μ are ζ_α, θ_α, $-\zeta_\alpha\theta_\beta$, -1. Regard the ζ_α as anticommuting roots of -1, the θ_α likewise; then the 'direct square' of the ζ-algebra, with numbers

$$\Sigma(\zeta_\alpha q_\alpha + \theta_\alpha q'_\alpha + \zeta_\alpha \theta_\beta q_{\alpha\beta}) + q_{16}$$

gives, for real q_α, the real E-numbers. The justification (1) is regarded as superficial compared with (2):

...we shall find the nature of observational measurement is such that a world constructed from the results of measures necessarily has an

E-frame for its basis. In this deduction the E-frame is introduced as the direct square of a ζ-frame, so that it has the specialized property which brings it under case (a)....

We can now distinguish the *real rotations* $q(...)q^{-1}$, namely, those which correspond to real E-numbers q....We do not lay down...that vectors must be represented by real E-numbers...[but] a complex E-algebra, unrestricted by reality conditions, would be unsuitable for application to physics [because of dimensionality of p.d.'s: cf. **F** § 61].... When a component $p_{\mu\nu}$ is not real, the p.d. is still one-dimensional; either it is restricted to imaginary values or to some other one-dimensional locus in the complex domain....The reality conditions, which are formally treated as specifying a domain of p.d., can be equivalently regarded as a criterion of physical reality.

46. **Distinction between space and time:** like **F** § 62. Although the position vector is written as $X = (E_{15}x_1) + E_{45}it$, it is stated that this (as against iX in **F**) is a matter of convention.

47. **Neutral space-time:** corresponding to parts of **F** §§ 63, 64. First, if the position vector is extended to $\sum_1^4 E_{\beta 5}x_\beta + E_{05}x_0$, the curvature of a *neutral* uranoid must be associated with the direction $E_{05}E_{16}$ (which is independent of the chirality of the frame). Thus '*Physical* space-time is not extended to five dimensions'. The original E-frame corresponds to a 'unichiral' uranoid, which is beyond experience in *molar* physics—and we have to begin with molar physics, since microscopic physics deals with entities which are designated by molar analogies.

Thus for practical purposes we must use a non-chiral (neutral) E-frame instead of the real E-frame. That is why we get the $\sqrt{-1}$ occurring so often in quantum theory. The neutral E-frame is obtained by substituting the imaginary directions $E_{01}E_{16}$, etc., for the real directions E_{01}, etc., for the 6 electrical symbols. The neutral frame then specifies 16 directions none of which is reversed by changing from a r.h. to a l.h. set; it is effectively two chirally opposite frames superposed. It has 10 relativity rotations corresponding to the mechanical symbols (though in molar physics these are not all effective); but the rotations corresponding to the 6 electrical symbols are barred, because they rotate the two superposed frames opposite ways.

The extended momentum vector P is separated into an electrical 5-vector with an invariant, and a mechanical 10-vector as in the 5-dimensional representation (..not the 5-space in which we picture hyperspherical space-time) in § 44. The 10 surviving relativity rotations transform these independently, so that as far as vector and tensor properties are concerned they are quite unconnected characteristics. There is, however, a particular reason for stringing them together into a single expression P, which will appear in § 50. The extended vector in a chiral frame is monothetic; but in a neutral frame we substitute imaginary for real electrical directions, or equivalently we substitute imaginary for real components p_μ in those directions. The domain of p.d. and therefore of physical reality is altered accordingly; and we have the important reality condition—

The electrical part of a vector is antithetic to its mechanical part. (47·2)

...The specialized role of the different E-symbols depends mainly on two considerations, *reality* and *neutrality*. Reality is all-important in respect to relativity rotations; neutrality..in application to the molar world of neutral matter and the corresponding Riemannian space.. they are not the same consideration in different guise. They respectively single out a triad-pair of real symbols and a pentad of electrical symbols; and other symbols become distinguished according to their commutative relations to these two sub-sets.

48. **Strain vectors:** like **F** § 59; the reality condition **F** (63·4) is included.

49. **Determinants and eigenvalues:** like **F** § 65. It is added:

..the real E_μ are antithetic to.., the imaginary E_μ homothetic with, their eigenvalues. In a position 4-vector the space components are homothetic with their eigenvalues and the time component antithetic. It is convenient to generalize this nomenclature:

Any symbolic expression is space-like if homothetic with its eigenvalues, and time-like if antithetic to them. (49·7)

50. **Idempotency:** like **F** § 66; a passage like that quoted in **11·2** § 3·1 occurs.

51. **Standard form of idempotent vectors:** close to **F** § 67.

52. **Spectral sets of particles:** like **F** § 68, but ending just before the definition of *scalar particle* in **F** (q.v.).

53. **Dictionary of symbolic coefficients:** like the first part of **F** § 69, with an interlude related to **F** § 77. This interlude on microscopic quantities is, however, not in agreement with the later part of **F** § 77. The section begins:

In four-dimensional representation the momentum vector includes two invariants, both electrical, $E_{16}p_{16}$ and $E_{05}p_{05}$...electric charge and ..magnetic pole-strength.... We now find the symbolic coefficients of the principal molar quantities....

...In microscopic theory we have an additional dimension of p.d., the variates being scale momentum and phase. This must be associated with E_{05}, which thus regains its interpretation as normal to space-time. In molar physics fluctuation of scale is replaced by curvature, and the constant scale is incorporated in the measure of the molar quantities.

Their symbolic coefficients therefore include that of the scale; and the latter must be removed to obtain the symbols of the corresponding microscopic quantities. The molar linear scale is fixed with respect to the radius of curvature of the uranoid, which has symbolic direction $E_{05}E_{16}$ (§47). Since $(E_{05}E_{16})^{-3} = E_{05}E_{16}$, the same coefficient will apply to the scale of volume reciprocals and of momenta. We therefore divide (or multiply) the molar symbols by $E_{05}E_{16}$ to obtain the symbols in microscopic theory... this rule may be stated in the form—

Quantities represented by vectors in molar theory are represented by the corresponding vector-densities in microscopic theory. (53·1)

..the relation is reciprocal...

Strain vectors are the same in molar and microscopic theory, because the factor $E_{05}E_{16}$ is brought in twice (once in the vector and once in the volume reciprocal) and $(E_{05}E_{16})^2 = 1$. The strain vector, however, is significant only in connexion with a momentum vector, and is inappropriate to positional quantities, forces, etc.

The scale, an algebraic real constant in molar, is a momentum.. $E_{05}E_{16}$ or $E_{05}i$ in microscopic theory. This refers to the direct equivalent of molar scale and momentum. But we have seen the quantum momentum is an *analogue* obtained by multiplying the classical by $-i$. The linear quantum momentum vector, including scale momentum, is therefore of the form

$$\sum_{1}^{5} E_{0\mu}p_{0\mu}. \qquad (53·2)$$

The multiplication by i, introduced by the inversion of energy, applies to all components of P; for otherwise the reality condition (47·2) would be violated. Thus, corresponding to the molar momentum vector P.. a quantum vector $-iP$.. strictly idempotent.

The table follows substantially as in **F** §69 (full text), the $E_{\mu\nu}$ being headed 'Molar Vector'. There are parallel columns (as far as Electric Moment) headed 'Strain Vector' and 'Microscopic Equivalent', obtained by multiplying the corresponding $E_{\mu\nu}$ by $-E_{45}$ and E_{05}, and interchanging i and E_{16} where necessary to make just one E-symbol appear.

A later comment is relevant to the spin 6-vector early in **F** §53:

...certain quantities...have no compact names because they vanish (in molar physics) for bodies at rest. The a.m. 6-vector has three partially temporal components $E_{(1)4}$. The magnetic..and electric moments have time components E_{04}..and $E_{45}i$...

12·2. Draft G *30* VI. Wave Vectors

54. **The linear wave equation:** like **F** §70 and part of §72; corresponding to **11·2** §3·3 (q.v.). In both these MSS. the 'wave equations' are established for a vector satisfying $P^2 = iP$ (as **F** §68,

not $P^2 = P$ as in **F** §70), so that the normalizing condition is $p_{16} = \frac{1}{4}$ (not qs $P = \frac{1}{4}$). In this connexion:

The condition $p_{16} = \frac{1}{4}$ is a matter of normalization...we shall often use 'idempotent vector'..to denote..$P^2 = \alpha P$...These vectors have the property stressed in §50 [**F** §66] that there is a determinate relation between mean linear and mean quadratic characteristics entirely independent of the p.d. The carrier will be called an idempotent or *pure* particle [contrast **F** §68]..the *germ* of an 'elementary' particle....

[*Ends*:] We shall now introduce a matrix representation, as a result of which the wave equations will be reduced to a more familiar though more specialized form.

55. **Matrix representation of *E*-numbers:** like **F** §71. As, however, the coefficients s_μ of a strain vector S are *real* in this draft, S is taken to be an anti-Hermitian matrix ($S = -S^\dagger$).

56. **Factorization of *E*-numbers:** like **F** §72 (omitting the 'star' notation).

57. **Wave vectors and tensors:** part of **F** §72 (the 'star' notation), the substance of §73 (operational form of momentum, the four classes of wave tensor) and §75 (phase space). The interpretation of phase-space transformations has a different emphasis, which bears on the varying definitions of (double frame) strain tensors in **F** §85 and elsewhere. The quotations are from this last part.

...*Thus the transformations* $[q(\,..\,)\bar{q}]$ *which transform phase space into itself are those of a covariant wave tensor.*

Accordingly the position vector $S = \Sigma E_\mu s_\mu$ of a point in phase space is taken to be a covariant wave tensor...a point can be displaced all over phase space; except that there are singular loci, corresponding to singular values of S, which the point cannot reach, though it may approach them asymptotically. (They cannot be reached by continuous transformations; but it is possible that every point may be reached by finite transformations, since there is no reason to exclude singular matrices q in the antiperpendicular rotations $q(\,..\,)q$.)

The foregoing result has wider generality when S is described as a covariant wave tensor than when it is described as a strain vector. For it is valid for all 16 rotations q, whereas the equivalence of covariant wave tensors and strain vectors holds only for the 10 mechanical rotations. Since S is homothetic, the rotations are subject to the reality condition that q is real and therefore $E_\mu \theta$ is real. Thus 6 rotations are circular and 10 hyperbolic. Phase space is not closed, and it would be meaningless to speak of a uniform p.d. over phase space. In practical applications we generally have pseudo-discrete distributions, and the openness of phase space creates no difficulty.

It is instructive to compare phase space, in which the coordinates are the mechanical components of S, with another 10-space with coordinates the mechanical components of P. In the former the 45 coordinate planes are all planes of rotation; in the latter 30 are planes of rotation, and 15 are not. In each case a q-rotation gives rotation in 3 coordinate planes; and the difference arises because there are 15 q-rotations which transform the S-space into itself, and only 10 which transform the P-space into itself. In addition, the S-space has an E_{16} rotation which expands or contracts it without affecting orientation. It is the existence of rotations in all planes, making the geometrical picture congruent with the physically possible displacements, that makes the covariant wave tensor the appropriate instrument for specifying a domain over which probability is continuously distributed.

58. **Space tensors of the second rank:** equivalent to **F** § 74 (compare **11·2** § 3·6 (a)). The four *space* tensors of **F** are first introduced (and it is observed that pairs of antisymmetrical suffixes, as in the R.C. tensor, count as one in the $E_{\mu\nu}$ notation); then one strain tensor (the T_5 of **F**):

Consider another permutation [of the wave vectors]:

$$T^5 = T_{\alpha\beta\underline{\gamma\delta}} = \psi_\alpha \phi_\beta \chi^\gamma \omega^\delta \tag{58·2}$$

..the outer product of a contra- and a covariant strain vector..a mixed strain tensor $\Sigma^\mu{}_\nu$. Comparing it with T^2 [T_2 in **F**],...

$$(\Sigma^\mu{}_\nu)_{\alpha\beta\gamma\delta} = (T^\mu{}_\nu)_{\alpha\gamma\delta\beta}. \tag{58·3}$$

We can similarly derive a strain tensor $\Sigma_\mu{}^\nu$ by permutation...an operation not essentially different from raising or lowering suffixes. We regard $T^{\mu\nu}$, $T^\mu{}_\nu$, etc., as forms of the same physical T, and this now extends to the strain tensors. In the usual application the carrier of T is a bi-particle; and if it is separable into the product of space vectors, the separated carriers are two particles. But if it is separable into strain vectors, the carriers of the strain vectors are betwixt the two primary particles, being composed of factors derived from both. We may describe the permutation (58·3) as..*crossing* two systems. Evidently the internal particle of the hydrogen atom is derived from the proton and electron by an operation of crossing.

...the rather artificial introduction of strain vectors by the definition $S = -PE_{45}$ is due to the abstract character of the systems considered in these preliminary investigations; and the strain vectors appear spontaneously (as covariant wave tensors of the second rank) as soon as we have advanced far enough to deal with the energy tensor.

59. **Angular momentum:** like **F** § 105. Eddington's procedure (of obtaining field a.m. as a correction to particle a.m.) was given, he observes, in *Proc. Roy. Soc.* A, **138**, 23 (1932).

60. **Symbolic coefficients in ξ-space:** corresponds to ideas in **F** §§ 77, 78. This earlier approach has many points of interest.

With..the current formulae for angular and linear momentum, our theory reaches the more typical part of quantum theory which I have distinguished as 'quantal theory'...study of internal particles, or of the electrical world....We...examine the anchorage of a frame in the electrical world...we include the scale and phase as an additional dimension of p.d.

If x_α, x'_α are..two particles, their internal particle is..$\xi_\alpha = x'_\alpha - x_\alpha$ with phase..$\alpha = 5$. When the scale is de-stabilized, the symbolic coefficients are modified, the scale-stabilized momentum $E_{15}(p_{15}/1)$ being recognized as the ratio $E_{10}p_{01}/E_{05}p_{05}$ of a momentum along x_1 to the standard furnished by the scale momentum along the x_5 axis.

The molar momentum with de-stabilized scale is

$$P = E_{16}\{(E_{01}p_{01}) + E_{04}ip'_{04} + E_{05}p_{05}\} \tag{60·1}$$

with $p_{0(1)}, p'_{04}, p_{05}$ real. For, measuring $E_{16}E_{01}p_{01}$ with the scale $E_{16}E_{05}p_{05}$, we obtain the ratio $E_{15}.p_{01}/p_{05}$, and p_{01}/p_{05} has to be real. The factor E_{16} has to be inserted to avoid introducing chirality. The position vector X has a similar form. The scale for X and P is defined by p_{05} (not by the phase x_{05}); so that the condition that p_{05} must be real does not immediately apply to x_{05}. However, we define x_{05} so that it is related to p_{05} in the same way as..other coordinates and momenta.

On passing to ξ-space we adopt the quantum reckoning of momentum, antithetic to the molar (classical) in (60·1). This is done by dropping the factor E_{16}, and the momentum vector is then

$$\Pi = (E_{01}\varpi_{01}) + E_{04}i\varpi'_{04} + E_{05}\varpi_{05}. \tag{60·2}$$

This is purely electrical (as we should expect in the 'electrical world') and changes sign if the chirality of the frame is reversed so as to change the sign of the charges. It is not the practice to make a corresponding change of coordinates, and the momentum vector in quantum theory is therefore antithetic to the position vector, as shown in the equations $\varpi_\alpha = -i\hbar\,\partial/\partial\xi_\alpha$.

A 'μ-space' (of electric moment $\mu_\alpha = \pm e\xi_\alpha$) is discussed. Returning to (60·1, 60·2), a system is to be contemplated at one instant, so that the time coordinate $\xi'_4 = t' - t$ is stabilized at zero, the ξ-space limited to four dimensions, and the momentum and position are

$$\Pi = (E_{01}\varpi_{01}) + E_{05}\varpi_{05}, \quad \Xi = E_{16}\{(E_{01}\xi_1) + E_{05}\xi_5\}. \tag{60·4}$$

Consider the reality conditions for (60·4)....In (60·1) we have.. de-stabilized the scale of a particle in x-space, i.e. associated with the external object-particle a scale-bearing comparison particle whose momentum is represented by the term $E_{16}E_{05}p_{05}$. This is the same momentum which appears as $E_{05}\varpi_{05}$ when the comparison particle is associated with the internal object-particle. Thus

$$E_{05}\varpi_{05} = E_{16}E_{05}p_{05} = E_{05}ip_{05}.$$

Since p_{05} is real, ϖ_{05} is imaginary, and we write

$$\varpi_{05} = i\varpi'_{05}. \tag{60·5}$$

If $\varpi_{0(1)}$ are real, Π is monothetic. The internal particle is then mechanically analogous to a classical particle with real momenta $\varpi_{0(1)}$ and real energy ϖ'_{05}. The fact that energy is now associated with suffix 5 instead of 4 makes no essential difference to the analogy.

The condition that ϖ_{05} is imaginary is *necessary*; because ϖ_{05} being a scale standard common to x- and ξ-space is directly connected with molar measurement. It can scarcely be said that the reality of $\varpi_{0(1)}$ is equally necessary, because they are purely domestic variates of the electrical world. These quantities never occur as constants (eigen-characteristics) of a quantal system; so there is no possibility of molar measurement of their sum over a large number of atoms. All we can say is that states for which the quantum momenta are real have a special importance in quantal mechanics, because they are the analogues of the physical real states of particles in classical mechanics.

..existing quantum theory rests on..the analogy between quantal and classical mechanics....Following the nomenclature [of analogy], a quantal quantity is physically real if it is the analogue of a physically real quantity. The reality conditions can then be...stated:

$$\Pi \textit{ is monothetic and imaginary}, \ \Xi \textit{ monothetic and real.} \qquad (60\cdot6)$$

An a.m. is obtained by multiplying a radial vector Ξ by a transverse vector Π. It is therefore imaginary. Thus

The complete momentum vector of a particle in ξ-space is monothetic and imaginary. (60·7)

61. **The differential wave equation:** a substantial section, bearing on **F** §§ 106, 107, 78. This begins:

Corresponding to (59·52) [**F** (105·5)] for the total a.m., we have for the total linear momentum

$$\mathbf{p}_{0\mu} = -i\hbar(\partial/\partial x_\mu - \tfrac{1}{2}\alpha E_{0\mu}), \qquad (61\cdot1)$$

..α of dimensions $(\text{length})^{-1}$, inserted in order not to restrict the unit of length....Unlike (59·52), (61·1) is not derived as a whole, but the two terms..have been obtained separately.

The meaning of 'rigid field' is now discussed, for the 'electrical world' at present under consideration, together with the wave functions of § 59 [**F** § 105]:

Occupation of a vector wave function means not only the insertion of a particle but also of a field energy tensor. The object of the differential wave equation is to secure the right proportion between these, so that after any change of occupation the field is self-consistent....These considerations are limited to the momentum 4-vector and the corresponding energy tensor with 10 components [and not applied to the a.m. which tends to cancel out, unlike gravitational energy]....

..(56·6) [**F** (72·81)] gives for the pentad ϖ_0,

$$(\Sigma_\mu E_{0\mu} p_{0\mu} - ip_{16})\psi = 0, \quad \chi^*(...) = 0. \tag{61·2}$$

Suppose that another pentadic expression $\Sigma E_{0\mu} p'_{0\mu}$ satisfies

$$(\Sigma_\mu E_{0\mu} p'_{0\mu} - inp_{16})\psi = 0 \quad \chi^*(...) = 0. \tag{61·3}$$

Multiplying these by..initial $\chi^* E_{01}$ and..final $E_{01}\psi$ and adding, we obtain $2\chi^*\{-p'_{01} - inp_{16}E_{01}\}\psi = 0$, or $p_{16}\{p'_{01} - np_{01}\} = 0$ by (57·3) [**F** (73·1)]. Similarly..$p'_{0\mu} = np_{0\mu}$. In particular, setting $p'_{0\mu} = -i\hbar\,\partial/\partial x_\mu$ in (61·3), the differential equation

$$\{-i\hbar\Sigma E_{0\mu}\,\partial/\partial x_\mu - inp_{16}\}\psi = 0, \tag{61·4}$$

with the corresponding equation for χ^*, expresses the condition that the field momentum 4-vector is n times the particle vector.

Applying this to the electrical particle in ξ-space we have the *differential wave equation*

$$\{-i\hbar[(E_{01}\,\partial/\partial\xi_1) + E_{05}\,\partial/\partial\xi_5] - m\}\psi = 0, \tag{61·5}$$

where

$$m = inp_{16}. \tag{61·6}$$

The energy tensor involves the square of the momentum (or root) vector, so for..$W_{\mu\nu} = -(k+1)E_{\mu\nu}$ [$E_{\mu\nu}$ particle energy as in **F** II, *not* an E-symbol], n must be imaginary. Since p_{16} is real by (59·53) [**F** (105·42)], m is a real constant....

...we change notation. The x- and ξ-space are intrinsically similar, each having 3 space-like and 1 time-like dimension....We accordingly write (61·5) as

$$\left\{-i\hbar \sum_1^4 E_{\mu 5}\,\partial/\partial x_\mu - m\right\}\psi = 0. \tag{61·8}$$

The imaginary x_4 will be replaced by a real coordinate $t = ix_4$, when required. This change of notation..[applies] the nomenclature..of molar theory to the analogues in quantal theory..[cf. **F** §78]...the original suffixes 0, 5, 4 become 5, 4, 0. If it becomes necessary to introduce as perturbing factors any of the quantities represented in x-space, their symbolic coefficients are obtained by making this substitution in the table in § 53 [**F** § 69]. We notice a.m. and magnetic force are unchanged....

The equation (61·8) for the internal particle of a hydrogen atom is usually known as the 'wave equation of the hydrogen atom'....An extra term representing Coulomb energy...is superfluous since the term spontaneously appears in the course of the solution of the equation.

62. **The eigenscale:** bearing on **F** § 108.

In (61·8)..the energy operator is

$$\boldsymbol{\epsilon} = -i\hbar\,\partial/\partial t = -\hbar\,\partial/\partial x_4 \tag{62·1}$$

..called the energy by analogy, actually..the scale momentum. If $\boldsymbol{\epsilon}$ reduces to an eigenvalue ϵ, the scale is exact....

It is well known that the wave equation for hydrogen (and other systems with negative energy) gives a discrete series of eigenvalues of $\boldsymbol{\epsilon}$.

...The wavelength of hydrogen light is uncertain, being dependent on the transition the atom happens to make; but..the H_α line [serves] as an exact standard of length. This change of attitude, which construes a spectrum as an exact pattern and not an uncertainty of..wavelength.., tends to obscure the role of ϵ as scale constant...it is the link which connects all measured characteristics..with the extraneous standard. It is, as it were, the Foreign Secretary in the Cabinet of characteristics.

...we are interested only in characteristics which are capable of simultaneous eigenvalues with $\boldsymbol{\epsilon}$....

63. **Perturbation theory:** a blank sheet, so headed; followed by:

CHAPTER VII. THE HYDROGEN ATOM AND THE NEUTRON

63. **Symmetric degeneracy:** an incomplete section, corresponding to **F** § 110 in its objects. To summarize by quotation:

..the internal wave equation (61·8)...

$$(\mathbf{W}-m)\psi=0, \quad \mathbf{W}=\sum_1^4 E_{\alpha 5}\mathbf{p}_\alpha, \tag{63·1}$$

where $\mathbf{p}_\alpha=-i\hbar\partial/\partial x_\alpha$,...has to be solved subject to the condition that the energy component of $\mathbf{p}_\alpha$ reduces to an eigenvalue. But how is the direction of this component to be defined?...we can only define *relative* orientations [in internal space]....The orientated characteristics of an eigenstate are energy and a.m. Thus the direction of the component which is to be distinguished as the energy...must be defined relatively to the plane or planes of a.m. The a.m. is a 6-vector, which..can be reduced to..$E'_{23}\omega'_{23}+E'_{14}\omega'_{14}$...the direction of the time [i.e. phase or 'energy'] axis x'_4 in the plane $x'_1x'_4$ is left undefined....Thus the condition..is not that $\mathbf{p}'_4$ shall reduce to an eigenvalue, but that the component *in some direction* in the plane $x'_1x'_4$ shall reduce to an eigenvalue. The states for which this is satisfied, being indistinguishable, merge into one multiple state; this is..*symmetric degeneracy*....

Equation (63·1) makes no reference to a.m., but the a.m. is contained implicitly in the circumstance that the axes employed have not been defined observationally, so that as it stands it contains unobservables. (We are now taking up the study of 'uncertainty of orientation', the complement of uncertainty of origin and of scale.) A.m. in a plane is always associated with symmetrical degeneracy in that plane. The dynamical aspect is that normally the a.m. of a microscopic system is rapidly dissipated, and has no cumulative statistical effect; but when the system has a plane of symmetry, the integral of a.m. in that plane persists, perturbing forces from outside having little or no resultant effect on it. Symmetric degeneracy is treated by transforming **W** so that the a.m. $\mathbf{\Omega}$ appears explicitly. The effect is that two definables (observables), namely, radial and angular momentum, are substituted for two indefinables (unobservables), namely, momentum components in two undefined directions in the plane....

The MS. ends abruptly after beginning the mathematics of **F** § 110.

12·3. Draft G *15* VII. Double Frames

61. **The *EF*-frame:** like **F** § 79. Concerning the energy tensor:

In a momentum vector p_{45} and p_{15} are antithetic; so in the outer square $p_{45,45}$ and $p_{15,15}$ are to be interpreted with opposite sign. This outer square is a co- or contravariant energy tensor $T_{\mu\nu}$ or $T^{\mu\nu}$. In the mixed tensor the relative sign is reversed, so that diagonal components of $T_\mu{}^\nu$ are incorporated in the symbolic form without modification of sign. It is therefore convenient to...[use]

$$T_0{}^0 = \Sigma E_{\mu 5} F_{\nu 5} T_\mu{}^\nu. \tag{61·44}$$

Then $T_\mu{}^\nu$ agrees with the ordinary notation as regards density and pressure components...a factor i in momentum components.... Corresponding to (61·44)

$$T_{00} = -\Sigma E_{\mu 5} \bar{F}_{\nu 5} T_\mu{}^\nu, \quad T^{00} = -\Sigma \bar{E}_{\mu 5} F_{\nu 5} T_\mu{}^\nu. \tag{61·45}$$

62. **Chirality of the double frame:** the basis of **F** § 80.

63. **The interchange operator:** like **F** § 81. Two amplifications: (i) for the end of **F** § 81 on the factors of I:

Distinguishing ζ and θ (or σ and τ) as the *spin* and *counterspin*, the counterspin remains attached to the suffix in Dirac's operation, so that it must be carried by the comparison particle as well as the suffix, the object particle being treated as having only the one spin σ.

(ii) on continuous interchange (a page pencilled '? delete'):

..The rotation operator which produces *continuous interchange* is

$$q = e^{\mathscr{I}\theta}, \qquad \mathscr{I} = \tfrac{1}{2} i(1+I), \tag{63·7}$$

which gives $q = -I$ when $\theta = \pi$...$\mathscr{I}$ is imaginary....This representation of a discontinuous transition by an imaginary rotation is the recognized method of treating 'quantum jumps'; and we shall use the same procedure in treating transition probabilities in perturbation theory. The corresponding real rotation $q = e^{\frac{1}{2}\mathscr{I} iu}$ is included in the real relativity rotations...; so we have two variates associated with one symbol $\mathscr{I}$.

...the interchange rotation constitutes a departure...[to avoid] more highly multiple frames...Reducing the quadruple to a double p.d. (so that it can be represented in a double instead of quadruple frame) has been explained in § 25;...to validate the reduction, provision must be made for representing interchange.

64. **Duals:** like **F** § 82; with the addition [cf. **F** § 83] that for the strain and space tensors

$$(Z_0{}^0)_{\alpha\beta\gamma\delta} = \psi_\alpha \phi_\beta \chi^\gamma \omega^\delta, \quad (T_0{}^0)_{\alpha\beta\gamma\delta} = \psi_\alpha \omega^\beta \chi^\gamma \phi_\delta, \tag{64·32}$$

then

$$Z_0{}^0 = \tilde{T}_0{}^0, \quad T_0{}^0 = \tilde{Z}_0{}^0. \tag{64·33}$$

65. **The *CD*-frame:** like **F** § 83, omitting the later discussion.

66. **Double phase space:** this corresponds to **F** § 85, without the benefit of the preliminary **F** § 84. The components z_μ of $Z=\Sigma' EF_\mu z_\mu$ (summed over space-like or *real* EF_μ) are taken as coordinates in a 136-phase-space. The transformations show Z to be a strain tensor $Z_0{}^0$ (i.e. as (64·32) above); as it is space-like,

$$(Z_0{}^0)_{\alpha\beta\gamma\delta}=(Z_0{}^0)_{\beta\alpha\delta\gamma},$$

and the associated $$(T_{00})_{\alpha\beta\gamma\delta}=(Z_0{}^0)_{\alpha\delta\beta\gamma}$$

satisfies $$T_{00}=IT_{00}I \quad (T_{00\alpha\beta\gamma\delta}=T_{00\gamma\delta\alpha\beta}),$$

and so is a *symmetrical* (space) tensor. This specification of T_{00} as symmetrical is, however, less fundamental (e.g. for elementary carriers) than the specification that the energy strain tensor is to be space-like.

67. **The uranoid and the aether:** corresponds to **F** § 86. Three extracts:

(i) ...in system A the particle is considered in association with empty space. For that very reason Z_a (rather than Z_u) is to be regarded as the *intrinsic* strain vector [tensor] of the particle; and $Z=-TI$ is the primary connexion of the space- and strain-tensor forms of the same physical characteristics.

(ii) ...the double frame..is suited for the representation of two vector particles...Taking one..as object-particle, one particle is available to represent the environment. By giving the latter particle exact momentum..we obtain a crude representation of the zero-temperature uranoid, which is easily developed into a precise representation by treating the particle as an unidentified member of an assemblage of N particles...the centroid provides a physical origin.. uncertainty σ..so it is possible to develop a rigorous positional theory. The alternative..application..represents the environment by one particle with exact position and uncertain momentum..a very inadequate substitute for a universe...we should have to take N particles with exact but *different* positions...also..'empty space' forms a non-static (de Sitter) universe....Thus the main development is concerned with uranoid, not aether, theory....

(iii) Since only time suffixes are involved, $E_{45}F_{45}$ is indifferently the mixed, co- or contravariant energy tensor of the uranoid; but $-I$ is, by its invariance, definitely the mixed energy tensor of the aether, the corresponding covariant tensor being

$$\bar{I}=\tfrac{1}{4}\Sigma E_\mu \bar{F}_\mu. \tag{67·3}$$

The following form of the relation between the energy tensors T^{ae}, T^{u} of aether and uranoid is of interest. If we write

$$(T_{00})^{ae}=(T_{00})^{u} I', \tag{67·41}$$

we have $$I'=E_{45}F_{45}\bar{I}=\tfrac{1}{4}\Sigma E_\mu F_\mu^\dagger, \tag{67·42}$$

where $F^{\dagger}$ denotes the symbols of the corresponding l.h. frame as in §43. By analogy I' might be called the 'antichiral interchange operator'; but it does not actually effect interchange.

68. **The tensor identities:** like **F** §89. It is added that **F** (89·7) applies to a tensor whose cross-dual consists of density components only. Equation [**F** (89·4)] applies when V is space-like and U time-like [with $T_{\gamma\alpha\delta\beta}=V_{\gamma\delta}U_{\alpha\beta}$ in **F** (89·35)] when $\mu=\nu$, but not when $\mu\neq\nu$. Equation [**F** (89·7)] applies when V is time-like and U space-like. Thus the case in which $T^{\times}$ has momentum components can be dealt with if necessary; but we shall find it not of much physical interest.

69. **The quantum-classical analogy:** a lengthy section, beginning with the mathematics of **F** §90, and then discussing the significance of the energy and R.C. tensors in terms of quantum 'analogy'. After reaching **F** (90·6), the section proceeds:

This result is subject to two limitations...First, it depends on the absence of density and momentum components in $T^{\times}$....Secondly, it is not invariant for change of the time axis...the change would introduce momentum components. Analytically the non-invariance is introduced in (69·1) [**F** (90·1)], where we form an EF-number T to represent the array of components of the R.C. tensor in a particular coordinate frame....

...what we have encountered is not a fundamental law, but a fundamental analogy...quantum quantities are analogues..of classical.... Such an analogy begins as a mistaken identification. There must be some genuine similarity to cause the mistake...the spontaneous part of the analogy...The foregoing result..exhibits the spontaneous part. ...Since..(69·5) [**F** (90·5)] is not a tensor relation, it cannot apply to the true (classical) energy tensor and R.C. tensor; i.e. the attempt to represent these as parts of the same extended tensor fails. But it may well apply to the quantum analogues...this may be the substance of the analogy. We shall accordingly take T to be the quantum energy and R.C. tensor.

...since T cannot be a space tensor it should be a strain tensor. The cross-dual $T^{\times}$ of a strain tensor is a covariant space tensor; and we shall provisionally assume this is the true (classical) R.C. tensor...The assumption enables us to define..the time axis to which the non-invariant formulae [**F** (90·5)] refer as the axis with respect to which the true R.C. tensor reduces to pressure components...

The proposed quantum-classical analogy is accordingly

$$(B_0{}^0)_{\text{quantum}}=T, \quad (B_{00})_{\text{classical}}=T^{\times}. \tag{69·6}$$

It would be plainer to call $T^{\times}$ and T the R.C. space and strain tensors. But plain language implies dropping the language of analogy...quantum theory talks this language, and..(69·6) is the key...

Since time-like components of a strain vector are physically meaningless, T is wholly space-like and generates the 136-space..as in §66. This makes $T_{\alpha\beta\gamma\delta}$ symmetrical for interchange of $\alpha\gamma$ with $\beta\delta$, $T^{\times}$..for $\alpha\beta$ with $\gamma\delta$; so that $B'_{\mu\epsilon\nu\sigma}$ [i.e. B_{00}] has the symmetry of an R.C. tensor for interchange of $\mu\epsilon$ with $\nu\sigma$. Of the 136 components of B', 55 are pressure components; and the conditions of the analogy restrict B' to these... Quantum mechanics arises primarily as a way of treating a pure pressure-system...

Accordingly our programme involves the separation of part of the molar tensor, consisting of pressure terms only, for intensive study by.. 'quantal' theory, leaving the rest to be dealt with by scale-free theory as in I–IV...this division has already appeared in various forms—a separation of transition from initial energy, of a pure electric from a mechanical world, of correlation wave functions from distribution functions, of self-normalizing quantized from pseudo-discrete states...

There is no intention of applying the theory to a system *constrained* to be a pure pressure system, having only 55 d.f.'s; it is to be applied to the pressure part of an ordinary system with 136 d.f.'s..the multiplicity is 136. In fact the vanishing of the density and momentum terms is not..exact but almost exact...the remaining 81 terms are dealt with in another part of the physical theory.

70. **Recoil rotations:** like most of **F** § 87. Reference is made to the treatment in **RTPE**, p. 191. From the final paragraph:

This reinterpretation of the classical R.C. tensor..is intended to be applied to the quantum R.C. tensor or energy tensor in accordance with the quantum-classical analogy...

71. **Transformation to a relative frame:** this corresponds to part of **F** § 111 (cf. also **RTPE**, § 10·8); but the later discussion here is of great interest.

Let E_μ, F_μ ($\mu = 1, \ldots, 4$) be corresponding tetrads, and

$$G_\mu = (E_\mu + iF_\mu)\, I/(1+i), \quad H_\mu = (E_\mu - iF_\mu)\, I/(1-i). \qquad (71{\cdot}11)$$

...we derive complete sets G_μ and H_μ..we take

$$G_{16} = H_{16} = -E_{16} = -F_{16}. \qquad (71{\cdot}3)$$

...The connexion between the GH- and EF-frames is based on a particular 5-space. We define the 'field of congruence' of the two frames to consist of (a) the corresponding 5-vectors, (b) quantities in the interchange direction...I, and (c) algebraic quantities (quarterspurs). Spin momenta are outside the field...

Consider two 5-vectors $\mathbf{p} = \Sigma E_\mu p_\mu$, $\mathbf{p}' = \Sigma F_\mu p'_\mu$ in the field, and let

$$\varpi_\mu = p_\mu - p'_\mu, \quad \varpi'_\mu = p_\mu + p'_\mu. \qquad (71{\cdot}41)$$

We associate ϖ, ϖ' with the H and G-frames so that

$$\boldsymbol{\varpi} = \Sigma H_\mu \varpi_\mu, \quad \boldsymbol{\varpi}' = \Sigma G_\mu \varpi'_\mu. \qquad (71{\cdot}42)$$

Hence..

$$(\mathbf{p} + i\mathbf{p}')\, I = (\boldsymbol{\varpi} + i\boldsymbol{\varpi}')/(1+i). \qquad (71{\cdot}5)$$

Let p_μ and p'_μ be the momentum vectors of two particles of equal mass. Then ϖ_μ and ϖ'_μ are the momentum vectors of the internal and external particles...the multiplication by I constitutes the quantum-classical analogy...introduced just at the right point, namely, when we separate the momentum vector of the internal particle with which quantal theory is concerned...(71·5) is the fundamental form of the two-particle transformation...

...the allocation of different frames E_μ, F_μ or G_μ, H_μ to two particles is a generalized..suffixing..(§63). The suffixing of the E- and F-particles introduces an interchange momentum or energy ϵ_i which should be included in the transformation. It has the symbolic direction Ii, since continuous interchange is an imaginary rotation. The two particles share equally in the interchange circulation, and their momentum vectors are therefore extended to $\mathbf{p}^0 = \mathbf{p} + \frac{1}{2}Ii\epsilon_i$, $\mathbf{p}'^0 = \mathbf{p}' + \frac{1}{2}Ii\epsilon_i$. The corresponding addition to the l.h.s. of (71·5) is $(1+i)\,\frac{1}{2}iI\epsilon_i I$, which equals $-\epsilon_i/(1+i)$; hence

$$(\mathbf{p}^0 + i\mathbf{p}'^0)\,I = (\boldsymbol{\varpi}^0 + i\boldsymbol{\varpi}'^0)/(1+i), \tag{71·6}$$

where

$$\boldsymbol{\varpi}^0 = \boldsymbol{\varpi} - \epsilon_i, \quad \boldsymbol{\varpi}'^0 = \boldsymbol{\varpi}'. \tag{71·7}$$

Thus in the GH-frame the interchange energy is a purely algebraic addition to the momentum vector of the internal particle....

In (63·7) [p. 223] the continuous interchange operator is given as $\mathscr{I} = \frac{1}{2}i(I+1)$, but here we have only $\frac{1}{2}iI$. The remainder would give an additional $-I\epsilon_i$ in $\boldsymbol{\varpi}^0 + i\boldsymbol{\varpi}'^0$, which is most appropriately included by setting $\boldsymbol{\varpi}'^0 = \boldsymbol{\varpi}' + Ii\epsilon_i$. Since I is also the interchange operator for the G and H-frames, the addition is interpreted as the energy of interchange of the external and internal particles (allotted unsymmetrically to the external particle). The treatment must be adjusted so that this term is suppressed; for we do not contemplate the external and internal particles as interchangeable. We can see in a general way how the suppression occurs. The vectors $\mathbf{p}$, $\mathbf{p}'$, $\boldsymbol{\varpi}$, $\boldsymbol{\varpi}'$ in the four frames have full freedom of relativity rotation, and therefore represent free particles. But..in the direct two-particle transformation the particles cannot all be free; even when interchange energy is included, the freedom of the internal particle is imperfect so that β-factors have to be introduced. It appears then that the algebraic part of the $\mathscr{I}$, which we find it necessary to suppress in (71·6), is the source of the β-factors; so that it will ultimately be taken into account in that form...

There is no way of extending the transformation to two particles of different mass...At this stage..all particles have the same mass; and it is not until multiplicity factors are introduced by the rigid-field treatment, and the masses modified to incorporate them, that diversity of mass appears....

...the a.m. are outside the 'field of congruence' and there is no simple cross-connexion....We might perhaps attribute the resultant a.m. of rotation of the sun and earth to a particle occupying the centre of mass; but it is hard to see any way of associating their difference of rotation with a particle having the relative coordinates of the sun and earth. We can only say the p.d. of spin momenta must be differently analysed according as it is assigned to the E- and F-particles or to the external and internal particles. The former analysis is remote from experimental application; we lose little by being unable to connect it with the a.m. of the external and internal particles of which the experimental importance is immediate.

12·4. Draft G *22* VII. Double Frames

This MS., dated April 1943, was rewritten in May as the MS. *15* discussed in **12·3**; and its sections, §§61–66, were expanded into *15* §§61–69. The sections here are

61. The EF-frame, 62—, 63 The dual frame, 64 Double phase-space, 65 The two strain tensors, 66 The R.C. tensor.

A slip attached to this MS. gives the scheme for continuation:

Other Chapters | Wave functions | Occupation vectors | Special problems | (Nuclear Magnetic moments) | Unitary Theory | [and the remark] Continuous interchange operator is idempotent.

Some brief quotations:

62 (cf. **F** §81):

the continuous interchange rotation $I'+i$ $[I'=-iI]$ being imaginary introduces an additional d.f. of the system; whereas I is a combination of rotations already recognized and involves no additional d.f. This is taken into account in the raising of the multiplicity from 136 to 137 in §26 [**F** §28].

63 (cf. **F** §83 on the relation of duals):

It is of special importance that space and strain tensors (in the same frame) are related as duals..; but the mixed space tensor must be used, since the dual of a covariant or contravariant tensor is another tensor of the same class...*The continuous interchange operator $I'+i$ is self-dual, and is common to both CD- and EF-frames.*

65 (cf. **F** §86, on uranoid and aether, and also **4·3** §1·2):

The energy tensor present at any point consists of

$$(-1/8\pi\kappa)(G_\mu{}^\nu-\tfrac{1}{2}g_\mu{}^\nu G)\ \text{(material)},\quad (-1/8\pi\kappa)\lambda g_\mu{}^\nu\ \text{(cosmical)},\qquad (65\cdot3)$$

the sum being $(-1/8\pi\kappa)\{G_\mu{}^\nu-\tfrac{1}{2}g_\mu{}^\nu(G-2\lambda)\}$. The cosmical tensor is that of empty space or aether: the sign agrees with the negative sign in $\Sigma[Z_a$ of **F** $(86\cdot1)]=-TI$...We recall also that the cosmical constant λ causes dispersal of the galaxies and has therefore the effect of a negative pressure. But all our energy tensors refer to steady states, since otherwise the formulae are modified by expansion energy (§20). Thus, out of the whole pressure present, a positive pressure must be set aside to neutralize the cosmical repulsion; and this is the pressure component of the cosmical energy tensor. A positive pressure corresponds to negative $T^{(1)}_{(1)}$. The aether is thus a system with positive pressure and negative density.

66 (on **F** §§ 88–91; the R.C. tensor and the tensor identities are here discussed together):

...the additional components of the R.C. tensor are $T_{\mu\nu}$, not $G_{\mu\nu}$. It has always been difficult to see any physical significance in the process of contraction by which $G_{\mu\nu}$ is derived from $B_{\mu\epsilon\nu\sigma}$. It has no physical significance, since it represents an unfinished calculation forming a stage in the derivation of $T_{\mu\nu}$ by means of the identity (66·43) [**F** (90·23)].

...The amalgamation of the R.C. and energy tensors in one complete tensor puts into symbolic form the usual representation of energy-density, momentum-density and pressure by curvature. But when U, V [$UV = T^{\times}$] are time-like, the energy and R.C. tensors are not amalgamated and..these cannot be represented by curvature of Riemannian space-time. If we have a passion for geometrization, we can no doubt meet this case by introducing a non-Riemannian geometry. But it is simpler to accept the distinction as it stands; so that one part of the energy tensor is incorporated in space-time, and the other part is superposed on space-time. The first part provides 100, the second part 36 independent components of the complete energy tensor; they are, however, mixed together in the energy tensor, and to separate them it is necessary to transform it into its dual. Broadly speaking the study of the first part is a formalization of what we have already treated by other methods; but the second part introduces altogether new problems, which the molar theory of the energy tensor does not foreshadow. It is therefore the 36 terms provided by the time-like U, V that are concerned in the more distinctive part of quantum theory. [*End.*]

12·5. DRAFT G *3* VIII. [No title: dated October 1943.]

This MS., aiming at the wave equation, has a more tentative feeling than most drafts,[a] and was left unfinished. The main quotations are from a useful survey of particle versus wave (or classical versus quantum) theory in § 79.

73 [changed to '76']. **Angular momentum:** like **F** § 105; it is mentioned that particle and field a.m. correspond to conventional spin and orbital a.m.

74 (*a*). **Polar coordinates:** this is related to **F** §§ 108–110, but falls into no clear scheme. The conclusion is:

If the extra-spatial a.m. reduces to an eigenvalue ω, it is taken into account by adding a term $E_{45}\,\omega/r$ to the momentum vector P operating on 3-dimensional wave functions. The 3-space covered by the wave functions

[a] The first Chapter VIII of **G** was a on Occupation Symbols; see **14** below. The 'tentative feeling' of **12·5**, **12·6** implies that a recasting of the argument was in process.

is mobile, having a Eulerian coordinate θ'_{14}; but since this has been allowed for in the calculation of $E_{45}\omega/r$, it is to be treated as though it were fixed.

74(*b*). **The differential wave equation** (short): like part of **F** § 107.

75. **The symbolic frame in relative space:** the basis of **F** §§ 76, 77.

76. **Reality conditions in relative space:** largely equivalent to **F** § 78.

77. **Relation of quantal and scale-free physics** (two versions, one fragmentary): Two points made are:

(i) Dimensionality of physical quantities is important in scale-free physics, because of the possibility of scale transformations; in quantal, scale-fixed, theory, magnitudes are expressed as pure numbers, as for idempotent quantities.

(ii) Nevertheless, classical systems are not quite 'scale-free', being limited by the finiteness of the universe; the relation of the 'classical' and 'quantal' scales corresponds to $\sqrt{N}$, or the relation of gravitation and electromagnetism.

78. **The metastable states of hydrogen:** corresponding to **F** § 93 in results, but not very clear. An earlier version is quoted in **12·6**.

79. **Particle and wave properties:** connected mainly with **F** §§ 76—78, but bearing on **F** X. Most of this interesting survey is quoted.

We have studied the junction of classical and quantum theory... time for a general survey. There are two consecutive junctions, first between molar theory and scale-free quantum theory, secondly between the latter and scale-fixed quantum theory or quantal theory.

... two extremes [are] the ideal classical particle with exact coordinates and the ideal quantum particle or simple wave... without specified coordinates and therefore carrying only an energy tensor or a momentum vector... In physical theory exact position is replaced by a small irreducible uncertainty.. σ, and infinitely uncertain position by even p.d. over a large volume V_0; these two physical extremes now become our ideal classical and quantum particles. By setting a limit V_0 to.. uncertainty of position, we.. define an ideal universe of volume V_0... The ratio of $V_0^{\frac{1}{3}}$ to σ determines.. N... We shall find ultimately that N has a fixed value... so that the element of arbitrariness in choosing the two limits is removed....

There is continuous gradation between the two extremes, so that what is counted as the same particle may on one occasion be a classical particle carrying a concentrated mass m and on another a quantum particle carrying a density ρ. Certain multiplicity factors are incorporated in the masses and densities as ordinarily defined; supposing these to be removed, we choose the fixed-scale units so that m and ρ are the same number. Then V_0 is the fixed-scale unit of volume, or natural normalization volume.

The gradation from quantum to classical particle is represented by a wave packet of continuously increasing concentration...The microscopic classical particle is a very unnatural construction; that is why in microscopic analysis we employ quantum particles, either scale-free or -fixed. And when we need..a scale-fixed particle characterized by a mass instead of a density, we do *not* employ a classical particle produced by concentrating the distribution..[but] the analogue of a classical particle (carrying the analogue of mass), produced by a change of frame from molar to micro-space. Apart from the change of stabilization..this analogue is just the quantum particle viewed from another aspect.

..To discuss the change of aspect: it is the inverse transformation that usually presents itself; we are strongly biased towards the classical conception of a particle...Strictly speaking, the quantum particle, carrier of the density ρ, has no uncertainty of position; it is everywhere that the uniform density ρ extends. Perhaps the reader finds it hard to concede this; in his view, the continuous density is just a device, and there is really a mass point somewhere if only we knew where it was. But if we knew where it was it would be a classical, not a quantum particle; and as for being a device, all representational theory is a device, and only the metaphysicist regards it as anything else. The continuous ρ must be accepted literally; but there is another way of regarding it by which it is equivalently described by a variate concentrated at some unknown point. The mistake is to suppose this variate is the mass; it is the mass-analogue, used as the mass in quantum dynamics. It is not formed by integrating ρ, though it can be made equal to the integral of ρ by choosing a domain V_0 of integration *ad hoc*, as we have done. Ignorance of the location of the point of concentration (relative to the adopted physical origin) is often of no consequence. When we investigate the internal energy levels of a hydrogen atom, we do not know or care where the extracule is. In such circumstances the second aspect naturally presents itself.

By this study of the point of convergence of the two theories we anchor scale-fixed to scale-free quantum theory, which in turn has been anchored to molar theory...The practical importance of the transformation begins to appear as we move away from the point of convergence. For extracules, the transition from molar to micro-space does not immediately open up any new development...The intracule is the focus of all our preparations; so that in general we couple the transition from molar to micro-space as well as the transition from wave vectors to wave functions with the transition from x- to ξ-space.

The vital difference between ξ- and x-space is that the origin is a physical singularity in the former. This makes possible in the ξ-space a series of structures which, because they involve a singularity, are impossible in x-space; so a treatment rather sterile in x- has abundant scope in ξ-space. The physical singularity is the coincidence of position of the two particles furnishing the extra- and intracule. Before we can speak of coincidence, or indeed define a ξ-coordinate, we must adopt the point of view represented by micro-space, in which the particles *have* positions though they are unknown; for it would be meaningless to speak of the distance between two superposed densities. Thus the two-particle transformation postulates micro-space. Results obtained in micro-space can, if necessary, be transformed back into molar space, using the relation between p'_μ and p_μ in (76·2) [**F** (78·2)]. Except in specially simple structures, such as metastable states, the micro-structures referred to molar space are not small-scale copies of real molar structures, since they fail to satisfy the molar reality conditions. 'Reality conditions' is only another name for the 'boundary conditions' corresponding to the standard environment. The micro-structures do not fit the standard environment, and consequently they radiate. If the standard environment is modified by a suitable radiation field, their proper boundary conditions are supplied, and they can exist permanently.

80. **The internal wave equation:** an unfinished section related to **F** §§ 108, 110. The opening is linked with the last paragraph of § 79 above.

We shall now treat the wave equation of a free intracule. We shall use analogous designation...so ostensibly the equations refer to..an extracule in molar space-time. But actually they have no application to an extracule, because the analysis turns on the existence of a singularity at the origin in relative space.

By § 74 the wave equation which gives isostatic compensation of the momentum 4-vector is

$$(\mathbf{W}-m)\,\psi=0,\quad \mathbf{W}=\sum_1^4 E_{\beta 5}\mathbf{p}_\beta,\quad (\mathbf{p}_\beta=-i\hbar\,\partial/\partial x_\beta).$$

The real fourth coordinate is $t=x_4/i$, and the corresponding real momentum is $\boldsymbol{\epsilon}=\mathbf{p}_4/i$...

We seek solutions in which certain quantities reduce to eigenvalues.. the scale..to an eigenscale,..in analogous designation the energy $\boldsymbol{\epsilon}$. ...The solutions must correspond..[also] to eigenvalues of the a.m. vector...The commutation condition [on $\boldsymbol{\Omega}$, $\mathbf{p}_4$ and $\mathbf{W}$] is satisfied by the total a.m. but not by the field a.m....

12·6. Draft G *18* VIII. [No title: written before G *3*.]

This is an earlier and fragmentary version of **12·5**. One section, the earliest MS. on the 'metastable states', is largely quoted.

73. **Angular momentum:** like **F** § 105.

74 (a). **The metastable states of hydrogen:** the basis of **F** § 93.

...The symbolic frame provides two axes normal to space, both time-like, with symbols E_{45}, E_{05}. We can transform to an equivalent pair E'_{45}, E'_{05} by rotation in the plane E_{04}. The equivalence is real, and the standard unaccented frame is only distinguished by its relation to the neutral uranoid. The suffixes have been defined so that unaccented... 0 is associated with electrical characteristics;..there is no difficulty in referring [a molar system] directly to the special axes E_{45}, E_{05}. But in microscopic theory we introduce new elementary concepts..; these are referred to axes E'_{45}, E'_{05} whose orientation...is determined in the course of the investigation.

The free intracule..has a linear momentum vector of length μ. If we take the momentum vector to be $E'_{45}\mu$,..the intracule will be at rest. Let it have in addition an integral number of quanta of a.m. $j\hbar$ in the plane E'_{23} $(=E_{23})$. The whole motion is then dynamically similar to that of a rigid body, whose rotation is necessarily in the space in which the linear velocity is zero (§41 [**F** §53]). The a.m. is represented by the vector $E'_{01}j\hbar$, E'_{01} being the symbol in a vector which corresponds to E'_{23} in a strain vector...appealing to the...analogy by which classical strain vectors become quantum space vectors.

Referred to the unaccented frame, $E'_{45}\mu$ and $E'_{01}j\hbar$ become

$$(E_{45}\cos\theta - E_{05}\sin\theta)\,\mu \quad \text{and} \quad (E_{41}\sin\theta + E_{01}\cos\theta)\,j\hbar.$$

We adopt the notation

$$E'_{45}\mu = E_{45}\epsilon - E_{05}f, \quad E'_{01}j\hbar = E_{41}\alpha\hbar + E_{01}g\hbar. \tag{74·1}$$

Then

$$f = (\mu^2 - \epsilon^2)^{\frac{1}{2}}, \quad g = (j^2 - \alpha^2)^{\frac{1}{2}}, \tag{74·2}$$

and

$$\cot\theta = \epsilon/(\mu^2 - \epsilon^2)^{\frac{1}{2}} = (j^2 - \alpha^2)^{\frac{1}{2}}/\alpha. \tag{74·3}$$

The unaccented frame corresponds to molar space-time determined by the neutral uranoid. Thus ϵ and $g\hbar$ are the ordinary energy and a.m., whilst f and $\alpha\hbar$ are extra-spatio-temporal linear and angular momentum. Thus by tilting the intracule in the E_{04} plane we have introduced an extra-spatial circulation with a.m. $\alpha\hbar$.

We have seen that the interchangeability of suffixes gives the intracule a quantum of extra-spatial a.m. in the rigid field, which (on account of the multiplicity factor) is reduced to $\hbar/137$ in the transformation to Galilean coordinates. By taking $\alpha = \frac{1}{137}$ in (74·3) we adjust the tilt so as to provide this extra-spatial circulation. Accordingly, the energy ϵ corresponding to a state of a.m. $j\hbar$ is given by (74·3) with

$$\alpha = \tfrac{1}{137}. \tag{74·4}$$

This is the Sommerfeld formula for the metastable states,...very accurately confirmed by experiment.

If we multiply $E_{41}\alpha + E_{01}g$ by E_{45}, we obtain the corresponding a.m. density $-E_{15}\alpha - E_{23}ig$; so that the plane of the interchange a.m. in the system considered is identified as E_{15}. It seems odd a circulation in this plane should be possible, since the ordinary relativity rotation is hyperbolic. The explanation is that continuous interchange is not one of the real relativity rotations of the double frame (§ 63); so that its reality condition is directly opposite to that of an ordinary rotation.

The general Sommerfeld formula which includes all states is (74·4′, 74·5) [**F** (93·81, 93·82)]. When $n \neq 0$, μ, the resultant of ϵ and f, cannot have the same tilt as j, the resultant of g and α, and the system does not become dynamically similar to a rigid body for any choice of axes...no elementary physical representation of the unstable eigenstates.

Reference to the analysis..shows no simple geometrical picture of the unstable states is possible...their wave functions are non-algebraic. An *algebraic wave function* is $\psi = \psi_0 . f(x_1, x_2, x_3, x_4)$, ψ_0 a constant wave vector; in a non-algebraic wave function the 4 components of ψ are independent functions of the coordinates. The momenta $-i\hbar\partial/\partial x_\alpha$ of a non-algebraic wave function have an E-symbolic character which twists them out of their nominal directions; e.g. we may find a momentum ω in the E_{12} plane represented as a symbolic momentum $E_{25}\omega$ in the E_{15} direction. There is no such distortion in an algebraic wave function.... Since the momenta of unstable states are subject to this directional distortion (presumably the source of instability), geometrical representation is inappropriate, and it is scarcely possible to investigate them otherwise than by blind analysis.

75. **The symbolic frame in relative space** [two versions], and

76. **Reality conditions in relative space** [unfinished]: these correspond to **12·5** §§ 75, 76.

74 (*b*). **The differential wave equation** [short, or unfinished]: like part of **F** § 107.

CHAPTER 13

DRAFTS OF F VI–XI: (iii) DRAFT H

13·0. Note

Chapters I–IV of **H** were summarized in **8**. The rest of **H**, treated here, covers not only the E-number theory (**13·1–13·3**) of **F** VI–VIII, but also (**13·4–13·6**) the developments of **F** IX–XI. In **13·1–13·3**, the only extensive quotations are from §§ 63 and 75, 76, corresponding to the difficult endings of **F** VII, VIII. **13·4** gives the only draft of 'Simple Applications'. An isolated fragment on affine field theory is appended to this chapter as **13·7.**

13·1. Draft H 7 V. The Complete Momentum Vector.

(Headed 'March 1943 revised December 1943'.)

This MS. is close to **F** VI; a few interesting variants are quoted.

41–43: close to **F** §§ 53–55 (same titles).

44. **Rotations:** like **F** §§ 56–57. There is a clear statement of the role of the E-frame:

We have been investigating the connexion between a symbolic E-frame and a geometrical coordinate frame. To make the connexion we study a physical system which can be referred to either...The result is very simple; components p_μ referred to the symbolic frame are also components of vectors in the geometrical frame, and it is only necessary to draw up a table of correspondences as in (44·7) [**F** (56·7)]...the basis of the procedure has been the concept of *relativistic equivalence*. This enables us to distinguish certain forms of transformation $p_\mu \to p'_\mu$ called rotation, which are such that the system described by p_μ is relativistically equivalent to that described by p'_μ. The connexion then results from the condition that the recognized system of transformations which gives relativistic equivalence in the ordinary geometrical representation must also give relativistic equivalence in the symbolic representation.

45. **Effective and ineffective relativity transformations:** close to **F** § 58, without the last remarks—strain vectors come later in this version than in **F**.

46. **Real and imaginary symbols:** equivalent to **F** §§ 60, 61, with considerable rearrangement. The real ζ- and E-numbers are introduced, and the initial argument of **F** § 60. Three reasons are given for choosing system (a) (6 real and 10 imaginary symbols): (1) it

corresponds to **F** (60·3) ('which was based on the real ζ-algebra') and (2) to the matrix representation; also (3):

If we substitute 'chiral' for 'imaginary' and 'achiral' for 'real', the same conditions of consistency must be satisfied in distributing chiral character among the symbols. Thus the two systems [(*a*) and (*b*)] provide for two different but homomorphic kinds of distinction of the symbols, of which one (chirality) has already been identified with the distinction of electric and mechanical characteristics.

...the reason (1) goes deepest...When..we investigate the ultimate origin of the E-frame, we shall find it is in fact introduced into physics *via* the ζ-algebra....

On E_{45}:

The chiral (electrical) symbols..are $E_{0\mu}$, E_{16}. Two symbols E_{04}, E_{05} are both real and chiral, and their product E_{45} may be regarded as the pivot of this scheme of distribution of suffixes. Also the real symbols $E_{(23),04,05,45}$ are converted into the electrical by multiplication by E_{45}. We shall find this pivotal symbol is associated with the time-axis....

47: close to **F** § 62.

48. **Neutral space-time:** like **F** § 63, with a modified argument on the antithesis of P_e and P_m which is equivalent in part to **F** § 64. Concerning the 'tilt' associated with a strong e.m. field:

..the radius of curvature is tilted through an angle 10^{-39} [cf. **F** text] ...interpreted as a strain due to the high potential. Our usual procedure is to treat an actual environment as the sum of the standard uranoid and a superposed disturbance; and the two components of the radius.. become separated. The large space-like radius $E_{05}E_{16}R_0$ is a characteristic..of neutral space-time; the small time-like component $E_{05}r_0$ is a characteristic of the superposed e.m. field.

Since P_e is dormant in a strictly neutral environment, an investigation of the extended vector P_m+P_e implicitly postulates a dual environment of the kind here described. The connexion of object with environment splits up into (1) a connexion of P_m with the neutral uranoid through the mechanical E-symbols, or equivalently through Riemannian space-time, and (2) a connexion of P_e with a superposed electric field through the chiral symbols. In the 5-dimensional representation of the uranoid the axes have been labelled $E_{(1)5}i$, E_{45}, $E_{05}E_{16}$. A homologous 5-space is given by axes labelled $E_{0(1)}i$, E_{04}, $E_{05}E_{16}$. In the dual treatment we adopt homologous Lorentz frames, one in each space, which provide realizations of the symbolic 4-vectors $\Sigma E_{\mu 5}p_{\mu 5}$ and $\Sigma E_{0\mu}p_{0\mu}$. For purposes of *analogy* between mechanics and electrodynamics, the two spaces are treated as congruent; but for purposes of *unification*, they must be superposed in such a way that lengths along the fifth axes have the symbolic relation $E_{05}E_{16}R_0$, $E_{05}r_0$ already found. That is to say, in transforming from dual to unified representation we must multiply measures in the second 5-space by $-E_{16}$; or equivalently the symbols associated with real measures must be changed to

$$-E_{0(1)}E_{16}i, \quad -E_{04}E_{16}, \quad E_{05}. \tag{48·3}$$

The electrical 4-vector is thus antithetic to the mechanical.

In order to give meaning to P_e we have had to superpose an electric potential on the standard uranoid; but P_e remains determinate as the potential is reduced. The reality conditions therefore apply to the neutral uranoid, on the understanding that it is defined as a *limit*....

49. **Strain vectors:** like **F** §59. The final paragraph is quoted on 'ordered volume' (cf. **F** full text).

A neater form is obtained if we define the ordered volume by

$$E_{\tau\lambda}V_{\tau\lambda}=E_{\mu\nu\sigma\tau\lambda}E_{\mu5}d^1x_{\mu5}E_{\nu5}d^2x_{\nu5}E_{\sigma5}d^3x_{\sigma5}, \tag{49·8}$$

where $E_{\mu\nu\sigma\tau\lambda}$ is $+E_{16}$ or $-E_{16}$ according as $\mu\nu\sigma\tau\lambda$ is an even or odd permutation of 0, 1, 2, 3, 4, and zero if the suffixes are not all different. When the three elements are along the space axes, so that $E_{\mu5}d^1x_{\mu5}=E_{15}i\,dx_1$, etc., this reduces to one component

$$E_{04}V_{04}=-E_{16}E_{15}E_{25}E_{35}i^3dx_1dx_2dx_3=E_{04}iV \quad (V=dx_1dx_2dx_3). \tag{49·9}$$

Thus the ordinary volume $E_{45}V$ and the ordered volume $E_{04}iV$ are respectively mechanical and electrical vectors along the time axis, and are antithetic as the reality conditions require.

50. **Determinants and eigenvalues:** close to **F** §65. The passage quoted in **12·1** §49 also occurs here.

13·2. Draft H *6* VI. Wave Vectors. (Headed 'March 1943 revised December 1943'.)

This is close to **F** VII; differences appear in the last two sections.

51. **Idempotency:** like **F** §66. It is observed that idempotency, by removing the indeterminacy of relations between linear and non-linear characteristics, enables us to *get beyond scale-free physics.*

52: very close to **F** §67.

53. **Spectral sets:** like **F** §68 and the later part of §69.

The result of spectral analysis of a scalar particle is shown conspicuously by the exclusion principle [**F** §41], where the two protons and two electrons occupying a unit cell are the equivalent of one scalar particle whose linear momentum vector p_{15}, p_{25}, p_{35} is used to define the cell.

54. **Table of symbolic coefficients:** close to the first part of **F** §69. The table is equivalent to that in **F**, save that the last line

(Radius of space curvature $E_{05}i$) is replaced by 'Magnetic pole density $E_{04}i$'. Also:

The ordered volume is to be regarded as an electrical characteristic; it will be seen from (49·9) that its reciprocal is a magnetic energy, just as the reciprocal of an unordered volume is a mechanical energy.

55, 56: close to **F** §§70, 71.

57: close to most of **F** §72.

58. **Wave tensors:** like **F** § 73 (after beginning with the 'star' notation of **F** §72).

59. **Phase Space:** like **F** §75.

..*Thus the position vector of a point in phase space is a strain vector.* ..Conversely..*when the space vectors undergo a real relativity rotation the representative points of the complete strain vectors remain within the* 10-*dimensional phase space*....

In **RTPE** § 7·4, phase space was wrongly stated to be closed. The error has been pointed out by A. J. Coleman [see Bibliography].

60. **Space tensors of the second rank:** like **F** § 74. The following detail should be compared with **F**:

The relation between an associated strain and space tensor may be written

$$(Z_0{}^0)_{\alpha\beta\gamma\delta} = (T_0{}^0)_{\alpha\delta\gamma\beta} = (-T_{00})_{\alpha\delta\beta\gamma}. \qquad (60\cdot4)$$

The suffix 0 is used to indicate the character of a tensor without referring to individual components. In the permutation $\alpha\beta\delta\gamma$ the pairing of the wave vectors to form space vectors is unaltered and (60·3)

$$[-(T_{\mu\nu})_{\alpha\beta\delta\gamma} = (T_\mu{}^\nu)_{\alpha\beta\gamma\delta}]$$

applies to each component separately; but in the permutation $\alpha\delta\gamma\beta$ the wave vectors are crossed, and (60·4) applies only to the tensor as a whole.

61. **The quantum-classical analogy:** like **F** §76 (*not* §78).

62. **The symbolic frame in relative space:** like **F** § 77.

The argument is fuller, and these extracts should be compared with the text of **F**. The four-dimensional *vector density* $\mathfrak{P} = PiE_{05}$ in **F** §57 is here introduced as $\mathfrak{P} = -PE_{05}$.

...Thus the result of the de-stabilization of scale is that P is converted into $\mathfrak{P}$. The de-stabilized scale momentum duly appears in $\mathfrak{P}$ as the term $-E_{05}ip_{16}$.

To put it another way: (62·21) [**F** (77·11)] has been arrived at as a vector of scale-free physics. Scale-free theory provides no way of normalizing it and the adopted scale is therefore an arbitrary stabilized constant. To fix the scale observationally we must make an extension either to cosmical physics which furnishes the linear standard R_0 or to quantal theory which furnishes quantum-specified standards. In either case the standard appears as a vector in a fifth dimension; so that (62·21) must be multiplied by $\pm E_{05}$ to transform its scale into this direction before de-stabilization.

...The dropping of the x_5 dimension and consequent stabilization turns $\mathfrak{P}$ into P, where $\mathfrak{P} = -PE_{05}$. Similarly, in relative space, the dropping of the x_4 dimension and consequent stabilization of the relative time turns $\mathfrak{P}$ into P_0 [P_q in **F**], where $\mathfrak{P} = -P_0 E_{04}$. The relation between the vectors P_0 in ξ-space and P in x-space is therefore

$$\mathfrak{P} = -PE_{05} = -P_0 E_{04}, \tag{62·3}$$

so that

$$P_0 = PE_{45} = -S \qquad (62·4)\ [\mathbf{F}\ (77·3)]$$

by (49·1) [**F** (59·1)]...therefore

The analogues in ξ-space of classical vectors are actually classical strain vectors. (62·5)

63. **Reality conditions in relative space:** like **F** § 78, with some differences in the argument. After the transposition **F** (78·1) and list **F** (78·2) of the components of P_q (here P_0), **F** (78·3) is replaced by the statement:

..writing $P_0 = \Sigma E'_\mu p'_\mu$ as an alternative notation, p'_μ is the designation of a component by analogy, whereas the expression in terms of p_μ given in (63·2) [**F** (78·2)] is its designation by identity. The E'_μ frame in ξ-space is the analogue of the E_μ-frame in x-space.

The discussion of *relative* reality conditions follows much as in **F** up to **F** (78·4). Then:

Among the results obtained by applying [**F** (78·4)] to [**F** (78·2)] we notice:

(1) $E'_{05} i p_{16}$ is homothetic with $E'_{16} i p_{45}$ or $E'_{16} p'_{16}$; hence p_{16}, p'_{16} are homothetic,

(2) $E'_{16} i p_{45}$ is antithetic to $E'_{15} p_{15}$; hence $p_{(1)5}$, p_{45} are homothetic.

In regard to..(2)..the intracule is an electric doublet, and cannot be expected to conform to the specification of a simple massive particle otherwise than by analogy.

The absolute reality condition is determined by the fact that we use a quantum-specified standard in molar physics...Quantum-specification implies an eigenscale...On de-stabilization the symbolic scale is multiplied by the real symbol $-E_{05}$..; but whereas the stabilized (scalar) scale is homothetic with its eigenvalues, the de-stabilized (vector) scale is antithetic to them, and the eigenvalues of the scale are reversed in character. Thus the condition that the molar scale shall be a real multiple of the eigenscale of a quantal system requires that the symbolic scales shall obey opposite reality conditions; so that $E_{16} p_{16}$ (or $E'_{16} p'_{16}$) for an intracule is antithetic to $E_{16} p_{16}$ for a molar system. Since p_{16} is real for a molar momentum vector, it is imaginary for a momentum vector in ξ-space. The general reality condition is

Using analogous designation throughout, the mechanical part of a momentum vector in ξ-space is imaginary and the electrical part is real. (63·4)

With p_{16} imaginary in [**F** (78·2)] it follows from [**F** (78·4)] that $p_{(1)5}$, p_{45} are real. Thus the spatial momenta $p_{(1)5}$ of the intracule are imaginary classical momenta. (This identification of quantal with imaginary classical momenta was introduced in [**F** (18·31)].) *It is a reality condition for the intracule that its momenta shall be imaginary in classical reckoning*...

Since density has dimensions (mass)^2 in natural units, the multiplication of the mass-scale by $-E_{05}$ multiplies the density scale by -1. Thus, in de-stabilizing the scale, the density and other components of the energy tensor are reversed in sign. This is the inversion of energy found in a more elementary way in §§ 17, 21 [**F** § 21].

The analogous or quantal designation is intended primarily for vectors in the relative space of the intracule; but just as we sometimes want to express these quantal vectors in classical designation, so we sometimes want to express molar vectors in quantal designation. It is useful to compare the reality conditions for the momentum vectors of molar systems and intracules by listing the real and imaginary components in both designations.

The results **F** (78·7, 78·8) follow. Finally:

For a strain vector the reality conditions have the simple form—
A 1 Molar particle in classical designation. All components s_μ imaginary.
B 2 Intracule in quantal designation. All components s_μ real... (63·8).

13·3. DRAFT H *5* VII. DOUBLE FRAMES. (Headed 'Jan. 1944'.)

This is mainly close to **F** VIII; the last two sections again are important.

64. **The *EF*-frame:** close to **F** § 79. On contragredient frames:

The asymmetry of [**F** (79·71)] is due to the fact that the ordinary notation uses real time, whereas the symmetrical development would require imaginary time. In a symbolic rotation $q(\,..\,)q^{-1}$,..(unless we use the transposes $\bar{E}$, $\bar{F}$) the vectors P, P' can only be made to transform contragrediently by making the F-frame transform contragrediently to the E-frame. Occasionally, e.g. in § 69, we introduce a contragrediently linked EF-frame...Equations [**F** (79·71, 79·72)] refer to a *directly linked EF-frame*...

65. **Chirality of a double frame:** close to **F** § 80.

...the dormant components in the EF-frame are more deeply asleep than those in the E-frame; 'dead' would perhaps be a fitter description.

66. **The interchange operator:** close to **F** § 81.

67. **Duals:** close to **F** § 82. The result **13·2** (60·4) reappears here as a 'dual', namely, $Z_0{}^0 = \tilde{T}_0{}^0$; this $Z_0{}^0$ is the $Z_0'^0$ of **F** (74·1); cf. also **F** (83·6).

68. **The *CD*-frame:** like **F** § 83, stopping before **F** (83·6).

69. **Double vectors:** like **F** § 84. The point is elaborated that the permutation correspondence between space and strain tensors is invariant only for contragredient linkage of the frames, which implies opposite time axes; but:

Usually we define our time axis in a particular way at the beginning of a problem and keep to it. Then in such names as energy, R.C. tensor, 'tensor' is only a courtesy title..we are only concerned with...a particular frame; and there is no harm in using the usual directly linked frame in which the correspondence is not invariant. The present investigation is exceptional because it supplies the proof that the 'courtesy' tensors really are tensors of the kind stated when tensor transformations are applied; for this we have to adopt a linkage which permits us to apply the transformations. But there is no further occasion to use contragredient linkage. In the next section tensor transformations of space and strain tensors...are the general transformations q of double wave tensors unrestricted by any linkage of rotation of the E- and F-frames.

70. **Double phase space:** like **F** § 85 in general.

...The eigenvalues of $E_\mu F_\nu$ are ± 1. The 136 real $[EF_\mu]$ symbols are homothetic with their eigenvalues and by the classification in (50·7) [cf. **12·1** §49] are classed as space-like; the 120 imaginary symbols are time-like...If EF_μ, EF_ν are space-like, the product is space-like if they commute...as for E-symbols.

Consider a space-like EF-number

$$Z = \Sigma' EF_\mu z_\mu. \tag{70·1}$$

..Z is the position vector in the phase space...

$$Z_{\alpha\beta\gamma\delta} = Z_{\beta\alpha\delta\gamma}. \tag{70·4}$$

..Z is a strain tensor in the most general sense..the '*phase tensor*'..the cross-dual of the covariant energy tensor...

71. **Uranoid and aether:** like **F** § 86 in general; also like *15* § 67 (cf. **12·3**).

72. **The R.C. tensor:** like **F** § 87, and part of § 88 (possibly left unfinished in this draft). On the de Sitter universe, **F** (88·2)

expresses the truism that, when there are no electrons and we consider spatial or temporal displacement of an electron that isn't there, the 'there' where it isn't keeps pace with the displacement. Yet from [**F** (88·2)] we derive a definite geometry of the realm of not-thereness!

The tensor [**F** (88·3)] applies only if the whole universe is empty. It is of the nature of a singular solution of the differential equations for $g_{\mu\nu}$ in empty space, and is not even the limiting form obtained by making $\rho \to 0$ in a material universe whose total mass remains finite. Its use is therefore for mathematical comparison rather than physical application.

73. **The tensor identities:** very close to **F** § 89. After **F** (89·7) it is observed that the $t^{\times}_{mn,mn}$ are all density components of $T^{\times}$. The line after **F** (89·7) (full text) should read 'For $\mu = 1, 2, 3, 4, 5, 0$', as in this MS.

74. **The contracted R.C. tensor:** close to **F** § 90.

75. **Interstates:** corresponds to most of **F** § 91; but the presentation differs, and most of this will be quoted.

..the tensor defined in (74·1) [**F** (90·1)] is the covariant space tensor T_{00}. Its cross-dual $T^{\times}$ is the phase tensor $Z_0'^{0}$ [$Z_0{}^0$ of **F** (85·6)]. The reality conditions limit $Z_0'^{0}$ to space-like $E_\mu F_\nu$; so that $T^{\times}$ consists of 100 pressure ..and 36 density terms,..momentum terms being absent. In obtaining (74·4) [**F** (90·4)] density terms have been excluded....

Thus the merging of the R.C. and energy tensors..is subject to a restriction..this is not the first time we have had to emphasize that.. relativity..and quantum theory normally have different fields...[but] the overlap is here very much wider than in previous investigations, where the two methods were connected only in uniform spherical space. Outside the overlap,..relativity continues to form the energy tensor by contracting the 4-dimensional R.C. tensor, setting

$$-8\pi\kappa T_\mu{}^\nu = G_\mu{}^\nu - \tfrac{1}{2}\delta_\mu{}^\nu(G - 2\lambda), \tag{75·1}$$

although it then includes components which have not the appropriate symbolic direction. Wave mechanics adheres to symbolic direction (in other words, group structure) as the defining characteristic of the energy tensor, so that

$$8\pi\kappa T_\mu{}^\nu = B_{\mu5}{}^{\nu5} + B_{\mu0}{}^{\nu0}. \tag{75·2}$$

..(75·1) satisfies the conservation law $(T_\mu{}^\nu)_\nu = 0$ identically..; [but] (75·2) is not self-conserved if $T^{\times}$ includes density terms...But..in quantum theory we are more concerned with conservation of the *analogues* of classical energy and momentum.

In wave mechanics..conservation does not enter..until we deal with transitions, i.e. with perturbation energy; and the energy tensor to which it applies is altogether different from that which determines the energy levels in an atom. We shall call it the *interstate energy tensor*. The 'motion' to which it refers is a flow of probability between states.... The ultimate purpose is..to treat perturbations..changes in probability..predictive theory. We are at present concerned only with.. structural theory...A steady transition circulation..[is] an *interstate*,

..the occupant..an *oscillator*. The transformation of description replaces states and particles..by interstates and oscillators. This is the change we make when we transform a double wave tensor into its cross-dual, or..the EF- into the $C'D'$-frame [**F** § 83 text]. A covariant $T_{\alpha\beta\gamma\delta}$ can be regarded either as the product of double wave vectors $\Psi_{\alpha\gamma} = \psi_\alpha \phi_\gamma$, $X_{\beta\delta} = \chi^\beta \omega^\delta$ or...space vectors $V_{\alpha\beta} = \psi_\alpha \chi^\beta$, $W_{\gamma\delta} = \phi_\gamma \omega^\delta$. In the cross-dual ϕ and χ are interchanged, so

$$\Psi^\times = V, \quad X^\times = W, \quad V^\times = \Psi, \quad W^\times = X, \tag{75·3}$$

and $V^\times$, $W^\times$ are strain vectors. If T is an energy tensor, V, W, $V^\times$, $W^\times$ are momentum vectors. Conceptually we associate a momentum vector with..'flow'; and in general two independent wave vectors, e.g. ψ, ϕ, are needed to specify the flow. We shall call $\psi_\alpha \phi_\beta$ the momentum vector of the interstate of ψ and ϕ...; ϕ may be $\psi^\dagger$..the interstate then degenerates into a pure state specified by one ψ. The reality conditions require $\psi\psi^\dagger$ to be a strain vector..so only $V^\times$ and $W^\times$ can degenerate....Setting $V^\times$, $W^\times$ to be momentum vectors of pure states, so that $\phi = \psi^\dagger$, $\omega = \chi^\dagger$, we have...$W = V^\dagger$. Thus W and V are momentum vectors of interstates of opposite flow.

Let us now make a fresh start, taking $V_{\alpha\beta} = \psi_\alpha \chi_\beta$ to be the momentum vector of an oscillator....The two [related] states must be the pure states specified by ψ and χ....If χ is changed continuously to $\psi^\dagger$, the flow between states is changed to a flow between the conjugate parts of the same pure state, so that it now forms an internal flow in the state, and $\psi\psi^\dagger$ is the momentum vector of this flow.

The cross-dual transformation accordingly transforms the state energy tensor $T^\times = V^\times \times W^\times$, contributed by the momentum vectors of particles in two states, into an interstate energy tensor $T = V \times V^\dagger$, contributed by the momentum vectors of oscillators in the two conjugate interstates....

The interstate momentum vector $\psi_\alpha \chi^\beta$, which is a space vector, does not in general satisfy the reality conditions for a momentum vector. In steady-state theory this means that conjugate oscillators are not separable from each other in the way that particles are separable. The opposite flows corresponding to V and $V^\dagger$ really form one interstate which is divided only symbolically, in the same way that the factors ψ, $\psi^\dagger$ constitute a symbolic division of a state corresponding to $\psi\psi^\dagger$. We can therefore consider only a joint occupation of V and $V^\dagger$; and it is sufficient that the energy tensor T of this joint occupation should satisfy the reality condition. In perturbation theory the fact that an oscillator does not satisfy the standard reality conditions implies that it is not simply superposable on a neutral uranoid; and the oscillator is accordingly recognized as an instrument of radiation.

76. **Antisymmetrical wave functions:** corresponds to **F** § 92, although the arguments meet only in the major points.

By (62·5) [**13·2**] the quantum momentum vector of an intracule..is actually (i.e. in classical designation) a strain vector. This is..generalized by the result (§ 75) that the state momentum vector $V^\times$ is a strain vector.

Classical designation is used throughout §§ 74, 75, since (74·4) [**F** (90·4)] is..a link with the molar energy tensor...

Thus..we may call the state momentum $V^\times$ and energy $T^\times$ the quantum momentum vector and energy tensor. The quantum energy tensor is as it stands the phase tensor Z..the position vector in phase space. We divide it into pressure and density terms Z_p, Z_d. The interstate energy T_p $[Z_p^\times]$ is self-conserved; but Z_d is recalcitrant and excluded from.. § 74. Z_d..is..the concern of quantum theory..introducing the fixed-scale characteristics. When Z_d is zero we do no more than translate the classical conception of physical change as produced by motion into the wave-mechanical conception..as transition between states...

Z_d corresponds to the time-like or antisymmetrical components of $V^\times$ and $W^\times$, suffixes (23), 05, 04, 45. This is the dormant part of a strain vector, composed of electrical characteristics....In classical theory the e.m. energy tensor is

$$E_\mu{}^\nu = -F_{\mu\alpha}F^{\nu\alpha} + \tfrac{1}{4}\delta_\mu{}^\nu F_{\alpha\beta}F^{\alpha\beta}, \tag{76·1}$$

which gives $(E_\mu{}^\nu)_\nu = F_{\mu\nu}J^\nu$; so $E_\mu{}^\nu$ is not self-conserved, and we have to introduce a non-Maxwellian energy tensor..if we wish to preserve conservation....There is nothing alarming in the breakdown of conservation for Z_d, and therefore for $Z_p + Z_d$. The new point is that the usual patching-up by a non-Maxwellian fiction is out of keeping with our chosen line of advance.

Although Z_d and T_d appear...as the e.m. parts of the energy tensors, this is not the most suitable way of defining the distinction. Two ways are important, one applying to the quantum Z, the other to the interstate or classical energy tensor T.

The antisymmetrical matrices (23) and 05, 04, 45 are the spin and co-spin matrices [**F** (81·5)]; the permutation [**F** (78·1)] does not affect this. Thus

The recalcitrant part of the energy tensor arises from the spin and co-spin terms in the quantum momentum vector (*analogous designation*). (76·2)

Since $V^\times, W^\times = \Upsilon, \mathrm{X}$ by (75·3), the double wave vectors of T_d are antisymmetrical. In later developments the double wave vectors become double wave functions. Thus

The self-conserved part of the interstate tensor corresponds to symmetrical, the recalcitrant part to antisymmetrical wave functions. (76·3)

It was in this form that the distinction first originated in § 73 [**F** § 89], where the symmetrical and antisymmetrical forms of Υ, X were treated separately.

The result (76·3) forms an important junction with current quantum theory...In current theory the separation of antisymmetrical and symmetrical wave functions..is extended to more highly multiple functions...But...[n-tuple wave functions] are only introduced as a step in deriving the symbolic wave functions representing 'second quantization'. The n-tuple functions are unnecessary in the present theory which arrives at second quantization more directly in Chapter [*sic*].

We are therefore in fundamental agreement..as to..separating symmetrical and antisymmetrical wave functions; but..for double wave functions..the equivalent separation (76·2) is more suitable.

13·4. Draft H *4* VIII. Simple Applications. (Headed 'Feb. 1944'.)

77. **The metastable states of hydrogen:** like **F** §93. The first quotations should be compared with the full text of **F**.

Consider an intracule..momentum vector $E_{45}p_{45}+E_{01}p_{01}$..energy $p_{45}=\epsilon$ and magnetic moment $p_{01}=g\hbar$. This might well be the intracule of a standard particle; but for a hydrocule it is too simple, because there is nothing to represent the extra-spatial interchange circulation. To remedy this we give the intracule a time-tilt....

[After **F** (93·24)]: The interchange circulation is now provided for by the term $E_{14}ig\hbar \sinh u$..an a.m. in the plane through a space axis E_{15} and the 'time' axis E_{45}; but the latter is the interchange direction used as analogue of time in the quantum-classical analogy. The tilt simultaneously introduces an E_{04} term representing magnetic energy...

[After **F** (93·5)]:..the tilt may be expected to have discrete eigenvalues determined by quantization of a.m...We shall suppose provisionally that $j\hbar$ is an integral number of quanta, and $\alpha\hbar$ is one 'small quantum', i.e. $\frac{1}{137}\hbar$ or e^2; so that [**F** (93·7)]..[hence] the Sommerfeld formula for..the metastable states (including the ground state..) of hydrogen.

More symmetrically, we may take..$\alpha\hbar$ an integral number Z of small quanta; so that $\alpha=\frac{1}{137}Z$. Then [**F** (93·5)] is the general formula for the metastable energies of hydrogen-like ions, and the quantum number Z is the atomic number of the nucleus.

..the introduction of the small quantum..is a consequence of the reduction from rigid coordinates to the observational (Galilean) system....

The energy ϵ of the level is the real classical energy of the intracule, but it is in a tilted time direction E'_{45}...it is the resultant of a mechanical and a magnetic energy (symbols E_{45}, E_{04})...[appropriate for radiation].

Comparing the formula $\epsilon^2=\mu^2-f^2$ with the ordinary Lorentz formula $\epsilon^2=\mu^2+p^2$, we see that the magnetic energy f (real in classical reckoning) is dynamically equivalent to an imaginary momentum ip...the intracule, as depicted in the quantum-classical analogy, has no mechanical energy p_{45} and only a small magnetic energy....[Formulae **F** (93·81, 93·82) are stated.]

78. **Deuterium and neutron:** equivalent to **F** § 94 in conclusions, but simpler, omitting 'modular equivalence'.

..a metastable hydrogen intracule..has a momentum vector.. $E_{16}i\mu+E_{23}ij\hbar$ in quantum designation. Considering the ground state $j=1$, and taking..$\mu=\frac{1}{2}$, $\hbar=\frac{1}{2}$..[this] is

$$P_1=\tfrac{1}{2}i\{E_{16}+E_{23}\}, \qquad (78·1)$$

and $(-P_1)^2 = -P_1$.... We can therefore represent the intracule by an idempotent..but of rank 2. In our analysis..rank 1 has been considered the more elementary; it gives a pure particle as the product of two wave vectors...According to this view the intracule is composite—a superposition of two pure particles...This..is repugnant to our conception of the elementary character of the intracule.

A way out of this difficulty has been found in § 53 [**F** § 68]. The momentum vector

$$P_2 = \tfrac{1}{2}i(E_{16} + E_{45}) \tag{78·2}$$

is also (when the sign is reversed) idempotent of rank 2; and

$$P = P_1 P_2 = -\tfrac{1}{4}i\{E_{16} + E_{45} + E_{23} + E_{01}\} \tag{78·3}$$

is idempotent of rank 1. We call P_1 a *spin intracule* in allusion to the term E_{23}. Then P_2 is a *co-spin intracule*; and the pure vector P represents a *double intracule*..one spin and one co-spin...

A wider view...the quantum momentum vector of the hydrogen intracule in an appropriately time-tilted frame has the general form $E_{16}p_{16} + (E_{23}p_{23})$. By (46·9) [**F** (60·5)] this is the general ζ-number. The θ-algebra is obtained by substituting co-spin for spin. Thus

The momentum vectors of the spin and co-spin intracules are ζ- and θ-numbers.

The E-algebra..is a multiplicative combination of ζ- and θ-algebras, so that in general we have to do with a multiplicative combination of spin and co-spin intracules, i.e. with double intracules. The hydrogen intracule provides the easiest introduction to quantal theory but it shows us only one half of the realm.

..P in (78·3) can also be expressed as P_1P_3, where $P_3 = \tfrac{1}{2}i\{E_{16} + E_{01}\}$, or as P_2P_3. Why choose P_2 rather than P_3 as the second type of intracule? The choice is strongly suggested by the fact that the spin and co-spin terms form the recalcitrant part Z_d of the quantum energy tensor,.. the part to which steady conditions must be applied, since otherwise it is not self-conserved. When we approach the problem from the subalgebras, the question is settled conclusively; P_3 is a mixture of ζ- and θ-symbols, and is not comparable with P_1, P_2....

Both kinds of intracule are to be combined with extracules and the double intracule with a double extracule...[i.e. P_1, P_2, P give hydrogen, neutron, deuterium respectively].

...a neutron is literally a hydrogen atom in a 0-quantum state. The contraction of the 'electron-orbit' almost to a point..is evidently an appropriate extrapolation to $j = 0$ of..the decrease of radius with j. But the usual wave equation..agrees with (77·6) [**F** (93·5)] in giving no solution for $j = 0$....The wave equation expresses a condition which.. is to eliminate co-spin terms...The neutron is given by the complementary condition which eliminates spin terms.

We shall confirm this identification of the neutron and deuterium (*a*)..by the theory of nuclear spin, (*b*)..by masses, (*c*) by the magnetic moment of the neutron.

79. **Mass of the neutron:** equivalent to **F** § 95. The β-factor at the very end of **F** § 95 is not mentioned, so that

$$n' - \mathrm{H}' [= 1{\cdot}5\mu] = 0{\cdot}000818;$$

but the β is pencilled in, giving 0·000824. The observational value 'is stated to be 0·00079 (Wigner) or 0·00081 (Barkas)'. These remarks follow:

..it seems clear *there can be only one eigenstate of the neutron.* For a linear series of eigenstates would be equivalent to at least one extra d.f.; and a multiplicity 5 would lead to $n' - \mathrm{H}' = 0{\cdot}00055$ in marked conflict with experiment...

We picture the neutron state as that of a hydrogen (V_{10}) intracule of rest mass μ which has been given the energy $1{\cdot}5\mu$ necessary to excite it from the usual spin state to the co-spin state. In terms of protons and electrons, we describe $1{\cdot}5\mu$ as the energy of neutron or nuclear binding of the electron, the energy of atomic binding at high quantum level being taken as the zero of reckoning.

80. **Atomic mass of deuterium:** equivalent in result to **F** § 98 (omitting β-factors), but differing in approach.

Consider an assemblage of protons and electrons in an arbitrary pseudo-discrete state specified by a complete energy tensor. It can be described indifferently as..hydrogen or deuterium...As deuterium, it is referred to an initial rest state corresponding to a D (deuterium) atom on the verge of ionization; this has much less energy than the corresponding rest state for H (hydrogen), so that a pair of H atoms in any ordinary state are equivalent to a D atom with extremely high transition energy. The difference between the two initial energies is the mass-defect of deuterium....

By separating symmetrical and antisymmetrical wave functions, we analyse the distribution into D atoms each consisting of a V_{100} and a V_{36} bi-particle, masses m_2, μ_2. These masses represent the energies present.. in a certain initial state..S_1. An arbitrary transition energy $\delta T_{\mu\nu}$ can be added; this must be separated into symmetrical and antisymmetrical parts....In H we have a similar separation (according to functional symmetry and antisymmetry) of two kinds of particle of masses m_0, μ; but in accordance with the rigid-field treatment the transition energy is assigned wholly to the intracule μ. To coordinate the two representations, we must treat the V_{36} particles as intracules and the V_{100}'s as extracules. To make the intracules capable of carrying transition energy they must be de-stabilized into V_{136}'s. The change of multiplicity changes the mass from μ_2 to

$$\mu_2' = \tfrac{36}{136}\mu_2. \qquad (80{\cdot}1)$$

This gives a representation of transition energies of the D atom homologous with..that of an H. This is in fact the customary representation, in which the D atom is reduced from a 4- to a 2-particle system of a deuteron and an electron. Accordingly μ_2' is the mass of the intracule in this 2-particle representation. The mass of the D intracule is not quite the same as the mass μ of a H intracule, but the difference is so small that it would be out of place (and probably misleading) to take it into account at this stage. We can therefore with sufficient accuracy set $\mu_2' = \mu$. Then the loss of energy in passing from the initial state S_1 with intracule energy μ_2 to initial state S_2, corresponding to the rest state of the deuteron and electron with intracule energy μ, is

$$\mu_2 - \mu = \tfrac{100}{36}\mu = 0{\cdot}001514 \text{ atomic weight units} \qquad (80{\cdot}2)$$

by (79·3) [cf. § **F** 95].

We have connected what appear to be the two most significant states of the system of four particles...The lower, S_2, is identified with the rest state of a deuteron and electron. The upper, S_1,...has very much greater energy; this can scarcely be other than the state of dissociation into protons and electrons at rest. Assuming this identification, which will be examined more fully in § 81, (80·2) is the mass defect of D; so that

$$2H^1 - H^2 = 0{\cdot}001514. \qquad (80{\cdot}3)$$

The observed values $H^1 = 1{\cdot}00813$, $H^2 = 2{\cdot}01473$ give $2H^1 - H^2 = 0{\cdot}00153$. [*End.*]

81. **Simple and double intracules:** like **F** § 96 (omitting the β-factor addendum), centring on the two analyses of the energy tensor into (*a*) simple and (*b*) double extra- and intracules. The interpretation of the states S_1, S_2 of D used in § 80 is now confirmed.

82. **Atomic mass of helium:** fairly close to **F** § 99; the main argument omits the explicit distinction of 'sub- and superthreshold analysis' which appears in **F** IX but not in this MS. chapter. In the 'assemblage of balanced particles':

We take two D atoms in state S_1 [cf. § 80] without mutual interaction as the initial state S of our 8-particle system. Introducing the rigid coupling [cf. **F**], we obtain a new state S'..a balanced particle. In state S the system is 4 free electrons and..protons at rest or equivalently 4 standard particles. In state S'..an unnatural balancing of the electrical energies..coherency of phase...; the electrical part of the momentum vector of one D atom cancels..the other..; the mass is then $4m_0/\beta$, since the purely mechanical part of the energy of a standard particle is represented by the mass of a hydrocule m_0/β...

When the separation of mechanical and electrical energy was previously dealt with in §§ 26, 27 [**8·3**], a rather elaborate procedure was required...All this is irrelevant when the electrical energy is not being reserved but destroyed—so that there is nothing left for quantal treatment. We employ only the result that, in a uniform distribution of protons and electrons at rest, $\frac{136}{137}$ of the energy is mechanical and $\frac{1}{137}$ electrical. Naturally it makes no difference whether the density is contained in V_{136} carriers (used in deriving the ratio) or V_{10} carriers (subsequently substituted). The final reckoning of mass refers to V_{10} carriers; and the masses in states S, S' are correspondingly $4M, 4M/\beta$ [$M = \frac{136}{10}m_0$]. Thus the mass defect of a balanced particle is

$$4M - 4M/\beta = 4M/137. \tag{82·1}$$

The argument on the 'two-legged intracule' follows as in **F**, yielding **F** (99·2):

$$[d=]4H^1 - He^4 = \tfrac{4}{137}M - \mu\sqrt{2} = 0·02866. \tag{82·2}$$

The observational value is 0·02863.

(82·2) is compared with the nuclear energy constant A as in **F**. Some comments:

The theoretical ratio d/A contains $\sqrt{\pi}$ as a factor; so it is clear the difference is not due to the omission of a minor correction...

The calculation for helium is frankly approximate....I think the method is exact for neutron, and that there is a good chance that it is exact for deuterium also. It is possible, however, that the definition of the mass of a deuteron is not precise enough to give the question a meaning. The deuterium nucleus is much less rigid than the helium nucleus; and, having regard to the difficulty of defining the mass of a single charged particle, it may well be that the resultant mass of three charged particles non-rigidly connected is not provided for by existing definitions.

83. **The separation constant of isobaric doublets:** like **F** § 100, with a paragraph representing **F** § 102. The argument proceeds to **F** (100·4); then a digression:

The exclusion energy here calculated is alternatively represented as interchange energy, which in turn has been identified with Coulomb energy. The energy μ_2 is contained in the four particles when they exist as free protons and electrons at rest, its partial release when a deuteron is formed being responsible for the mass defect of deuterium. Thus the Coulomb energy as ordinarily reckoned is $\mathfrak{E}_0 = \mathfrak{E} - \frac{1}{2}\mu_2$. Since the mean exclusion energy is $\frac{3}{5}\mathfrak{E}$, the whole Coulomb energy of the nucleus is $\frac{1}{2}n(\frac{3}{5}\mathfrak{E} - \frac{1}{2}\mu_2)$. From this we can derive a general idea of the radius of the nucleus; but otherwise the result is not helpful, because the nucleus also contains non-Coulombian energy for which we have no parallel method of calculation. It has been found that the non-Coulombian energy arises from coincidences of two protons or near-coincidences of their coordinates (§ 39) [**8·4**]; presumably the integration of the non-Coulombian energy must be attacked in an altogether different way, since coordinates are not introduced in the foregoing treatment.

The **F** argument is then carried to **F** (100·62); but as the value $d=0·001514$ (**F** (98·2)) is used for the mass defect of deuterium, without the factor β of **F** (98·3), **F** (100·63) is replaced by

$$M(n,-1) - M(n,0) = \tfrac{1}{2}.1·250n^{\frac{2}{3}} - 0·757, \tag{83·62}$$

and the same factor persists in **F** (100·7), which is here

$$M(n, -\tfrac{1}{2}) - M(n, \tfrac{1}{2}) = \tfrac{1}{2} . 1{\cdot}250(n-1)\, n^{-\frac{1}{3}} - 0{\cdot}757. \qquad (83{\cdot}7)$$

Wigner's formula (**F** (100·8)) is here quoted as (83·8), and the section ends with the following comments:

The greatest difference [of **F** (100·8)] from (83·7) is 0·09 mMu...well within the uncertainty due to raggedness. Thus the theoretical calculation is fully confirmed....

Wigner's investigation of the isobaric doublets was based on the spin theory of the nucleus, which has a natural place in the present fundamental theory, and will be treated in the next section. We have thus far used his formula only as a convenient embodiment of observational data; but we may now refer to its theoretical setting. The agreement between (83·7) and (83·8) is a bridge between two legitimate ways of approaching the problems of the nucleus, which supplement one another. Wigner's general formula is

$$M(n, T_u) - M(n, 0) = aT_u - \tfrac{1}{2} b T_u (n - 1 - T_u)/n^{\frac{1}{3}},$$

which includes (83·62) for n even and (83·7) for n odd as special cases. The near agreement of his constant $a = 0{\cdot}79$ with ours is fortuitous, since he took it to be the mass excess of the neutron—an interpretation which must certainly be rejected. His constant $b = 1{\cdot}27$ was purely empirical; and it is by providing a theoretical determination of this separation constant that our method supplements the previously existing theory. On the other hand, the nuclear-spin theory supplements our method, first by including isobaric doublets in a very general theory of isobars; and secondly it makes much clearer the nature of the approximations involved in developing the mathematical analogy between isobaric multiplets and multiplets in atomic spectra.

There is a paragraph (before the one just quoted) on *radii of nuclei* (cf. **F** § 102):

...we follow the current convention, although this, being based on the neutron theory of the nucleus, does not fit very logically. The conventional radius is the radius r of a uniform spherical distribution of N_p free protons giving the same Coulomb energy E; so that

$$E = \tfrac{3}{5} N_p (N_p - 1)\, e^2/r = \tfrac{3}{5}(\tfrac{1}{2}n - T_u)\,(\tfrac{1}{2}n - T_u - 1)\, e^2/r. \qquad (83{\cdot}91)$$

In general r will be a function of n and T_u; but, remembering that r determines the non-Coulombian energy which is unchanged in the transition, we must take it to be an even function of T_u. Thus in the transition $(n, -\tfrac{1}{2})$ to $(n, \tfrac{1}{2})$, r is unchanged, and we have

$$\delta E = \tfrac{3}{5}(n-1)\, e^2/r. \qquad (83{\cdot}92)$$

In adapting the current definition to our theory, the constant term in (83·7) must be omitted from the Coulomb energy (because current theory, which supplies the definition, treats it as binding energy of neutrons). Thus δE is to be equated to $\frac{1}{2}.1{\cdot}250(n-1)/n^{\frac{1}{3}}$ mMu. This gives

$$r = 1{\cdot}485 \times 10^{-13}.n^{\frac{1}{3}}\,\text{cm}. \tag{83·93}$$

84. **Nuclear spin:** an incomplete section, close to the early part of **F** § 101.

85. **Mass of the mesotron:** like **F** § 104. After the definition of a *forbidden oscillator* ('*f.o.*'):

..the interaction in a complex system may be partly composed of f.o.'s. The state energy tensor of the complex system is $\sum_r Z^r$, and consists of two parts Z_p, Z_d represented by symmetrical and antisymmetrical double wave functions. In general the interaction will include oscillators representing transition circulation between states comprised in Z_d. A f.o. represents similar transition circulation between a symmetric and an antisymmetric state.

It is commonly said that there are no transitions between symmetrical and antisymmetrical states; but that must be understood to mean that the simpler kinds of perturbation are unable to provoke them. Our own conclusion is that a tensor of the third rank at least is involved [cf. **F** § 92]. The necessary degree of complication is unlikely to occur except in nuclei where many particles are in simultaneous interaction. According to our theory of deuterium and helium there are no Z_m terms in their structure. It would appear that the only stable sources of forbidden oscillators are the nuclei of elements heavier than helium.

The argument follows closely that in **F**, but the β-factor is not introduced in the same way:

...Thus

$$\mathfrak{m}_0 = m_e + \tfrac{76}{60}(m_0 + m_e) \tag{85·3}$$

$$= 174{\cdot}440 m_e \quad (\text{standard } m_e),$$

or, in terms of current electron mass

$$\mathfrak{m}_0 = 173{\cdot}23 m_e. \tag{85·4}$$

It will be noticed that $\mathfrak{m}_0 = \mathfrak{m} - m_0$; this is because the energies of the V_{60} and V_{136} bi-particles are reckoned from a zero level m_0 below the level of ordinary reckoning for single particles which does not take account of the mass of the comparison particle. The mass (85·4) is at any rate not inconsistent with the scanty experimental data. The time of decay of the mesotron will be calculated in [*sic*].

...heavy mesotrons..mass..substituting m_p for m_e in (85·3); the result is $2{\cdot}36 m_p$.

...we infer that nitrogen and oxygen (either or both) contain forbidden oscillators. This will be verified from their atomic weights in § 86....

The next extant section is § 87 below.

13·5. DRAFT **H** *2* IX. WAVE FUNCTIONS. (Headed 'April 1944'.)

87. **Angular momentum:** close to **F** § 105. The distinction **F** (105·6) between field and particle a.m. is

the appropriate extension to quantal physics of the distinction between field and particle characteristics introduced in scale-free physics in Chapter II.

88. **The gradient operator:** close to **F** §§ 106 and 107, omitting the final statement on the purpose of the wave equation.

89. **Wave equation of the hydrogen intracule:** fairly close to **F** § 108. At the end:

Common eigensymbols of the four operators..constitute the wave functions of the intracule.... Actually the wave functions ψ_s of hydrogen do not form a complete set; i.e. it is not possible to expand an arbitrary function ψ in a series $\Sigma a_s \psi_s$. All the functions ψ_s have an unnecessarily high degree of convergence at the origin, and cannot be adapted to represent a highly concentrated density; this is due to the discontinuity between the lowest hydrogen state and the solitary neutron state of the proton and electron.

90. **Solution of the wave equation:** close to **F** § 109.

91. **The Coulomb energy:** like **F** § 110. An amplification on the a.m. associated with phase-to-space orientation:

This [the a.m.] has a plane of symmetry (*Footnote*: Since the components of a.m. do not reduce to eigenvalues simultaneously, there is no plane that can be described as the plane of the resultant a.m. in the classical sense. But there is a plane of symmetry, which is the principal plane of the spherical harmonic contained in the wave function.) which we can adopt as the plane x_2x_3. The antiperpendicular plane x_1x_4 then becomes fixed...the axis x_4 remains undefined..[as in **F** full text, p. 227].

The final paragraph is:

Although **W** refers to field momentum, the correcting Coulomb term includes both the field and particle momentum of the interchange circulation; if the particle momentum were not included, the Coulomb term would be halved. This means that the wave functions of hydrogen derived from $\mathbf{W} - i\hbar/r$ represent the interchange (electrical energy) as though it were wholly field energy, although properly it is half field and half particle. This simplification extends throughout electrical theory; so that, whereas we divide the mechanical energy into particle and gravitational-field energy, we have normally only one kind of electrical energy, which is treated as field energy. This difference of treatment corresponds to the difference of character of gravitational and electrical fields in practical conditions; the gravitational field of a particle extends throughout the universe, but its electrical field is quenched in a short distance by the polarization it induces in the environment. Another equivalent view of the apparent doubling of the Coulomb term has been explained elsewhere.

92. **Fixed-scale units:** this is related to the later part of **F** § 69, but applies these units explicitly to the intracule.

The extended momentum vector strings together 16 quantities of all sorts of dimensions both in ordinary and natural units. The components are proportional to the characteristics they are said to represent, but are not their actual measures unless the units are specially chosen. These special units will be called *fixed-scale units*.

In an eigenstate of the hydrogen atom the particle momentum vector of the intracule is the product of wave vectors ψ, χ. Thus, referred to suitable axes, it has the standard form of a pure momentum vector

$$P = -im\{E_{45} + E_{23} + E_{01} + E_{16}\}, \qquad (92\cdot1)$$

where m is real by B 2 in (63·5) [**F** (78·7) in **9·2**]. We can eliminate time-tilt by considering a metastable state of very high quantum number, and (92·1) then corresponds to the usual reference frame.

Since P refers to a free intracule the measure system is the theoretical system B [**F** § 27 in **3·3**]. Also, since the wave functions represent transition energy added to initial energy of a distribution of V_{136} particles (hydrocules), the coordinates are rigid coordinates in which the time is $\frac{1}{136}$ of the Galilean time (§ 19). By § 91, this also applies to the time analogue. In these coordinates the mass p_4 of the free intracule is $136\mu_B$.

The 4 components represent mass, a.m., magnetic moment and electric charge. Since (92·1) is an extension of the ordinary momentum 4-vector, m is always the measure of the mass; but m will not be the measure of the other 3 quantities unless fixed-scale units are used. For practical identification..fixed-scale units are a system in which the mass, a.m., magnetic moment and charge of an intracule (in the foregoing conditions) all have the same value. We shall show the momentum vector of the intracule is

$$P = -i\{E_{45}(136\mu_B) + E_{23}(\tfrac{1}{2}\hbar) + E_{01}(\tfrac{1}{2}e\hbar/\mu_B) + E_{16}(136e)\}, \qquad (92\cdot2)$$

so that by (92·1) $\quad m = 136\mu_B = \frac{1}{2}\hbar = \frac{1}{2}e\hbar/\mu_B = 136e. \qquad (92\cdot3)$

The coefficients in (92·2) are quantities whose values in c.g.s. units are known, having been derived in a roundabout way from observation. We have to show that (92·2) accords with the theory which gives this observational connexion. The first two coefficients are verified immediately; the time-transformation factor 136 does not affect the spatial a.m. $\frac{1}{2}\hbar$. The magnetic moment..is connected with observational measurement through the Zeeman effect. Our development has joined up with current theory in the analysis of the eigenstates of hydrogen, and the agreement extends to the magnetic splitting of the eigenstates. We can therefore take over from current theory the expression $(\frac{1}{2}e\hbar/\mu)\,H$ for the energy in a molar magnetic field H. Since H is a spatial tensor $F_{(23)}$, it is unaffected by the transformation to rigid coordinates; but the question arises whether the measure H_B required in the theory is the same as the measure H_A in the observational system. We shall find in § [*sic*], rather surprisingly, that it is. We can therefore leave H out of consideration; and the theory of the Zeeman effect then confirms that the magnetic moment is $(\frac{1}{2}e\hbar/\mu)_B$ in the conditions postulated by the wave equation.

The equations $m=136\mu_B=\frac{1}{2}\hbar=\frac{1}{2}\hbar e/\mu_B$ give $m=136e$; so that the fourth coefficient is confirmed without more ado. It would have been rather difficult to confirm it directly, because the simpler observational manifestations of e come from the interchange (Coulomb) energy, and not from the E_{16} term in the momentum vector.

By (92·3), $136e=\frac{1}{2}\hbar=\frac{1}{2}.137e^2$; so that $e=2/\beta$, and

$$m=2.136/\beta, \quad \mu_B=2/\beta. \tag{92·4}$$

Accordingly the fixed-scale unit of mass is $\frac{1}{2}\mu_B\beta$. By (27·22, 27·23) [**F** (29·4, 29·5)] the mass of a hydrogen atom in fixed-scale units in the observational system [A] is

$$M=\frac{136^2}{10}\frac{2}{\beta^2}. \tag{92·5}$$

13·6. Draft H *1* XI. Molar Electromagnetic Fields.

(Originally numbered 'X'; dated 'June 1944'.)

This draft is unfinished and has few equations numbered. It corresponds to the first four sections of **F** XI.

96. **The gauge transformation (molar application)**: close to **F** §113.

97. **Action invariants:** like **F** § 114.

...The mechanical action as ordinarily defined, which yields the energy tensor $-8\pi T_{\mu\nu}=G_{\mu\nu}-\frac{1}{2}g_{\mu\nu}G$, is $\int G\sqrt{(-g)}\,d\tau$. This is an adaptation to *localized* systems of the more fundamental action $\int G^2\sqrt{(-g)}\,d\tau$ here employed....The scale-free action density excludes cosmical..as well as quantal physics; and neither the cosmical nor the quantum constant are lurking in it....

$\mathfrak{A}_2=\sqrt{\{-|{}^*G_{\mu\nu}|\}}$, called the 'generalized volume'....

98. **The gauge transformation (microscopic application):** related to **F** §115 and part of §116. This begins:

When γ is a constant, the gauge transformation reduces to a simple scale transformation and $G_{\mu\nu}$ (like $^*G_{\mu\nu}$) is invariant. Then the usual mechanical action-density $\mathfrak{G} = G\sqrt{-g}$ varies as γ^2, whereas the electrical action density $\mathfrak{E} = F_{\mu\nu}F^{\mu\nu}\sqrt{-g}$ is invariant. This difference..forbids any simple association of $\mathfrak{G}$ and $\mathfrak{E}$....But T and G have the same gauge dimensions. A convenient form of the comparison of mechanical and electrical energy tensors is

$$T_{00} \text{ and } \mathfrak{E}_0^{\ 0} \text{ are invariant for scale transformation.} \qquad (98·1)$$

This result..is required by §§74–76 [**F** §§90, 91]. It means that T is an interstate..E a state energy tensor. The connexion between these is the cross-dual. According to our previous results, the cross-dual of T_{00} is a strain tensor $Z_0^{\ 0}$..a space tensor in quantum designation. But this ..refers to Galilean coordinates, in which tensors and tensor-densities are not discriminated. When attention is paid to the $\sqrt{-g}$ factor, the relation between interstate and state tensors is

$$T^{\times}{}_{00} = \mathfrak{Z}_0^{\ 0}. \qquad (98·2)$$

This result is **F** (115·41), and is established as in **F**. Eddington now moves to the material of **F** §116:

An energy tensor $T_{\mu\nu}$ is treated as the sum of elements $\Delta T_{\mu\nu} = p_\mu p'_\nu \cdots$; p'_ν the momentum vector of the carrier of the normalization volume V, which is a comparison particle (for a mutual energy tensor) or a fictitious duplication of the object particle (for a self-energy tensor)....[as in **F** full text] p'_ν varies as γ^{-2}; and, by (98·1), p_μ varies as γ^2...Thus $T_{\mu\nu}$ is the product of p_μ (momentum vector of the object-particle) of index 2, and of p'_ν (of the normalization system) of index -2.

The gauge transformation gives the object-particle a complex momentum vector p_μ; but..p^μ has index 0 and is real....

We have noticed (§19), in the only other investigation in which non-Galilean coordinates were used, that the energy and momentum density referred to coordinate volume are components of $\mathfrak{T}_\mu^{\ \nu}$; so that, when precise definition is required, this is the form normally intended in physical application. Thus for the quantum energy tensor the gauge-invariant form $\mathfrak{Z}_\mu^{\ \nu}$ is also the form on which our practical calculations of multiplicity factors, etc., have been based.

The section then ends with the last two paragraphs of **F** §115.

99. **Complementary electromagnetic fields:** equivalent to later parts of F §116.

...a field of e.m. potential $\kappa'_\mu = \partial\phi/\partial x_\mu$ can be created artificially by a gauge transformation..[contrast 'irreducible' fields with $F_{\mu\nu} \neq 0$].

The effect of the gauge transformation in quantum mechanics is to turn the wave vectors ψ_0, χ_0 into wave functions $\psi = \psi_0 e^{i\phi}$, $\chi = \chi_0 e^{-i\phi}$. We then have

$$\left(-i\hbar\frac{\partial}{\partial x_\mu}\right)\psi = \left(\hbar\frac{\partial\phi}{\partial x_\mu}\right)\psi, \qquad (99·1)$$

so that (apart from a difference of unit represented by the factor $\hbar$) the field momentum vector is identified with the potential vector of the artificial e.m. field. We now extend this identification to irreducible fields. The generalized field momentum vector is then $\hbar(\kappa_\mu + \partial\phi/\partial x_\mu)$ or, in operational form,

$$-i\hbar\,\partial/\partial x_\mu + \hbar\kappa_\mu. \tag{99·2}$$

Here κ_μ is the extraneous molar potential, because the artificial field which yields the first term occurs only in the system described by the wave functions.

The formula (99·2) is generally written as $-i\hbar\,\partial/\partial x_\mu - e\kappa_\mu$; so that the potential $(\kappa_\mu)_c$ in current units is related to the potential $(\kappa_\mu)_g$ in the theoretical units introduced in gauge theory by

$$(\kappa_\mu)_c = -(\hbar/e)\,(\kappa_\mu)_g = -137e(\kappa_\mu)_g. \tag{99·3}$$

The foregoing extension to irreducible fields is wholly a matter of nomenclature. In introducing the field momentum vector in [**F** § 105] we assumed the absence of an extraneous e.m. field; indeed, all our investigations up to now have necessarily been subject to this restriction since we have had no means of calculating its effect on the system. Thus it turns out that, as used in quantum mechanics up to now, the term 'field momentum vector' has been synonymous with 'e.m. potential vector'. All that we now do is to continue this alternative nomenclature, remembering that e.m. potential (natural or artificial) refers to the actual identification, and field momentum vector refers to an analogy (more or less close) with classical mechanics.

In § 19 we introduced an artificial gravitational field by a transformation to rigid coordinates in order to satisfy the conditions of rigid field treatment. Evidently the gauge transformation is here used with the same object in dealing with the e.m. field. The gauge determined by the wave functions may be described as the rigid gauge. [*End.*]

13·7. Draft *14* IX. The Molar Electromagnetic Field

This MS. (of unknown date) has a single section headed '*Affine field theory*'. In the following (almost complete) quotation equation numbers have been inserted.

The section begins with two paragraphs like the text of **F** § 113, and then proceeds with a summary of affine field theory (cf. **MTR** VII):

We set up a one-to-one correspondence between covariant vectors at neighbouring points, so that $A_\mu + \delta A_\mu$ at $x_\nu + dx_\nu$ corresponds to A_μ at x_ν, ..and the sum of two vectors at $x_\nu + dx_\nu$ corresponds to the sum..at x_ν. The most general continuous formula specifying such a correspondence is

$$\delta A_\mu = \Gamma_{\mu\nu}{}^{\alpha}\,dx_\nu A_\alpha, \tag{1}$$

where the $\Gamma_{\mu\nu}{}^{\alpha}$ are arbitrary functions of position. We impose the restriction $\Gamma_{\nu\mu}{}^{\alpha} = \Gamma_{\mu\nu}{}^{\alpha}$. The correspondence is then called an *affine connexion*. In general the correspondence is non-integrable, so that correspondence between vectors at distant points depends on the connecting route. The $\Gamma_{\mu\nu}{}^{\alpha}$ do not form a tensor....

Let A_μ be a vector field having the value A_μ at x_ν, and $A_\mu + dA_\mu$ at $x_\nu + dx_\nu$. Then $A_\mu + dA_\mu$ and $A_\mu + \delta A_\mu$ are vectors at the same point, and their difference $dA_\mu - \Gamma_{\mu\nu}{}^{\alpha} A_\alpha dx_\nu$ is a vector. It follows that

$$A_{\mu,\nu} = \partial A_\mu / \partial x_\nu - \Gamma_{\mu\nu}{}^{\alpha} A_\alpha \tag{2}$$

is a second-rank tensor. This is called the *affine derivative* of A_μ...the affine derivative of a product is formed by the distributive rule

$$(A_\mu B_\nu)_{,\sigma} = A_{\mu,\sigma} B_\nu + A_\mu B_{\nu,\sigma};$$

the affine derivative of an invariant is its ordinary derivative. The general formula...

$$A^{\sigma\tau..}_{\mu\nu..,\lambda} = \frac{\partial}{\partial x_\lambda} A^{\sigma\tau..}_{\mu\nu..} - \Gamma_{\mu\lambda}{}^{\alpha} A^{\sigma\tau..}_{\alpha\nu..} - \Gamma_{\nu\lambda}{}^{\alpha} A^{\sigma\tau..}_{\mu\alpha..} + \Gamma_{\alpha\lambda}{}^{\sigma} A^{\alpha\tau..}_{\mu\nu..} + \ldots. \tag{3}$$

The affine derivatives of tensors are tensors.

Forming second derivatives, $A_{\mu,\nu,\sigma} - A_{\mu,\sigma,\nu} = {}^*B_\mu{}^\epsilon{}_{\nu\sigma} A_\epsilon$, where

$${}^*B_\mu{}^\epsilon{}_{\nu\sigma} = -\frac{\partial}{\partial x_\sigma} \Gamma_{\mu\nu}{}^{\epsilon} + \frac{\partial}{\partial x_\nu} \Gamma_{\mu\sigma}{}^{\epsilon} + \Gamma_{\mu\sigma}{}^{\alpha} \Gamma_{\alpha\nu}{}^{\epsilon} - \Gamma_{\mu\nu}{}^{\alpha} \Gamma_{\alpha\sigma}{}^{\epsilon}. \tag{4}$$

It follows that ${}^*B_\mu{}^\epsilon{}_{\nu\sigma}$ is a tensor,..wholly defined by the affine connexion,..the *curvature tensor* of the connexion. Setting $\epsilon = \mu$ and $\epsilon = \sigma$ in (4), we obtain two contracted tensors

$${}^*F_{\sigma\nu} = \partial \Gamma_\sigma / \partial x_\nu - \partial \Gamma_\nu / \partial x_\sigma \quad (\Gamma_\sigma = \Gamma_{\sigma\alpha}{}^{\alpha}), \tag{5}$$

$${}^*G_{\mu\nu} = -\frac{\partial}{\partial x_\alpha} \Gamma_{\mu\nu}{}^{\alpha} + \frac{\partial}{\partial x_\nu} \Gamma_\mu + \Gamma_{\mu\sigma}{}^{\alpha} \Gamma_{\nu\alpha}{}^{\sigma} - \Gamma_{\mu\nu}{}^{\alpha} \Gamma_\alpha. \tag{6}$$

We see that ${}^*G_{\mu\nu} - {}^*G_{\nu\mu} = {}^*F_{\mu\nu}$, and that the tensor

$${}^*R_{\mu\nu} = {}^*G_{\mu\nu} - \tfrac{1}{2} {}^*F_{\mu\nu} \tag{7}$$

is symmetrical.

Thus the affine connexion defines a symmetrical and an antisymmetrical second-rank tensor; since this is just what is needed to specify the combined gravitational-e.m. field, we may take an affine connexion as the basis of molar physics, ${}^*R_{\mu\nu}$ and ${}^*F_{\mu\nu}$ being identified more or less directly with the metrical tensor $g_{\mu\nu}$ and the e.m. field tensor $F_{\mu\nu}$. If ${}^*F_{\mu\nu} = 0$, the identification ${}^*R_{\mu\nu} = \lambda g_{\mu\nu}$, where λ is a constant depending on the unit of length, is unique. Also, if ${}^*F_{\mu\nu}$ is small so that its square can be neglected, the identification

$${}^*R_{\mu\nu} = \lambda g_{\mu\nu}, \quad {}^*F_{\mu\nu} = \lambda' F_{\mu\nu} \tag{8}$$

is unique. When the square of ${}^*F_{\mu\nu}$ is not negligible there are many possible alternatives. For example, we may define ${}^*G^{\mu\nu}$ as the minor of ${}^*G_{\mu\nu}$ divided by the determinant of ${}^*G_{\mu\nu}$, and identify $\lambda g^{\mu\nu}$, $\lambda' F^{\mu\nu}$ with the symmetrical and antisymmetrical parts of ${}^*G^{\mu\nu}$.

Much attention was formerly given to the question which is the right alternative. They are all right! When the e.m. field is not negligible the definition of length by a quantum-specified standard breaks down, because the eigenstate which the standard system is required to occupy is non-existent owing to electric or magnetic splitting of the eigenstates. To the first order in $F_{\mu\nu}$ we may doubtless take the mean of the split states as equivalent to the unsplit state, but this clearly will not be true to the second order. Thus, to the accuracy required to differentiate between the alternatives, there is no recognized definition of length and metric in an e.m. field. The investigator is left to introduce his own definition; and, provided it converges to the accepted definition as the field tends to zero, no one can gainsay him. One identification of $g_{\mu\nu}$ in terms of the affine tensors is as good as another.

Early investigations on these lines were inspired by the hope that the 'right solution' would explain the structure of the electron and proton, and perhaps quantization, as arising from quadratic terms in $^*F_{\mu\nu}$; but we now know the intense intra-atomic fields must be treated by altogether different methods, essential features such as spin, correlation of probability, and interchange being lost in the molar averaging and therefore unrepresented in the molar energy tensor. This leaves very little to be accomplished by molar unified theory. (Formally, I think the most successful development..is due to Einstein, given in **MTR**, 2nd ed., p. 257....)

At present we need not commit ourselves to the identification (8); if not by this, by some direct definition, we introduce the tensor $g_{\mu\nu}$. This will enable us to raise and lower suffixes...The affine derivative of $g_{\mu\nu}$ is

$$g_{\mu\nu,\sigma} = \partial g_{\mu\nu}/\partial x_\sigma - \Gamma_{\mu\sigma}{}^\alpha g_{\alpha\nu} - \Gamma_{\nu\sigma}{}^\alpha g_{\mu\alpha}$$

$$= \partial g_{\mu\nu}/\partial x_\sigma - \Gamma_{\mu\sigma\nu} - \Gamma_{\nu\sigma\mu}. \tag{9}$$

Setting
$$S_{\mu\nu\sigma} = \tfrac{1}{2}(g_{\mu\nu,\sigma} - g_{\mu\sigma,\nu} - g_{\nu\sigma,\mu}), \tag{10}$$

we obtain $S_{\mu\nu\sigma} = -[\mu\nu,\sigma] + \Gamma_{\mu\nu\sigma}$, where [...] is the usual Christoffel bracket. Hence

$$\Gamma_{\mu\nu}{}^\sigma = \{\mu\nu,\sigma\} + S_{\mu\nu}{}^\sigma. \tag{11}$$

Contracting, and setting $S_{\mu\alpha}{}^\alpha = S_\mu$, we have

$$S_\mu = \Gamma_\mu - \frac{\partial}{\partial x_\mu}\log\sqrt{-g}, \tag{12}$$

so that
$$\operatorname{curl} S_\mu = \operatorname{curl}\Gamma_\mu = {}^*F_{\mu\nu}. \tag{13}$$

By (10), $S_{\mu\nu}{}^\sigma$ is a tensor, so S_μ is a vector, whereas Γ_μ is not. Thus if we adopt the identification $\lambda' F_{\mu\nu} = {}^*F_{\mu\nu}$, we have

$$\lambda'\kappa_\mu = S_\mu. \tag{14}$$

By (10), $S_\mu = -\tfrac{1}{2}g^{\nu\sigma}g_{\nu\sigma,\mu} = -\dfrac{1}{2g}g,_\mu$; so that

The e.m. potential vector is a multiple of the affine derivative of $\log\sqrt{-g}$. (15)

Independently of the identification (8), we obtain by inserting (11) in (4)

$$ {}^*B_{\mu}{}^{\epsilon}{}_{\nu\sigma} = B_{\mu}{}^{\epsilon}{}_{\nu\sigma} - (S_{\mu\nu}{}^{\epsilon})_{\sigma} + (S_{\mu\sigma}{}^{\epsilon})_{\nu} + S_{\mu\sigma}{}^{\alpha} S_{\nu\alpha}{}^{\epsilon} - S_{\mu\nu}{}^{\alpha} S_{\sigma\alpha}{}^{\epsilon}, \tag{16} $$

where $B_{\mu}{}^{\epsilon}{}_{\nu\sigma}$ is the ordinary R.C. tensor, and $(\ldots)_{\sigma}$ denotes covariant (not affine) differentiation. By contraction, we obtain for ${}^*F_{\nu\sigma}$ the value curl S already given in (13), and for ${}^*R_{\mu\nu}$

$$ {}^*R_{\mu\nu} = G_{\mu\nu} - (S_{\mu\nu}{}^{\alpha})_{\alpha} + \tfrac{1}{2}(S_{\mu})_{\nu} + \tfrac{1}{2}(S_{\nu})_{\mu} + S_{\mu\beta}{}^{\alpha} S_{\nu\alpha}{}^{\beta} - S_{\mu\nu}{}^{\alpha} S_{\alpha}. \tag{17} $$

Since S_{μ} represents the e.m. potential vector which we picture as an addition to empty space, it suggests itself that the tensor $S_{\mu\nu}{}^{\alpha}$ represents the most general molar physical system which can occupy space. Then the vanishing of $S_{\mu\nu}{}^{\alpha}$, or equivalently of the affine derivatives of $g_{\mu\nu}$, is the condition for empty space. Accordingly, in empty space (17) gives ${}^*R_{\mu\nu} = G_{\mu\nu} = \lambda g_{\mu\nu}$, by the usual law of gravitation. Thus the law of gravitation in empty space results from and requires the identification of ${}^*R_{\mu\nu}$ with the metrical tensor.

With the identification (8) we can calculate the total energy $T_{\mu\nu}$ and the e.m. energy tensor $E_{\mu\nu}$ in terms of $S_{\mu\nu}{}^{\alpha}$, and form their difference which is the non-e.m. part of the energy tensor. The resulting formula is cumbrous, and does not appear to give additional insight. We take this as an indication that the usefulness of the affine connexion, at any rate in this particular mode of application, is exhausted; and that the key to the structure of matter is to be found rather in microscopic theory.

Summarizing, we have found that affine geometry, which from the axiomatic point of view is much less specialized than metrical geometry, provides the basal tensors required in molar physics. Empty space corresponds to the vanishing of the affine derivatives of $g_{\mu\nu}$; and the e.m. potential κ_{μ} is the affine derivative of $\log\sqrt{-g}$. It will appear from the special case (Weyl's theory) treated in the next section that the constant λ' is imaginary; so that the separation of ${}^*G_{\mu\nu}$ into symmetrical and antisymmetrical parts is also a separation into real and imaginary parts.

CHAPTER 14

EPISTEMOLOGY AND THE COSMICAL NUMBER: DRAFT Ga

It was mentioned in **2** that Eddington filed with the final MS. of Chapters I–XII of **F** a Chapter VIII (Draft **Ga**) entitled 'Occupation Symbols'. Parts of this provided the basis for his paper 'The evaluation of the cosmical number', *Proc. Camb. Phil. Soc.* **40**, 37 (1944); this paper (**θ** in **2·1** above), which will be referred to as **FA**, was printed as an Appendix in **F**, to represent the projected Chapter XIII on Epistemological Theory.

It seems worth while here to reproduce Draft **Ga** *in toto*, rather than to give a full explanatory summary of **FA** (which would have to be very long) with only extracts from **Ga**; the order of development of **Ga** could not be seen very clearly from extracts. Some introductory remarks on the nature of the theory in **FA** and **Ga** will be given, however, in **14·1** and **14·2** before the text of **Ga** is presented in **14·3**.

14·1. Notes on 'The Evaluation of the Cosmical Number'

(i) The 'cosmical number' N is found in **FA** as the product of four factors

$$N = abcd, \quad a=2, \quad b=\tfrac{3}{4}, \quad c=136, \quad d=2^{256};$$

its significance and formulation will be briefly reviewed here in terms of these factors. N plays the part of the 'number of particles in the universe' in formulae which involve this conception; for example, in fundamental uncertainty of position (as for 'nuclear' forces), in the exclusion representation of gravitation and in galactic recession (involving the effective total mass). The factor $a=2$ occurs if the 'particles' are protons *or* electrons, but not if they are 'standard particles' equivalent to hydrogen atoms.

(ii) The main part of the calculation of N (strictly, of bcd) is, however, the enumeration of conceptually distinguishable measurements. A measurement is a comparison of two observables, and each of these is a relation of two primitive entities which themselves each have two 'eigenvalues' 1 and 0 for existence and non-existence. As entities, and likewise observables, are combined multiplicatively, a measurement has effectively 16 eigenstates, which relate it to the double or EF-frame. This frame is used in physics to represent the basic energy-tensor (of which the element represents a 'standard particle'), which has 136 independent components; so that the factor $c=136$ enters for this fundamental multiplicity.

(iii) The energy tensor itself, however, is regarded in physics as an *observable* (not a pure 'measurable'), being expressed in a dimensional form which implies a comparison background of similar physical dimensionality. Thus the basis of systematic physical measurement is in a quadruple or $EFGH$-frame.

(iv) In a simple E-frame a 'pure' measure (representing a measurement of a pure measurable, with 'pure' defined as in **F** VII) can be expressed as a sum of four commuting E-symbols associated with four 'anti-perpendicular' axes (e.g. $E_{01,23,45,16}$). By reversing the directions of any of these axes (a process of 'reflexion'), we associate 2^4 measurables with the standard form of measure. Similarly, in the *quadruple* frame there are $2^{4^4}=2^{256}$ measurables of the standard form. This 'grid constant' is the factor $d=2^{256}$ of N.

(v) There remains the factor $b=\frac{3}{4}$. This arises in **FA** § 11 by deliberately considering a volume distribution of particles in which the mass density is controlled by Einstein's general relativity law and the number density by the element of extension; so that the behaviour of these two densities can be contrasted when a test fluctuation is applied to the standard of length. The result is that, in an experimental test-addition of one additional particle, the standard scale (which is determined by the whole background of existing particles) will behave as if there were present $\frac{3}{4}n$ particles, where n is the number of measurables or particles calculated absolutely, without reference to the effect of scale on measurement. Thus for all purposes of (real or conceptual) observation, a factor $b=\frac{3}{4}$ appears in N. It is seen that in the course of this investigation the ultimate 'measurables' are identified in number with 'standard particles' or equivalent hydrogen atoms; this is largely on the grounds that these are the structural units that are conceptually countable (see also **Ga** below).

(vi) In this brief survey, the main omission is that of the concept of 'occupational operands' relating measures with measurables; this concept is allied to that of abstract angles and rotations, and so brings in the EF-frame in terms of rotation groups.

14·2. Notes on the MS. 'Occupation Symbols'

This Draft **Ga** differs from **FA** in one obvious way, namely, in being designed as a chapter of the book and not as a self-contained research paper. It also differs in arrangement, driving more directly and 'physically' at the evaluation of N, and then concluding with a discussion on the epistemological aspects. Materials of the sections §§ 2–11 of **FA** appear in **Ga** §§ 73–79 mainly as follows:

Ga § 74 is related to **FA** §§ 4, 11; § 76 to §§ 6, 7; § 79 to §§ 2, 3, 4.

To summarize: in §73, 'F.D. particles'[a] are defined as having states occupied by at most one particle, so that their occupation operators, J_r for the rth state, have eigenvalues 1 and 0. In §74 the total number of states (which will develop into the cosmical number) is taken to be $N = mn$, made up of m groups each of a large number n of detailed or nearly indistinguishable states; if these are *not* distinguished the occupants are 'E.B. particles'. The normal occupation of a group is $\frac{1}{2}n$, the occupation operator of a detailed state now being taken as $Y_r = J_r - \frac{1}{2}$. Scale considerations (equivalent in effect to those involving general relativity in **FA** §11, but based here on **F** §3 as in **3·1** above) applied to standard particles with 136 d.f.'s now give N the form $\frac{3}{2}.136.n$ (compare factors a, b, c in **14·1**). In §75, E.B. particles are characterized by the partial merging of occupation operands into the phase of wave functions. In §76 the n 'detailed states' are enumerated as the quadruple frame 'grid constant' 2^{256} (factor d in **14·1**), by considerations similar to those in **FA**. In §77 the cosmical number N is identified.

The remaining two sections are a review of the significance of the calculation. §78, 'Epistemological foundations', discusses physics as 'structure', the process of repeated 'E.-B.-ing' (so spelt by Eddington), and hence the nature of the identification of N, found by counting measurables, with the number of standard particles. §79, 'The primitive measurement', covers much of the material of **FA** §§2, 3, 4 and an element of §10 on the quadruple frame. It will be realized by comparison with the full text of **FA** that the concepts of 'observes' and of 'graduation' are very little developed in **Ga**, but the argument as a whole is clearer.

14·3. Draft Ga. Chapter VIII. Occupation Symbols. (Headed 'July 1943'.)

73. Fermi-Dirac particles

The trend of progress in microscopic physics has been to replace numerical quantities by symbolic operators. Accordingly, we introduce symbolic occupation factors J_r in place of the numerical factors j_r used in §14 and subsequently. Let S be any additive characteristic; then the total characteristic S^0 of the system will have the form

$$S^0 = \Sigma J_r S_r, \tag{73·11}$$

S_r being the value of the characteristic in the rth state when the state has unit occupation. Usually the additivity of characteristics is only approximate; and (73·11) will in practical application be supplemented by

[a] The contractions 'F.D.' and 'E.B.' are used for 'Fermi-Dirac' and 'Einstein-Bose' throughout this chapter.

interaction terms, or in scale-free problems the non-linearity will be allowed for by multiplicity factors.

We first consider states capable of occupation by one and only one particle. The corresponding particles are called *Fermi-Dirac particles*. Since $J_r S_r$ must reduce to S_r when the particle is definitely present, and to 0 when the particle is definitely absent, J_r must have eigenvalues 1 and 0. The essence of the new treatment is that when the particle is not definitely present or absent the degree of occupation shall be represented by a non-numerical symbol. Thus J_r has no other eigenvalues, and its characteristic equation is $J_r(J_r - 1) = 0$; that is to say, it is idempotent.

The symbolism corresponds to the generally accepted conception of probability. 'Probability' is a term which, as applied to an individual event, is not capable of numerical definition; but, as applied to an unidentified member of a class of events, it is used synonymously with 'frequency in the class'. The two usages are distinguished in mathematical theory, the non-numerical individual probability being represented by the symbol J; the numerical probability j, assigned to the same individual as a member of a class, is represented by the expectation value of J with respect to symbols defining the class.

If T is a mutual characteristic of pairs of particles, the total characteristic of the system will have the form

$$T^0 = \Sigma J_r J_s T_{rs}, \tag{73·12}$$

T_{rs} being the amount contributed by the rth and sth states when each is definitely occupied by a particle. We cannot make any distinction between $J_r J_s$ and $J_s J_r$ as combined occupation factors for the two states. The conception of occupation therefore implies that J_r and J_s commute. The properties of F.D. occupation symbols are therefore summarized as

$$J_r^2 = J_r, \quad J_s J_r = J_r J_s. \tag{73·2}$$

When J_r does not reduce to an eigenvalue the state is said to be *partially occupied*.

Since S_r is a fixed characteristic of a state which may be occupied, unoccupied or partially occupied, the reduction of J_r to an eigenvalue does not depend on S_r. Both sides of (73·11) are symbolic operators, and it is implied that in actual application there will be an operand q for them to act on. This operand symbolizes the whole state of occupation of the system, and will be called the *occupation operand*. The J_r are understood to act directly on q, so that they commute with any symbols (E-symbols, EF-symbols, differential operators) contained in S_r; and they reduce to eigenvalues only when q is an eigensymbol.

Let $e_{r\alpha}$ be a particular set (the 'α-set') of eigenvalues of the n operators J_r. Since $e_{r\alpha}$ is either 1 or 0, there are 2^n such sets. (We can conveniently regard the set of eigenvalues as a number α expressed in the binary scale.) Let

$$q_\alpha = \prod_1^n (J_r - 1 + e_{r\alpha}). \tag{73·3}$$

By (73·2), we obtain $J_s q_\alpha = e_{s\alpha} q_\alpha$; so that q_α is an eigensymbol corresponding to the α-set, and may be taken to be the occupation operand representing the α-state of occupation. For this state

$$S^0 q_\alpha = \Sigma S_r J_r q_\alpha = \Sigma S_r e_{r\alpha} q_\alpha = (\Sigma_e S_r)\, q_\alpha, \qquad (73\cdot4)$$

where Σ_e denotes summation over the *occupied* states. Thus by the transformation $S' = Sq_\alpha$, (73·11) is reduced to the simple summation $S'^0 = \Sigma_e S'_r$. It may be remarked that q_α is singular, so that there is no reciprocal transformation.

Usually there will be partial occupation and the occupation operand q will not have the special form q_α. The possible occupation operands must form a closed algebra which includes the 2^n special symbols q_α. I think that a rigorous development of this q algebra by a qualified algebraist would give considerable insight into the ultimate foundations of physics. But it is not a problem I can myself attempt, and I can only indicate the way I would expect it to develop. Presumably the conditions of closure and of containing all the symbols (73·3), with n given, are not sufficient to define the algebra uniquely; but it is likely that, discarding 'fancy algebras', one will easily be distinguished as of the type to which physicists are more or less accustomed. The question then is, what are the minimum axioms necessary to rule out all other solutions? When these are stated we can examine whether, as applied to the physical conception of occupation, they are mere truisms, or whether they involve a further deepening of the epistemological foundations of physical theory, or—as a last resort—whether their significance is associated with objective characteristics of the external world.

Symbols which, like J_r, have just two eigenvalues can be represented by twofold matrices. If S_r is a momentum vector or momentum strain vector, it is represented by a fourfold matrix. Thus $J_r S_r$, represented as the outer product of a twofold and a fourfold matrix, is half-way between a momentum vector and an energy tensor; and the symbolic sum in (73·11) is half-way between the simple addition of momentum vectors in Newtonian and elementary quantum theory and the simple addition of energy tensors in relativity theory. The hybrid quantity $J_r S_r$ is the result of the abstraction introduced when we allot mechanical characteristics as self properties to single particles, although they are properly mutual properties. It may be noticed that, although we try to manage with one particle as carrier of $J_r S_r$, we cannot avoid duplicating the particle in conception; it has to be at the same time the carrier of S_r, and also a possible occupant (or even a non-occupant!) of S_r.

The F.D. occupation symbols J_r can also be described as 'existence symbols', the eigenvalue 1 indicating that a particle with the characteristics S_r exists, and the eigenvalue 0 indicating that no such particle exists. Then q is an expression determining which of the particles that might exist have actual existence in the system contemplated. Existence symbols provide a formal distinction between

(*a*) entities, whose existence or non-existence is an unanalysable concept, (*b*) relations, whose existence is contingent on the existence of two entities (relata).

The existence symbol of a relation is the product $J_r J_s$ of the simple existence symbols of the two entities which it relates. This 'double-existence symbol' has the eigenvalue 1 if both entities exist, and 0 if either or both are absent. Being the outer product of two twofold matrices, it is a fourfold matrix, or equivalently an E-number. Allowing for repetition, its four eigenvalues are 1, 0, 0, 0. We may appropriately denote it by S^j, because it is an idempotent E-number with quarterspur $\frac{1}{4}$, just like the strain vector of a pure particle.

When matter is treated as an assemblage of protons and electrons, these are taken to be entities with simple existence symbols. An intracule is then the carrier of relations, and its existence is contingent on the existence of both a proton and an electron. Thus if the characteristic considered is the strain vector S of the intracules, we have

$$S^0 = \Sigma S^j_r S_r. \tag{73·5}$$

The product $S^j_r S_r$, being an outer product, is a strain tensor. This shows how a strain tensor can be generated by summation of strain vectors, when attention is paid to occupation symbols. It is in this way that a molar object characterized by an energy tensor is analysed into particles characterized by momentum vectors.

It follows from the symmetry of (73·5) that we might alternatively treat the S_r as occupation factors of the S^j_r.

Returning to the simple existence symbols, the algebra of twofold matrices is the ζ-algebra [**12·1** §45 or **F** §60 as in **9·1**], so that J_r may be represented as a vector in a ζ-frame. If we set

$$J_r = \tfrac{1}{2}(1 + i\zeta_r), \tag{73·6}$$

ζ_r is a square root of -1, and its most general form is $l\zeta_1 + m\zeta_2 + n\zeta_3$, where $l^2 + m^2 + n^2 = 1$. We shall later make use of the operator

$$\begin{aligned} Y_r &= J_r - \tfrac{1}{2} = \tfrac{1}{2}i\zeta_r \\ &= \tfrac{1}{2}i(l\zeta_1 + m\zeta_2 + n\zeta_3). \end{aligned} \tag{73·7}$$

By (45·1) a ζ-space is a 'spin space'; that is to say, vectors in it are homomorphic with the rotations in an ordinary 3-space. Thus Y_r is homomorphic with a spin of magnitude $\frac{1}{2}$.

There is little occasion to use the ζ-representation of J_r or Y_r, because no other quantity that concerns us is representable in the same ζ-frame. Any other occupation symbol J_s commutes with J_r, and must be represented in a distinct ζ-frame—distinct in the same way that the E- and F-frames are distinct. It is necessary to refer to it because certain highly important developments of current theory have (in effect) been expressed in this representation. Their purpose is to factorize J_r into two ζ-numbers which can suitably be associated with the two factors ψ, $\psi^\dagger$ of S_r, and then replace the symbolic addition of strain vectors by a symbolic addition of wave vectors or wave functions. This leads to a total wave function of the form $\psi^0 = \Sigma a_r \psi_r$, the a_r being symbolic coefficients. Setting $\phi_r = a_r \psi_r$, ψ^0 is the sum $\Sigma\phi_r$ of wave functions of a new type which express the state of partial occupation as well as the mechanical characteristics. These are the F.D. wave functions. By following this mode of derivation, we obtain the Jordan-Wigner commutation rules. The full derivation is given in **RTPE** § 16·2. The introduction of an equivalent of symbolic occupation was an important advance; but, so far as I can see, it is a needlessly complicated formulation of properties which are more easily handled in the form (73·11).

74. Multiple occupation symbols

It is usual to recognize another class of particles, namely, Einstein-Bose particles, the distinction being that any number (up to a high limit) of E.B. particles can occupy the same state, whereas only one F.D. particle can occupy a state. This distinction is artificial, because states must be defined by classificatory characteristics, and the number of recognized states will depend on the minuteness of the classification. For example, we may have a set of N states ψ_{rt} ($r = 1, 2, \ldots, m$; $t = 1, 2, \ldots, n$; $N = mn$), where r refers to certain major characteristics and t to minor characteristics; assuming that these are states capable of occupation by one particle only, a classification which ignores minor characteristics will give a set of m states ψ_r each capable of occupation by any number of particles up to n. E.B. particles are the result of the evanescence or neglect of minor characteristics.

A second distinction is that, whereas the total S-characteristic of a system of F.D. particles is defined as $S_f^0 = \Sigma J_r S_r$, the total characteristic of a system of E.B. particles is defined as

$$S_e^0 = \Sigma Y_r S_r, \tag{74·1}$$

where $Y_r = J_r - \frac{1}{2}$ as in (73·7). Logically the second distinction has no connexion with the first; and in general theory we can scarcely avoid referring to 'E.B. characteristics of F.D. particles' and vice versa. However, they generally become associated in practical applications.

If S_r is a major characteristic, we have

$$S_e^0 = \Sigma_{r,t} Y_{rt} S_r = \Sigma_r K_r S_r, \tag{74·21}$$

where
$$K_r = \Sigma_t Y_{rt} = \Sigma_t J_{rt} - \tfrac{1}{2}n. \tag{74·22}$$

We call K_r a *multiple occupation symbol*. Since the Y_{rt} commute, and their eigenvalues are $\frac{1}{2}$, $-\frac{1}{2}$, the eigenvalues of K_r are $\frac{1}{2}n$, $\frac{1}{2}n-1$, ..., $-\frac{1}{2}n$. Thus K_r satisfies the characteristic equation

$$(K - \tfrac{1}{2}n)(K - \tfrac{1}{2}n + 1)\ldots(K + \tfrac{1}{2}n) = 0. \tag{74·31}$$

Taking n even, this gives
$$K \prod_{s=-\frac{1}{2}n}^{\frac{1}{2}n} \left(1 + \frac{K}{s}\right) = 0. \tag{74·32}$$

As $n \to \infty$, this tends to
$$\sin \pi K = 0. \qquad (74·4)\ [\textbf{FA}(3)]$$

Similarly, if n is odd, the characteristic equation is $\cos \pi K = 0$.

We shall confine attention to one multiple state, and drop the suffix r. It will be seen that the secondary difference between F.D. and E.B. particles is that S_f^0 is measured from an initial state of zero occupation and S_e^0 from half occupation.

Denoting the particle density by

$$s = s_0(1+\zeta),$$

as in [F] § 3, with s_0 corresponding to half occupation, we have

$$\zeta = K/\tfrac{1}{2}n. \qquad (74\cdot51)$$

As in § 3, the fluctuations of particle density are equivalent to fluctuations of a linear scale constant $1+\epsilon$ determined by $1+\zeta=(1+\epsilon)^3$. If the fluctuation is small

$$\epsilon = \tfrac{1}{3}\zeta = 2K/3n. \qquad (74\cdot52)$$

Here ζ, ϵ are symbolic quantities which only become numerical when K reduces to an eigenvalue. It is only by substituting the expectation value of ϵ that we can treat the scale as continuously distributed over numerical ranges $d\epsilon$. The relation $\epsilon = \frac{1}{2}\zeta$ in (3·43) [see footnote, **3·1** § 3] refers to expectation values, and differs from (74·52) by an averaging factor introduced in relating the mean values of non-linearly related variates.

The whole investigation has assumed that the characteristics S_r combine linearly. Just as in the treatment of numerical occupation factors, it can be adapted to non-linear combination by inserting a multiplicity factor, provided that the fluctuations considered are small. For particles of multiplicity k, occupying (as here) a single state S_r, the energy varies as $j^{-1/k}$ [F § 16, **3·2** above]; so that an added particle has $-1/k$ of the efficacy of the particles already present. The sign is allowed for by the inversion of energy, and can here be disregarded. Thus the non-linearity is allowed for by giving the $\frac{1}{2}n$ particles present in the initial state a weight k, or equivalently replacing them by $\frac{1}{2}nk$ linearly combining particles. The only change needed is that (74·51) and (74·52) become

$$\zeta = 2K/nk, \quad \epsilon = 2K/3nk. \qquad (74\cdot53)$$

In particular, if the particles are standard particles,

$$\epsilon = K/N, \quad N = \tfrac{3}{2}.136n. \qquad (74\cdot6)$$

To solve the operational equation (74·4), we introduce a symbolic quantity θ defined by

$$K = -i\partial/\partial\theta. \qquad (74\cdot7) \text{ [FA(6)]}$$

Then for *any* function $f(\theta)$

$$e^{\pi\partial/\partial\theta}f(\theta) - e^{-\pi\partial/\partial\theta}f(\theta) = 2i\sin\pi K.f(\theta) = 0,$$

so that

$$f(\theta+\pi) - f(\theta-\pi) = 0. \qquad (74\cdot8) \text{ [FA(7)]}$$

Thus θ is representable by a geometrical angle, which has the property that no distinction exists between $\theta+\pi$ and $\theta-\pi$. It should be noticed that θ is a symbol, not a number, for no number has the property (74·8) for all functions f.

In the usual terminology (74·7) states that $K/\hbar$ and θ are a conjugate momentum and coordinate. This is the first time that conjugation has appeared spontaneously in the development of our theory. Previously the formula $p = -i\hbar\,\partial/\partial x$ has either been borrowed from current theory, or more non-committally has been adopted in order to make our nomenclature conform to current usage but without accepting responsibility for that usage. We shall therefore proceed as though we had no previous knowledge of this relation.

Since $K/\hbar$ is proportional to ϵ we describe it as the *scale momentum*. We notice that the scale for this purpose is the differential scale ϵ, not $1+\epsilon$. In accordance with the terminology already employed, the conjugate coordinate θ is the *phase coordinate*. If K reduces to an eigenvalue κ, the operand must by (74·7) contain the periodic factor $e^{i\kappa\theta}$.

The factor $e^{i\kappa\theta}$ is now the occupation operand q_κ. It is the eigensymbol of the multiple occupation factor K satisfying the characteristic equation (74·4), just as q_α was the common eigensymbol of the individual occupation factors J_r satisfying (73·2). The condition that K, or equivalently ΣJ_t, reduces to an eigenvalue is much less stringent than the condition that the J_t individually reduce to eigenvalues.

When $\kappa = 1$, S_e^0 corresponds to the addition of one particle to the initial half-occupied state. This is a 'built-up' particle composed of surpluses drawn from the partially occupied F.D. states; but conceptually we endow it with individuality, distinguishing it as an E.B. particle. Individuality is a matter only of conception; what is physically important is that E.B., like F.D. particles, have atomic nature. The number K of particles present is either integral or symbolic, never fractional, though the expectation number may be fractional.

The states capable of occupation by a large number of particles have previously appeared as pseudo-discrete states in scale-free theory. They were then divided into an arbitrary number of particles, the unit of occupation being defined by an arbitrary normalization density. Scale is introduced by including a periodic factor $e^{i\kappa\theta}$ in the wave function. Since $\theta - \pi$ and $\theta + \pi$ are identified, the distribution is represented as having a thickness 2π in the phase dimension as in § 24. The representation is in flat space since the scale fluctuation is taken into account directly; but the continuous variation of scale treated earlier is replaced by a distribution over discrete eigenscales. The eigenscale defines the number of particles; so that when we select a particular eigenstate for consideration, the selection automatically eliminates the extraordinary fluctuation which would be present in any unselected distribution.

Although not usually stated explicitly, it is implied in the definition of an eigenstate that the scale is an eigenscale; for a symbolic scale implies that the characteristics of a system, although numerically comparable with one another, are not numerically comparable with the extraneous standard.

75. Wave functions

If the occupation operand is non-singular, we can set

$$S^0 q = q S'^0, \quad S'^0 = q^{-1} S^0 q. \tag{75·1}$$

The main importance of this transformation (applied to all symbols which contain occupation factors) is that when q is an eigensymbol, S'^0 is the eigenvalue of S^0. But even when q is not an eigensymbol, so that S'^0 is symbolic, we must still regard it as the *special value* of S^0 for the state of occupation q.

To make this clear we must refer to the principle by which the structure of a world, whose intrinsic nature is unknowable, is rendered capable of mathematical formulation. The rule that physical quantities are represented by operators holds universally; the fact that they are sometimes represented by numbers presents no difficulty if the numbers are interpreted as a particular form of operator, namely, 'multipliers'. The operands of the operators are themselves operators. By the use of closed groups and algebras it is possible to formulate all knowledge derived from physical measurement without specifying any ultimate operand. Thus physics is presented as purely structural knowledge involving no conjecture as to what it is that has the structure described.

When the notation is such that the operators act from left to right, the ultimate operand on the right is left blank. The last terms that concern us may be written as qU, where U represents a skeleton framework, and q adapts it to a particular problem by stating how the framework is occupied in that problem. Behind these we may have a chain of operations of the type S^0. The purpose of the transformation (75·1) is to pass q backwards to the beginning of the expression; so that it can be left aside as an operation to be performed after all other calculations have been completed. Actually it is never performed. The symbolic treatment is used to avoid averaging prematurely, not to dispense with it altogether; and for the final comparison with observation, symbols are replaced by expectation values. After the final averaging the expression has become $q(\text{number})U$; so that q can pass back again over the number; and we have a number applied to the original operand qU.

Accordingly, instead of representing the state of occupation by an operand q, we represent it by a transformation $q^{-1}(\ldots)q$. This has the advantage that such symbols as have eigenvalues in the state of occupation are directly transformed into their eigenvalues. This method is applicable to the K and Y occupation factors, but not to the J factors, because the eigenvalues of the J_r correspond to singular occupation operands q_α.

We now apply the occupation transformation to $S^0 = KS$. If the state is pure, the strain vector S is the outer product of conjugate complex wave vectors ψ, $\psi^\dagger$. It is understood that symbols describing the characteristics of the state are independent of (commute with) those describing the occupation of the state; so that $q^{-1}Kq$ can be placed between the factors. We then have

$$S'^0 = (\psi^\dagger q^{-1}) K(q\psi) = (\psi^\dagger)' K\psi'.$$

The attachment of the occupation operand transforms the wave vectors ψ, $\psi^\dagger$ into *wave functions* ψ', $(\psi^\dagger)'$. They are functions because (74·8) shows that, in virtue of the characteristic equation of K, anything on which it operates is *ipso facto* a periodic function (numerical or symbolic) of its conjugate coordinate θ.

When K reduces to an eigenvalue κ, we have

$$\psi' = e^{i\kappa\theta}\psi, \quad \psi'^\dagger = \psi^\dagger e^{-i\kappa\theta}, \tag{75·2}$$

the wave functions, like the wave vectors, being complex conjugates. We may also give (74·7) the alternative forms

$$K = -i\frac{\partial}{\partial\theta} = i\frac{\delta}{\delta\theta}, \tag{75·3}$$

where $\delta/\delta\theta$ denotes the differential operator operating from right to left.

The occupation operand is contained in the wave functions; but the occupation factor K (which is now the scale) remains separate as a momentum operator to be applied to the functions. When a number of E.B. states are considered, the occupation factors are sometimes absorbed into the wave functions in order that they may be added to form a total wave function. The wave functions so formed are the 'E.B. wave functions'; they obey certain well-known commutation rules. We have already stated our view that the corresponding procedure for F.D. particles is a needless complication, and this applies also to the E.B. case.

The result (75·2) brings us to the ordinary starting point of wave mechanics—the wave function of a pseudo-discrete state. The argument of the periodic factor in the wave function is the phase coordinate—not, as is commonly assumed, the time coordinate or for a system in motion the proper time. The scale momentum turns out to be the analogue of the classical energy, and therefore *in terms of analogy* the phase coordinate may be described as the 'time'; but it is dangerous to confuse analogy and fact. In the description of steady states we have no concern with time at all; but in perturbation theory the occupation of the states is changing, and the occupation operand is a genuine function of the time. Thus a genuine time factor may be combined with the pseudo-time factors characteristic of the states.

76. The wave representation of phase

We consider a pure state specified by a strain vector $S = \psi\psi^\dagger = \Sigma E_\mu s_\mu$, or equivalently by the space vector $P = SE_{45}$. By [**F** § 68, in **9·2**] these have the antitetradic form

$$\left.\begin{aligned} S &= -\tfrac{1}{4}i(E_{01} + E_{23} + E_{45} + E_{16}), \\ P &= \tfrac{1}{4}(E_{01} + E_{23} + E_{45} + E_{16}). \end{aligned}\right\} \qquad (76·1)\ [\mathbf{FA}(11)]$$

We have found that the wave function of one E.B. particle in this state is $\psi' = e^{i\theta}\psi$, and that the phase coordinate θ is representable as a geometrical angle. Let us now examine the consequences of assuming that θ is representable as *an angle in the E-frame*—the frame in which S is represented. In vector notation an angle has a symbolic coefficient which defines its plane; and the assumption is that this coefficient is an E-number.

A representation of the strain vector and occupation operand in the same frame raises the question of commutation. The definition of occupation implies that there is no interference between the occupation variates and the state variates; and we have already made use of this condition, which requires that the two sets of variates commute. The symbols which commute with (76·1) are E_{01}, E_{23}, E_{45}, E_{16}; and these give four possible planes of θ which satisfy the condition of commutation with the characteristics of the state.

Alternatively, we can proceed as follows. Considered as a vector, the phase angle is $\Theta = i_\theta \theta$, where i_θ is a symbolic square root of -1 representing the plane of θ. The wave function $e^{i\theta}\psi$ is representable as $e^{\Theta}\psi$, if either (1) i_θ commutes with all the E-symbols so that, relatively to the E-frame, it is an algebraic square root of -1, or (2) ψ is an eigensymbol of i_θ so that $e^{i_\theta \theta}\psi$ reduces to $e^{i\theta}\psi$. By [F § 70] a factor of S is an eigensymbol of every pentadic part of S; so that ψ is an eigensymbol of E_{01}, E_{23}, E_{45}, E_{16}, and these are the possible planes i_θ. The alternative (1) is included in $i_\theta = E_{16}$.

In each plane there are two representations of θ, because we can take either $\Theta = E_{01}\theta$ or $\Theta = E_{10}\theta$, etc. These are distinct representations, because in one case θ is in the same direction as the component $\frac{1}{4}E_{01}$ of the vector P which specifies the state, and in the other case in the opposite direction. It is true that the alternative Θ's have opposite eigenvalues $E_{16}\theta$ and $-E_{16}\theta$; but until a representation of the occupation operand in the E-frame has been chosen, there is no reason for identifying the i which occurs in it with E_{16} rather than with $-E_{16}$.

The most general representation of θ in the E-frame is a combination of angles θ_1, θ_2, θ_3, θ_4 in the four antiperpendicular planes, namely

$$\Theta = \pm E_{01}\theta_1 \pm E_{23}\theta_2 \pm E_{45}\theta_3 \pm E_{16}\theta_4. \qquad (76·2)$$

Then $e^{\Theta}\psi = e^{i(\pm\theta_1 \pm\theta_2 \pm\theta_3 \pm\theta_4)}\psi = e^{i\theta}\psi$, where

$$\theta = \pm\theta_1 \pm\theta_2 \pm\theta_3 \pm\theta_4. \qquad (76·3)\ [\textbf{FA}\,(12)]$$

We call θ_1, θ_2, θ_3, θ_4 *phase sub-coordinates*. The positive directions of the sub-coordinate axes are defined to be the directions of the components of the state characteristic P. Adopting the + signs in (76·3) a *wave front*, or locus of constant phase, is a three-dimensional plane

$$\theta_1 + \theta_2 + \theta_3 + \theta_4 = \text{constant},$$

in the sub-coordinate system. This gives the usual picture of the wave function $e^{i\theta}$ as a set of plane waves—except that it is in non-Euclidean (antiperpendicular) space.

By varying the combination of signs in (76·3) we obtain 16 distinct wave systems. A particular combination of signs will be called a *reflexion*, and the set of 16 reflexions will be called a *grid*. Numbering the reflexions according to any plan, we can regard the serial number of the reflexion as an additional characteristic—the 'grid characteristic'—introduced into the state by its occupant.

If a relativity rotation is applied, the planes of the phase angle rotate along with the components of P, so that it is possible to identify corresponding reflexions in different states P_1, P_2. We shall therefore sometimes change our point of view and regard the grid characteristic as a characteristic of the state—a 'minor characteristic', according to the terminology of § 74—instead of as a characteristic of the occupation. The invariance of the grid characteristic for rotation enables us to apply it to pseudo-discrete states, and indeed all inexact states likely to be considered in practice, as well as to exact pure states. It is, however, desirable not to extend the rotation beyond 45°, in order to avoid confusion through the sub-coordinate axes being transformed into one another.

The phase angles are not angles of rotation; for the wave function is $\psi' = e^{i\theta}\psi$, whereas rotation through an angle θ gives $\psi' = e^{\frac{1}{2}i\theta}\psi$. Formally a rotation is obtained by setting $K = \frac{1}{2}$, and might be described as the wave function of 'half an E.B. particle'. There is, of course, no such eigenvalue of K; but we can regard the E.B. particle as a formation which is half rotatable and half immobilized. This behaviour is evidently associated with the change of the occupation factors from J to $J - \frac{1}{2}$.

We treat in the same way a state specified by a strain tensor

$$Z_0{}^0 = \psi\psi^\dagger\chi\chi^\dagger,$$

or by the associated space tensor

$$T_{00} = \psi\chi\psi^\dagger\chi^\dagger.$$

Setting $\Psi_{\alpha\beta} = \psi_\alpha\chi_\beta$, $\Psi^\dagger{}_{\alpha\beta} = \psi^\dagger{}_\alpha\chi^\dagger{}_\beta$, the space vectors Ψ, $\Psi^\dagger$ take the place of the simple wave vectors ψ, $\psi^\dagger$. Attaching the occupation operand, we have now double wave functions

$$\Psi' = e^{i\theta}\Psi, \quad \Psi'^\dagger = \Psi^\dagger e^{-i\theta}. \tag{76·4}$$

The vectors Ψ, $\Psi^\dagger$, being pure, have antitetradic form; and their outer product T has the form

$$T = \tfrac{1}{16}(E_{\mu\nu} + E_{\sigma\tau} + E_{\lambda\rho} + E_{16})\,(F^\dagger_{\mu\nu} + F^\dagger_{\sigma\tau} + F^\dagger_{\lambda\rho} + F^\dagger_{16}). \tag{76·5}$$

All the symbols in (76·5) commute, and the terms $E_{\mu\nu}F^\dagger_{\mu\nu}$, etc., form a set of 16 EF-symbols which commute with T. As in the previous discussion these are the possible symbolic coefficients of θ if it is representable as an angle in the EF-frame. There are now 16 phase sub-coordinates, and the grid consists of 2^{16} reflexions. It may be noticed that the double frame is rather simpler to treat than the single frame; that is because the corresponding system is less abstract.

Continuing the same treatment, we pass on to a state specified by a space tensor or strain tensor of the fourth rank, or equivalently by four wave vectors ψ, χ, ϕ, ω with their complex conjugates. This involves a quadruple frame $EFGH$, and quadruple wave functions

$$\Psi'_{\alpha\beta\gamma\delta} = e^{i\theta}\,\psi_\alpha \chi_\beta \phi_\gamma \omega_\delta.$$

The fourth rank tensor U is the outer product of $\psi\psi^\dagger$, $\chi\chi^\dagger$, $\phi\phi^\dagger$, $\omega\omega^\dagger$ (or we may substitute the associated tensor which is the outer product of $\psi_\alpha\chi_\beta$, $\phi_\alpha\omega_\beta$, $\psi^\dagger{}_\alpha\chi^\dagger{}_\beta$, $\phi^\dagger{}_\alpha\omega^\dagger{}_\beta$). This tensor is the product of four anti-tetrads in different frames E, F, G, H, and consists of 256 commuting $EFGH$-symbols. As before these are the possible symbolic coefficients of θ. There are now 256 sub-coordinates and the grid consists of 2^{256} reflexions.

The results for single and double frames are lemmas having no immediate application to physical problems. For the vectors and tensors P, S, T, Σ specifying the state are measured by comparison with an extraneous standard. If therefore we want to make the mathematical treatment conform to the principles of physical measurement, it is *wrong* to represent the scale and phase as contained in the E-frame or EF-frame of the system. The representation of the scale is inseparable from the representation of the phase, since the scale $K = -i\partial/\partial\theta$ is furnished by the gradient in the θ direction and has the same plane as θ. If the scale is represented in the EF-frame it is bound to the system in such a way that no rotation can be applied to the system without being applied to the scale; this is shown explicitly in the foregoing analysis, since, if the scale does not rotate with the system, K ceases to be numerical and the occupant of the state is no longer one E.B. particle. Clearly this is not the scale referred to in physical theory; the various states of the system have been defined by rotating it relatively to its environment, and the scale is determined by comparison with a standard contained in the environment.

This objection does not apply to the representation of the scale and phase with symbolic coefficient $E_{16}F_{16}$; because $E_{16}F_{16}$ is simply the algebraic number -1 common to all double frames, and forms, as it were, a no man's land not exclusively annexed to any frame. This algebraic coefficient, written alternatively as -1 or $E_{16}F_{16}$, reconciles the two aspects of the scale as a characteristic primarily outside the EF-frame, and yet comparable with quantities in the EF-frame and imported into the EF-frame for the purpose of such comparisons. In our earlier work this has been provided for by embodying the scale in a comparison particle with one (algebraic) degree of freedom.

It is not until we come to the quadruple $EFGH$-frame that we are able to make physical application of the grid. Since the data of physics are measures, the primary 'particles' are the carriers of measures; it will, however, be clearer to call these carriers 'quadriparticles' since they must be conceived as having fourfold constitution. This is expressed formally by saying that they have quadruple existence symbols $J_r J_s J_t J_u$, since the existence of the measure is contingent on the existence of four elementary entities with independent existence symbols. A measure $\boldsymbol{\alpha}$ is the ratio of an observable $\boldsymbol{\beta}_1$, furnished by two elements of the quadriparticle, to a comparable observable $\boldsymbol{\beta}_2$ furnished by the other two elements. It is a pure number, or array of pure numbers, or in operational theory a dimensionless operator whose eigenvalues and expectation values are pure numbers. Being dimensionless it is independent of scale as ordinarily defined; but, just as we have a variety of representations of the unit of mass and its conjugate coordinate, so we have a variety of representations of the unit number 1 and its conjugate coordinate. It is to be remembered that, in symbolic calculus, 1 stands for anything that behaves like the algebraic number 1 in respect to the operations of the calculus. The measure $\boldsymbol{\alpha}$ will have as many representations as there are representations of the unit 1 in which it is expressed. By treating the unit of number as a scale with a conjugate phase coordinate, the foregoing theory of representation is extended to dimensionless quantities such as measures.

The observable $\boldsymbol{\beta}$, being provided by a twofold system (bi-particle), is a tensor of the second rank; and $\boldsymbol{\alpha}$ is a fourth rank tensor $\boldsymbol{\beta}_1 \times \boldsymbol{\beta}_2^{-1}$. Our investigation of the quadruple frame therefore applies to a state specified by measures if, for example, ψ, χ represent an energy tensor $\boldsymbol{\beta}_1$, and ϕ, ω represent the reciprocal of an energy tensor which is also a tensor of the second rank. The measure $\boldsymbol{\alpha}$ describes a purely internal relation in the quadriparticle, and has no reference to any extraneous standard. The unit of numerical measure is therefore representable in the *EFGH*-frame; and the state specified by a measure possesses the grid-characteristic of the representation. We have seen that in this case the grid-characteristic has 2^{256} possible values.

77. The cosmical number

In the last section we referred to a change in the point of view, by which the grid characteristic is associated with the state instead of with the occupation or scale. The scale is then algebraic, and can be provided by an extraneous standard. This is an important step in unifying physics, because an extraneous standard is used throughout the molar part of physics.

We have distinguished in § 25 [cf. **F** § 35, full text] between casual and systematic measurement. The totality of our observational knowledge is a jumble of casual measurements. *Physical knowledge* is transformed into *physical science* by systematizing measurement. The scientific description of the physical universe is based, not on casual measures in which an observable is compared with any other observable of the same kind, but on standard measures in which the observable is compared with a standard; and the physical quantities used in the description are not pure numbers, but have a dimension index in terms of the standard. We have seen that only one standard is needed, and that the standard must be reproducible at any place or time. It is useful to distinguish between the standard, which is outside the frame of the object-system considered, and the particular reproduction of it used in measuring the object-system, which is formally brought into the frame by identifying its symbolic coefficient with the algebraic symbol $E_{16}F_{16}$ of the frame. Standard measures originated in the study of molar systems; but microscopic systems come into line provided that their scale is algebraic, and this is secured by relieving the scale of its grid characteristics as noted above.

It should be remarked that all experimental measures are molar: but for the description of atomic systems we employ ideal measurables which are only remotely connected with actual measurement. They arise by an analogy in which the analytical elements of actual objects are conceived to be measured by the analytical elements of actual measuring apparatus. It is understood that in idealizing a measure its essential characteristics are preserved; in particular, measures, observables and entities are related to one another in the same way in the analogy as in actuality. The prototype is a casual measure; for it would be an irrelevant restriction to insist that they must have the specialized form of standard measures when there is no question of actually comparing them with the standard. On the other hand, it is a great step towards unification if we can without loss of generality employ standard measures throughout microscopic as well as molar physics; and we have seen that this can be done by transferring the grid characteristic to the state.

We have then to consider the occupation of states with the usual major classificatory characteristics and also a minor grid characteristic. The latter does not affect the multiplicity factor since it is discrete. When it is ignored 2^{256} states coalesce. The analysis in § 74 is applicable with

$$n = 2^{256}. \tag{77·1}$$

We call n the *grid number*. The result (74·6) for standard particles becomes

$$\epsilon = K/N, \quad N = \tfrac{3}{2} \times 136 \times 2^{256}. \tag{77·2}$$

We call N the *cosmical number*.

To interpret ϵ, let y be the dimension index of the energy tensor (i.e. the major classificatory characteristic of standard particles) in terms of a linear standard. If T_0 is the energy tensor corresponding to scale 1, $T_0(1+y\epsilon)$ is the energy tensor in a state of occupation corresponding to scale $1+\epsilon$. Setting $K=1$, the addition of one E.B. standard particle changes the energy tensor from T_0 to $T_0(1+y/N)$. Hence the number of particles initially present is N/y. The energy tensors here considered are simply additive, because the correction for non-linearity is contained in the multiplicity factor 136 in (77·2).

In scale-free physics we have employed natural units, which are such that $y = -6$. In these units $\hbar$ and κ [gravitational constant] vary with the scale so as to preserve homology. We have now left scale-free physics behind, and are in fact using the grid number to fix a definite scale of particle structure. We are no longer concerned with deliberate transformations of scale to adapt the same analysis to widely different densities. The changes of scale contemplated are those formally described as 'fluctuations'; but instead of treating the Gaussian fluctuation of ϵ, we treat eigenfluctuations corresponding to the addition of 1, 2, 3, ... particles to the system. It was not intended that the variation of $\hbar$ with the choice of unit should extend to fluctuations of the unit; and it would be contrary to the practice of quantum theory to define $\hbar$ as a function of the occupation. Thus $\hbar$ is a fixed constant, and the momentum vector $-i\hbar\,\partial/\partial x$ has a dimension index -1 in terms of the linear standard. The quantum momentum vector has been identified with the root vector of the energy tensor, so that the energy tensor has dimension index -2. As usual the minus sign is compensated by the inversion of energy, and we have $y=2$. Accordingly the number of particles initially present is $\frac{1}{2}N$.

By definition the standard particle is characterized solely by an energy tensor, so that the grid distinction is disregarded. It follows that when we treat a standard particle as object-system and assign it a periodic wave function, it is added to a distribution of $\frac{1}{2}N$ standard particles; in other words the environment or uranoid consists of $\frac{1}{2}N$ standard particles. The standard particle was the starting point of our investigations of transition particles, V_{10} particles, hydrocules, etc., all referred to the same uranoid. It is therefore a general result that the uranoid consists of $\frac{1}{2}N$ standard particles or equivalently N protons and electrons. Thus (77·2) evaluates the natural constant N which has appeared in so many formulae. This value has been used in anticipation in the comparison of theory and observation in § 40 [**F** § 51].

The latter part of the calculation can be checked as follows. The uranoid has two characteristic linear extensions σ and R_0. In quantum theory R_0 must be regarded as fixed, because the coordinate system is rigid—unlike relativity theory where the curvature changes with the occupation. The physical (quantum-specified) standard is determined by σ, and is subject to fluctuation. The linear scale $1+\epsilon$ is proportional to R/σ, since this represents the result of measuring a fixed extension with the standard. But R/σ varies as $\sqrt{N}$; so that, if N is changed to $N+2$ by the addition of one standard particle, the scale is changed to $(1+2/N)^{\frac{1}{2}}$. Thus the value of ϵ for $K=1$ is $1/N$, showing that the N in (77·2) is the number of protons and electrons in the uranoid.

78. Epistemological foundations

The calculation of N in the last section is confirmed by an observational test which is accurate to about 1 part in 500. Various steps in the deduction have taken us through previously untraversed territory, and we shall now seek to consolidate this advance. To understand it fully we need to probe rather deeply into the foundations of physics.

Physical science might be defined as 'the systematization of knowledge derived by measurement'. The generalization that 'the data of physics are measures' is rather ambiguous. The measure number by itself would be of no use; it has to be the measure of some indicated quality of some indicated thing. Indication of the quality and the thing is needed to supply a *connectivity* of the measures. The search for a mathematical formulation of this connectivity is one task of physical science; development of the knowledge conveyed by the measures when so connected is another task.

In my early writings I tended to over-emphasize the numerical character ('pointer-readings') of the subject-matter of physics, implying perhaps that the setting of the numbers was altogether outside its scope.

The recent application of the methods of group theory and symbolic algebra to theoretical physics has enabled us to lay down more definitely the boundary between the physical and the extra-physical. The setting of the numbers, in so far as it contains their connectivity, is within physics; and the symbolic methods enable us to isolate the connectivity from the extra-physical aspects of the setting.

Since actual measurement is limited by practical difficulties there are large gaps in our knowledge, and the data of measurement are commonly supplemented by hypotheses—hypotheses as to what would be the result of the measurements we are unable to make. This may include hypotheses as to the correct interpretation of measurements, if the gap is caused by inability to perform the crucial measurement which would have decided between two interpretations. But it is desirable to distinguish these physical hypotheses from logical hypotheses covering gaps (recognized or inadvertent) in deductive reasoning. The term will here be confined to physical hypotheses. Our knowledge of the universe would be very meagre if all hypothesis were rigidly excluded. But there is one part of physics entirely free from hypothesis; this part deals with the properties possessed by measures simply because they *are* measures—not conditioned by the object to which measurement applies. This may be called *epistemological physics*; but it appears from the investigations in this book that epistemological physics is co-extensive with *fundamental physics*; that is to say, it comprises all the general 'laws of nature' and the constants contained in, or deducible from, them.

We have to indicate the quality and the thing associated with the measure-number, not as in experimental physics by just pointing at the thing or at the sense organ which responds to the quality, but by a communicable description. Communicable description is limited to *structure*; and the treatment of structure, abstracted from whatever it is that has the structure, was first made possible by the methods of symbolic algebra. In our theory the numbers have been associated with two sets of symbols, concerned respectively with the state and the occupation. The state symbols correspond to the quality, and the occupation symbols to the thing. The unification of physics has greatly reduced the variety of qualities to be provided for; and the table of symbolic coefficients in [**F** § 69, **9·2**] illustrates the way in which different qualities are brought together in a symbolic frame. The initial separation of state symbols and occupation symbols (qualities and things) becomes less definite as we proceed, as, for example, when we transfer the grid characteristic from the occupation to the state; and the whole symbolism becomes welded into a representation of what we have called the 'connectivity' of the measure-numbers.

Recognition that we are dealing only with structure, and that all our concepts must be structural concepts, removes what has seemed to be an illogicality in our distinction between observables and unobservables, measurables and unmeasurables. The distinction has been of immense importance in relativity theory and in quantum theory; but, as we remarked in § 1, modern physics is not over-scrupulous in postulating measurements of a highly impracticable character. In particular, microscopic analysis introduces so-called measures which are the elements of actual measures, and presumably supposed to be made with the elements of actual apparatus. Reference is made to actual observation or measurement for the purpose of ascertaining the structural characteristics of an observable or measurable; thereafter the term stands for a defined structural concept which may be used in analysis without implying that measurement is possible.

Our present starting point is a 'simple entity' carrying an extended vector or E-number; there is nothing to show what quality the vector represents. (In § 79 we shall go still farther back to show how this concept arises.) It seems an easy step to pass from this to an entity characterized by an extended momentum vector, and so reach the starting point of Chapter V [**F** VI]. Actually it is a long step, because a momentum vector is by definition a measurable, and a measurable is a structural concept involving four simple entities and a corresponding quadruple frame. Thus when we specialize the extended vector as a momentum vector we introduce a quadruple frame. So far as state characteristics are concerned, this is a nominal change, since $\Sigma E_\mu p_\mu$ is turned into a quadruple-frame vector by multiplying it by $1 = E_{16} F_{16} G_{16} H_{16}$. But the occupation characteristics are profoundly affected.

We have previously had occasion [e.g. **F** § 22] to study the conceptual transfer by which mutual characteristics of two or four entities are apportioned as self characteristics to the entities separately. But we are not now assuming that $\Sigma E_\mu p_\mu$ has been found in this particular way. It may have been deduced from measurement of a much less elementary system; it may be a stabilized characteristic; it may be simply a guess. It is just a momentum vector introduced for some reason in our analysis; and as such it has the structure of a measure—a structure which can only be represented in a quadruple symbolic frame.

If the measurement of $\Sigma E_\mu p_\mu$ had actually been made, we should have made some note of the circumstances that distinguish it from other measurements on our programme. We may therefore regard the measure as having a reference number r. The fact that the measurement has been made is incontrovertible evidence that the measurable 'exists', and accordingly its existence symbol or occupation factor has the value 1. But when the term is extended to quantities which have the structure of measures although the measurement may not actually have been made, this evidence is lacking and we can only assign a symbolic probability J_r with eigenvalues 1 and 0. As already explained, the concept of probability as applied to a single event is non-numerical; and, although the observational interest is confined to the numerical value which results from it by averaging, the strict symbolic representation must be used in the analysis in order to avoid premature averaging. To put it another way: the structural concept of a measure must be associated with the structural concept of existence, and the latter is representable by a symbol with just two eigenvalues corresponding respectively to existence and non-existence.

Initially we may suppose that there is no upper limit to the reference number r. A number s unwanted in actual application can be sterilized by adjusting the occupation operand so as to give J_s the eigenvalue 0.

In physical application it is always understood that the measure of the momentum vector is standard measure. We have not yet introduced this condition, which is evidently of great importance in establishing a connectivity of measures. To introduce standard measure we proceed to form an E.B. particle and substitute it as the carrier of $\Sigma E_\mu p_\mu$. The difference is

(1) The state characteristics and occupation operand are merged in a wave function containing $e^{i\theta}$;

(2) The p_μ are standard measures;

(3) The momentum vector has a grid characteristic which we may regard as a new reference number r' with an upper limit $n = 2^{256}$.

Unlike the previous reference number, r' has a representation in the symbolic frame and is contained in the structure of the measurable.

E.B. states may be occupied by more than one particle, but the number is indicated by the periodic factor. By incorporating $e^{i\theta}$ in the state characteristics we limit the occupation to one particle. This particle is definitely present. But we can now introduce *an occupation factor of the second order* $J_{r'}$.

Accordingly we recommence with an entity, carrying a momentum vector in standard measure and represented by wave functions, distinguishable by a grid characteristic r' instead of an arbitrary reference number r without structural interpretation. If the momentum vector has been found by actual measurement its occupation factor is 1; but if it is simply an element in our analysis of phenomena, identified as a momentum vector by its structure, it must be given a symbolic occupation factor $J_{r'}$ with eigenvalues 0 and 1. Treating r' as a minor characteristic ignored in the classification of states, we proceed to form an E.B. particle of the second order—an E.-B.^{2} particle. This will have a grid characteristic r'', also running from 1 to n. Structurally the E.-B.^{2} particle is just like an E.B. particle. We can go on to E.-B.^{3}, E.-B.^{4}, ..., particles without making any difference. In other words, the operation of E.-B.-ing is idempotent. The operation consists in forming a new particle out of bits of probability of other particles which add up to a probability 1. The 'bits' are symbolic; so that the new particle emerges discontinuously through the symbolic probability reducing to an eigenvalue. The process is in fact a transformation from one set of eigenstates to another; but the abnormal feature—abnormal for discrete eigenstates—is that the two sets are equivalent.

In practical application the particle considered is the carrier of an energy tensor, rather than the more abstract carrier of a momentum vector which can only be formed by introducing stabilization. Then, as has been shown, the particle always appears in an environment of $\frac{1}{2}N$ other particles of the same kind.

We have not shown directly that there are $\frac{1}{2}N$ standard particles in the universe; what we have shown is that quantum theory implicitly assumes that there are, by applying to physics a mathematical scheme in which the environment is represented in this way. But this comes to the same thing; for, if the quantum specification of a hydrogen atom includes the assumption that it is $2/N$ of the whole universe, and we develop the observational properties of a system having this specification, a system recognized experimentally as having precisely these properties is necessarily $2/N$ of the universe. The complete chain of proof extends throughout this book; for it includes the investigations that identify the standard particles with the systems recognized observationally as hydrogen atoms, and it must be made clear that the particles so identified are the E.B. carriers treated in this section. The latter identification is, I think, conclusively established by the fact that an extraneous standard has been postulated throughout the previous work; but a further check will be provided when we develop more fully the wave functions introduced by condition (1) above.

79. The primitive measurement[a]

We come now to the very beginning of the theory. In making the sequence of deduction as clear as possible, I cannot avoid repetition of some points already treated; but this will perhaps be excused because of the importance of presenting the argument in a complete and connected form.

The most primitive entity contemplated resembles a geometrical point in that it has 'no parts and no magnitude'; but instead of 'position only'—a far from primitive attribute—it has existence (or non-existence) only. It has no magnitude, because magnitude is an attribute of a measurable which, as we have seen, involves four entities. That it has no parts means that its existence is an unanalysable concept, not resolvable into the simultaneous existence of several parts any one of which can be conceived to exist without the others. We are not concerned with any metaphysical or absolute significance of existence—whether the entity exists in the sense in which *I* exist. The term here refers to a structural concept. The question is whether the entity exists (or, more simply, is) in the structure under discussion. This sole attribute of the primitive entity will be represented by a symbol J. It is clear that this symbol belongs to a certain category, which can be described as the 'yes-no' category. Such symbols have two eigenvalues, a yes-eigenvalue a_1 and a no-eigenvalue a_0.

The justification for this symbolism is as follows. Our mode of acquaintance, through sense organs, with a world external to the individual mind is such that our knowledge of it is necessarily structural knowledge; and our only means of describing abstract structure, when that which possesses the structure is unknown, is provided by the methods of symbolic algebra. These represent a structural pattern by the interrelations of a

[a] Although much of paragraphs 2–6 here appears in **FA** §§2, 3 and subsequent ideas in **FA** §4, the text has not been compressed because divergences of treatment soon develop.

closed group of symbolic operations. The entity and its attribute, being elements of this structure, cannot be represented otherwise than by symbols. Self properties of a symbol—those whose description contains no reference to other symbols of the structure—are contained in its characteristic equation, or equivalently in its possible eigenvalues. But strictly the field of numbers, supplying the eigenvalues and coefficients of the characteristic equation, must be counted as symbols of the structure, since they behave as algebraic numbers only with respect to the operations comprised in the structure. Thus the only genuine self property of a symbol is the category of its characteristic equation; unless the number of eigenvalues is infinite, this property is the order of the characteristic equation, or equivalently the number of eigenvalues. We have to choose a symbol with self properties that correspond to our conception of existence and non-existence as alternatives like yes and no. The only possible correspondence is that the characteristic equation shall be a quadratic offering a choice of two roots. Moreover the representation of the choice between the existence alternatives as a choice between eigenvalues agrees with the relative character of the existence concept. The existence-attribute of the entity is not identified with either alternative until we consider a structure in which the entity exists or does not exist; similarly the symbol is not identified with either eigenvalue until we consider an eigensymbol.

The most primitive measure is provided by four entities whose existence-attributes are independent. The corresponding measurable will not exist unless all four entities exist. It has therefore an existence symbol of the form $M = J_r J_s J_t J_u$. This has 16 eigenvalues, of which $a_{r1} a_{s1} a_{t1} a_{u1}$ is the yes-eigenvalue and the other 15 are no-eigenvalues. A measurable has 15 different ways of not existing. We must notice that M applies to all measurables provided by the four entities; we shall regard the whole array as a tensor measurable with the measurables corresponding to particular measurements as components.

Since the characteristic equation of J_r is $(J_r - a_{r1})(J_r - a_{r0}) = 0$ [**FA** (1)], the entity r exists in $J_r - a_{r0}$. Similarly the measurable exists in

$$M' = (J_r - a_{r0})(J_s - a_{s0})(J_t - a_{t0})(J_u - a_{u0}). \quad (79·1) \quad [\mathbf{FA}\,(2)]$$

Clearly no other measurable exists in M'; and we may therefore identify M' as the tensor measurable itself. M', like M, has 16 eigenvalues. Since the structure contemplated will contain other tensor measurables, we denote a particular tensor measurable by M'_p.

We have defined physics as the systematization of knowledge derived by measurement. This knowledge is formulated as a description of a structure which is called (in science, at any rate) the physical universe. In this structure the measurable (not the entity) is the unit. The scaffolding of entities, which we have just been using to ascertain the symbolic properties of a 'measurable', should be forgotten, and any entities subsequently introduced in describing the physical universe are to be defined in terms of measurables; these are not primitive entities because their existence is contingent on the existence of the measurables by which they are defined. Practical measurements involve four of the new entities; and the corresponding measurables reproduce the structure of the primitive measurables, only, as it were, at one remove. This cyclic return of structure into itself is a leading feature of the mathematical representation of structure by closed groups and algebras. Mathematically, there is no one point rather than another at which it can be said the cycle starts. But in physical application the structure is linked to human experience, and we use the concept of existence to formulate that linkage; this gives us a point of entry into the cycle, which forms the starting point for the purposes of physics.

Thus far we have said nothing about the result of measurement; all that we have discussed is the existence of something to which measurement can be applied, and all that we have established is that it has 15 different ways of not existing. The value of the measure will be denoted by X. The symbol X stands for an array of numbers, each number being the measure of one of the measurables comprised in the tensor measurable. A measurable M'_p has a unique measure X_q; but more than one measurable may have the same measure X_q.

We now invert our point of view and, instead of regarding the measure as something carried by the measurable, we regard the measurable as something occupying the measure. To express this we introduce an occupation symbol K_q and a corresponding occupation operand K'_q, such that the eigenvalue of K_q is the number of measurables occupying X_q. We shall call the product $W_q = K'_q X_q$ a *wave measure*. It is a symbol containing both the array of measures and the occupation. It is a simple commutative product, since the concept of occupation implies that the measure is unaffected by changes of occupation. It will be recalled that the concept of flexible occupation is a distinctive feature of the analysis in wave mechanics as opposed to that of molar relativity theory [**F** § 13].

The eigenvalues of K_q are the positive integers (including 0). This makes the theory of K and K' homomorphic with the theory of the harmonic oscillator. When this line of treatment is followed the occupants are F.D. particles. (Or more strictly, E.B. particles with F.D. characteristics. As pointed out in § 74, there are two kinds of distinction which have no logical connexion.) Its importance comes later in the theory in the study of small systems artificially isolated from the rest of the universe. We have not yet developed an analytical machinery for dividing the structure into isolable systems, and must deal with the universe as a whole. An enumeration of the measurables throughout the universe which have the particular measure X_q can scarcely be relevant to the 'knowledge derived by measurement' which we are endeavouring to systematize. We are concerned only with changes of K brought about by the phenomena we study. We therefore modify the definition of K_q so that it represents the excess of the occupation of X_q above a large, but at present unstated, number $\frac{1}{2}n$. The eigenvalues of K_q are then the positive and negative integers including 0. This property of K_q is expressed by a characteristic equation $\sin \pi K_q = 0$. We have seen in § 74 that the solution of this equation is $K = -i\partial/\partial\theta$, where θ is a symbol which satisfies

$f(\theta+\pi)=f(\theta-\pi)$ for all functions f. A symbol with this property introduces a new conception—a geometrical angle. For the first time in the present sequence of developments a geometrical concept appears, and we envisage the possibility of representing the physical universe as a *structure in space*.

Before proceeding further we have to consider *synthetic measurables*. It is the practice to make a number of measures $X_1, X_2, \ldots$, and, after performing various algebraic calculations on them, produce a quantity $Y=f(X_1, X_2, \ldots)$ which is also supposed to be the measure of something. Sometimes there is theoretical proof that Y is the measure of a measurable; and the procedure is merely a practically convenient way of obtaining a result which could ideally have been found by a single measurement. But in general there is no guarantee that Y corresponds to a measurable. For example, the determination of the mass m_e of an electron, however much it is idealized, involves at least two measurements; and the measure $9{\cdot}109\times10^{-28}$ g. does not correspond to anything having the structure of a measurable. In the present investigation we must distinguish the pseudo-measurables corresponding to synthetic measures from genuine measurables. In general they do not concern us; but there is one type of synthetic measurable which plays a vital part in the foundations of the theory, namely *additive measurables*. These are classed as momenta (as opposed to coordinates). Conceptually momenta are simply additive; that is to say, non-linearity of combination is treated by introducing extra momenta which are said to arise from 'interaction'. The synthetic measure is therefore simply ΣX_q summed for the measurables that exist in the system contemplated. More generally, we consider a synthetic measure $X=\Sigma J_{rq}X_q$ summed over all measurables, their existence or non-existence being left to be decided by an 'occupation operand' which will be included in the synthetic wave-measure. The J-symbols introduced earlier were the existence symbols of simple entities, and their eigenvalues were arbitrary numbers a_1, a_0. The new J-symbols refer to measurables, and decide in the same way whether the measurable exists in, or occupies, the measure considered; but they are more specialized because measurables (of a certain class) admit of addition, whereas entities do not. For this purpose we use specialized J-symbols (F.D. occupation symbols) with eigenvalues 1 and 0, so that a non-existent measurable contributes nothing to the addition.

Additive measures are necessarily standard measures. For there would be no sense in adding casual measures, e.g. a mass reckoned in pounds to a mass reckoned in kilograms.

We have said that there is no guarantee that a synthetic measure corresponds to a measurable; and, even when the synthesis is simple addition, it cannot be assumed as self-evident that the resultant is a measurable. But, unless additivity is actually inconsistent with the definition of measurables, we can investigate the class of measurables to which it applies. Their property is formally defined as follows: Let J_{rq} be the occupation factor of a measure X_q by a measurable r, and let $J_q=\Sigma_r J_{rq}$ summed over all the measurables; then if the occupation of the measures is described by total occupation factors J_q, the corresponding occupants are measurables (not pseudo-measurables).

The importance of employing additive measurables is that the undefined initial distinction indicated by the suffix r is eliminated. Thereafter, if we wish to distinguish between different measurables which might occupy X_q, we must define the distinction, e.g. by a grid characteristic. But the grid characteristic is associated with the symbols K, θ obtained from total occupation factors. If the occupants are still measurables we can identify r as the grid characteristic, and so set up a structural cycle in which measurables with grid characteristics result from the summation of measurables with grid characteristics. Such a cycle has no beginning, and we do not have to inquire how the grid characteristic first got into it. But if summation yields pseudo-measurables, so that the grid characteristic is a property of pseudo-measurables, we can never get a start; because we can never fulfil the implied promise to define structurally the distinction between measurables to which r refers. It is, of course, essential to the theory of structural knowledge, that no distinctions can be admitted except those that have a structural equivalent. It follows that structural theory must be based on additive measurables.

The normal measurable is now to be conceived as a unit of occupation. This leads to a conception familiar in quantum theory, namely a measurable with uncertain measure; that is to say, the unit (regarded as one measurable definitely existing) may be composed of elements occupying different measures. This gives a generalized interpretation of the additivity of measurables, not limited to additions in the same measure state.

Besides the normal measurables which correspond to the eigenvalue $K=1$, there are 'measurables of weight m' corresponding to $K=m$, and composed of m units of occupation. Their physical interpretation is as follows. We have seen (§ 74) that $K/\hbar$ is the scale momentum; so that a mixture of values of K corresponds to a fluctuating scale. When the small uncertainty σ_ϵ of the continuous scale obtained by averaging is replaced by a probability distribution over eigenscales the individual fluctuations are large. We have to provide for measures which are on $\frac{1}{2}$, $\frac{1}{3}$, ... of the normal scale. Effectively these are reduced to the normal scale by giving them double, triple, etc., occupation; so that they are equivalent to a normal measure with normal occupation in the addition. The simple addition of measures assumes, of course, that they are all on the same scale; so that in the present theoretical study of additivity of measurables, we postulate normal measurables occupying measures on the normal scale $K=1$; and we may at once replace K in the occupation operand by this eigenvalue.

We have now reached the point at which the investigation in § 78 fits on. For additive measurables the occupation factor is K, and the occupation operand (the symbol for the measurable itself) is $e^{iK\theta}$; the wave-measure is $X_q e^{i\theta}$, the eigenvalue $K=1$ being inserted as explained above. But, whereas in § 78 the form of X_q was given to be that of an *EFGH*-number, here we know nothing about its form. We have the meagre information that $e^{i\theta}$ is a measurable, and that X_q is a standard measure. We know also that θ is representable as a geometrical angle.

It will be remembered that θ is a symbol, no number having the property (74·8). We can write $i\theta = i_\theta \theta'$, where i_θ is a symbolic square root of -1, and θ' is the ordinary measure of the angle. If $q = e^{i\theta}$, q is the operator for rotation through an angle $2\theta'$ in a plane symbolized by i_θ. In order that q may have structural significance this rotation must be a member of a rotation group. We can easily determine this group. Since q is a measurable it has 16 eigenvalues. The group of rotations with 16 eigenvalues is the group of rotations of an EF-frame. Thus $e^{i\theta}$, and hence θ, are EF-numbers.

We can now distinguish the normal measurables by their grid characteristics in the EF-frame, and so justify the suffixing of the J_r at the beginning of the investigation. We now turn attention to the measures. Scalar measures X_q, $X_{q'}$ are distinct if X_q, $X_{q'}$ are different numbers; but the tensor measures of tensor measurables are arrays of numbers, and it is necessary to define structurally the distinction between different permutations, e.g. the distinction between the different transposes of a double matrix [F VIII]. Just as there was an implied promise to define the distinction indicated by the suffix r of the measurables initially considered, so there is an implied promise to define the distinction indicated by the suffix q of the measures initially considered; and our problem is to find measure arrays for which this promise can be fulfilled. In the structure thus far developed the most general type of number-array structurally defined is the EF-number; and the X_q must accordingly be EF-numbers. It need not be assumed that this initial system embraces all possible measures; for measures, unlike measurables, are not safeguarded against complexity by a limitation of the number of eigenvalues. But, if later we introduce more complicated measures, we must make sure that they are structurally defined. The point is that at the beginning we cannot help running into definitionary debt; we must first discharge this debt, and afterwards pay as we go.

Accordingly, the condition of additivity leads to a closed structure representable in an EF-frame, in which the measures are EF-numbers, and the measurables are wave systems in spaces with antiperpendicular axes. The additive measure is therefore an array having the form of an extended second rank tensor. This tensor, under the name of 'extended energy-tensor', has been the basis of all our previous investigations; so that we can now pass straight on to the developments in Chapters I–VII; and the whole theory, including the calculation of the natural constants, results solely from the principles of measurement without any arbitrary assumption at any stage. The wave systems in antiperpendicular space, representing measurables, provide similarly the basis of the further investigations which begin in the next chapter. For all these applications the theory of structure terminates with the EF-frame. The initial definitionary debt has been paid; and we start square, defining systematically any further measures which we introduce in working out the observational consequences for comparison with experiment. There is just one investigation which stands apart from all the rest, namely the calculation of the cosmical number. This takes us one step farther in the theory of structure.

Since perfect additivity requires that the measures shall be referred to an exact (and therefore unobservable) scale, the foregoing structure exists only in theory; but the requirement is so nearly fulfilled by the observable scale whose fluctuation σ_ϵ is of order 10^{-39}, that we usually accept it as the actual structure in physics. Even the small fluctuation σ_ϵ is not neglected but is taken into account statistically as a Gaussian scatter; and the approximation is good enough for the calculation of quantities depending on $\sqrt{N}$. The one problem outside its range is the calculation of N itself. Without this approximation a standard measure is the ratio of two exact-scale measures, namely an object-measure and a measure of the fluctuating standard both referred to an exact (stabilized) scale. It is therefore a fourth rank tensor or $EFGH$-number.[a] The rest of the calculation of N (depending on the properties of an $EFGH$-frame) has been given already.

It should be noticed that, although the measure is now an $EFGH$-number with 256 eigenvalues, the measurable by definition can have only 16 eigenvalues. The existence of the measurable is contingent only on the existence of the object-entities and of the scale employed. The exact scale, used as intermediary, has no existence symbol since its existence is impossible anyway. In the approximation, we do not replace the existent scale by an exact scale; we simply neglect its inexactitude. The approximation reduces the complexity of the measure; but it does not affect the measurable—to which indeed the conception of approximation is wholly inapplicable.[b]

[a] Denoting the two exact-scale measures by $T_{\mu\nu}$ and $S_{\mu\nu}$, their ratio $W_{\mu\nu\sigma\tau}$ is (in the notation of ordinary tensor calculus) defined by $T_{\mu\nu} = W_{\mu\nu\sigma\tau} S^{\sigma\tau}$ [**FA** (18)].

[b] If James and John are almost indistinguishable twins, one does not say that James is approximately John; but the measures of James are approximately those of John. [A.S.E.]

BIBLIOGRAPHY

The following list is of papers and books, published mainly after 1944, in which I have found material of value in the study of **F** (*Fundamental Theory*). Some indication is added of the content of the research papers. The earlier papers of McCrea and Lemaître are mentioned as valuable preliminaries to E-number theory as used respectively by Eddington and Kilmister. The most important current developments appear to be the studies of Kilmister and Bastin, on the algebraic and logical foundations. I am grateful to Dr Kilmister for summaries (used freely here) of this work, and for an indication of forthcoming investigations[a] concerning particularly the 'statistical theory' in **F**.

E. W. Bastin and C. W. Kilmister (1952). The analysis of observations. *Proc. Roy. Soc.* A, **212**, 559.

An 'observable' is defined as a set of n real numbers, which must themselves provide a standard in terms of which they are expressed. This 'scale' requirement is shown to lead, under certain restrictions, to even Clifford algebras and so to E-number theory.

E. W. Bastin and C. W. Kilmister (1954*a*). The concept of order. I. The space-time structure. *Proc. Camb. Phil. Soc.* **50**, 278.

E. W. Bastin and C. W. Kilmister (1954*b*). Eddington's theory in terms of the concept of order. *Proc. Camb. Phil. Soc.* **50**, 439.

In (*a*), an attempt is made to determine the fundamental characteristics of physics by considering to what extent physical theories are independent of measurement. There is assumed to be a hierarchy of 'theory languages', each with its appropriate experiments, instead of a series of layers of complexity with the more complex 'closer to reality'. The essence of an intelligible 'theory-language' is that its elements should be ordered; permutable elements introduce the concept of simultaneity, so that the simplest language leads to the $(3+1)$ structure of space-time.

In (*b*), Eddington's use of the E-frame with its characteristic *pentads* [**F** VI or **9·1** §53] is contrasted unfavourably with the quaternary basis of the languages of (*a*) above, in which the $(3+1)$ structure appears naturally in mechanics and electromagnetism; Eddington's device of transforming from molar to micro-space [**F** VI or **9·2** §76] illuminates and overcomes this structural flaw. The kernel 2^{256} of Eddington's 'cosmical number' is said to owe its importance to its association with the fourth-order 'theory-language'.

E. W. Bastin and C. W. Kilmister (1955). The concept of order. II. Measurements. *Proc. Camb. Phil. Soc.* **51**, 454.

The analysis of (1954*a*) is extended to include physical theories as continually developing in unpredictable ways. The necessary restrictions to make measurements possible in such theories are discussed.

A. J. Coleman (1945). Phase space in Eddington's theory. *Phil. Mag.* (7), **36**, 269.

This proof of the infinity of phase space is referred to by Eddington, **13·2** §59 above.

D. W. J. Cruickshank (1947). Scientific method and Eddington's fundamental theory. *Sci. Progr.* p. 652.

A simple but useful examination of the epistemology.

[a] To appear in *Rend. Circ. Mat. di Palermo.*

A. Deprit (1955). A. S. Eddington's E-numbers. *Ann. de la Soc. scientifique de Bruxelles*, **69**, 50.

An important analysis of the basic algebra of E-numbers.

H. Dingle (1954). *The Sources of Eddington's Philosophy* (Cambridge).

This (like Ritchie and Whittaker (1951) below) is the text of one of the annual Eddington Memorial Lectures; these three are particularly relevant.

P. A. M. Dirac, R. Peierls and M. H. L. Pryce (1942). On Lorentz invariance in the quantum theory. *Proc. Camb. Phil. Soc.* **38**, 193.

This was a reply to the first paper of γ of **2·1** above and was answered by Eddington in the second paper of γ.

G. Haenzel (1941). Die Diracsche Wellengleichung und das Ikosaeder. *J. für Math.* (Crelle), **183**, 232.

B. Higman (1955). *Applied Group-Theoretic and Matrix Methods* (Oxford).

This contains a lengthy chapter describing Eddington's E-number theory, with particular reference to its physical identification; there is also a useful discussion of the inverse relation of energy, etc., to number of degrees of freedom.

H. Jeffreys (1941). Epistemology and modern physics. *Phil. Mag.* (7), **32**, 177.

H. Jeffreys (1948). *Theory of Probability*, 2nd ed. (Oxford).

On p. 283 of the book, Jeffreys finds the numerical agreements in **F** to be too good, and suggests reasons. On p. 386 he points out that Eddington did not consider Einstein's theory a mathematical necessity *until* the 1919 eclipse results confirmed it. The 1941 paper is a useful commentary on Eddington's general position. To quote one main point: 'Eddington's starting point is not purely epistemological because it assumes the considerable amount of observational evidence that is needed before we can establish a measuring system at all.' Jeffreys's analysis of the role of the E-frame in physics should be compared with Bastin and Kilmister's later work.

C. W. Kilmister (1949*a*). The use of quaternions in wave-tensor calculus. *Proc. Roy. Soc.* A, **199**, 517.

Hamilton's quaternions are used to express Eddington's algebraic work in a general space in which no metric is defined. It is shown that Eddington's derivation of Dirac's equation is possible in an affine space of distant parallelism.

C. W. Kilmister (1949*b*). Two-component wave equations. *Phys. Rev.* **76**, 568.

A note on the simple quaternion form of wave equation.

C. W. Kilmister (1951). Tensor identities in wave-tensor calculus. *Proc. Roy. Soc.* A, **207**, 402.

E-numbers expressed as matrices are linear transformations of a space V_4; they also represent a set of two four-vectors, a six-vector and two scalars in a space S_4 [cf. **F** VI, or **9·1** (56·7) above]. 'Tensor identities' are equations (in E-numbers) which are Lorentz-invariant both in V_4 and S_4. Corresponding equations in a double frame include Eddington's identity [**F** VIII, or **9·3** (89·4)].

C. W. Kilmister (1953*a*). A new quaternion approach to meson theory. *Proc. R. Irish Acad.* **55**, 73.

Eddington's derivation of Dirac's equation is generalized to the double frame. The resulting equations provide a new particle formulation of meson theory. The concept of 'field-free conditions' is analysed.

C. W. Kilmister (1953*b*). A note on Milner's $\mathscr{E}$-numbers. *Proc. Roy. Soc.* A, **218**, 144.

It is shown that Milner's 'new type' of tensor transformation (see Milner below) is algebraically equivalent to the usual simple rotation transformations.

C. W. Kilmister (1955*a*). The application of certain linear quaternion functions of quaternions to tensor analysis. *Proc. R. Irish Acad.* **57**, 37.

C. W. Kilmister (1955*b*). The analysis of observations. II. *Quart. J. Math.* (*Oxford*), (2) **6**, 161.

Eddington's proof of his fundamental result [**F** II or **3·2** (15·51)] on the division of characteristics between scale-free particles and their field is elaborated in terms of an extension of the concepts of Bastin and Kilmister (1952) to variable observables.

G. Lemaître (1937). Sur l'interprétation d'Eddington de l'équation de Dirac. *Ann. de la Soc. scientifique de Bruxelles*, **57**, 165.

An elegant quaternionic version of E-numbers and their factorization, anticipating Kilmister's approach.

W. H. McCrea (1938). A theorem concerning Eddington's E-numbers. *J. Lond. Math. Soc.* **13**, 283.

The theorem is that an E-number T satisfies its characteristic equation $\det(T-\lambda)=0$, where $\det T$ is the function defined *directly* as in **F** VI (**9·1** here) equation (65·1).

W. H. McCrea (1939*a*). On the representation of Eddington's E-numbers by matrices. *Proc. Camb. Phil. Soc.* **35**, 123.

A 16 by 16 matrix, enabling the product of two E-numbers to be read off.

W. H. McCrea (1939*b*). On matrices of quaternions and the representation of Eddington's E-numbers. *Proc. R. Irish Acad.* **45**, 65.

W. H. McCrea (1940). Quaternion analogy of wave-tensor calculus. *Phil. Mag.* (7), **30**, 261.

The calculus of two-component wave-vectors on a quaternion basis; a beginner's handbook for Eddington's presentation.

E. A. Milne (1947). Last testament of a physicist. *Nature, Lond.*, **159**, 486.

A review, both critical and appreciative, of **F**. In the identification **F** §§41–43 (**3·4** above) of exclusion and gravitation, formula (41·3) is condemned as a 'palpable fudge' designed to give (41·41); but the argument in §43 shows this criticism is unwarranted.

S. R. Milner (1952). The relation of Eddington's E-numbers to the tensor calculus. I. The matrix form of E-numbers. II. An extension of tensor transformation theory. *Proc. Roy. Soc.* A, **214**, 292, 312.

In I an E-number $\sum_1^{16} E_\mu t_\mu$ is relabelled as a double sum $\sum\sum_1^4 \mathscr{E}_{\alpha\beta} t'_{\alpha\beta}$ where the $\mathscr{E}_{\alpha\beta}$ are products of quaternionic sets S_α, R_β, and from this form is converted into a 4 by 4 matrix. The (ik) matrix element of $\mathscr{E}_{\alpha\beta}$ satisfies $(\mathscr{E}_{\alpha\beta})_{ik} = (\mathscr{E}_{ik})_{\alpha\beta}$ so that the transformation can be reversed. The transformation from $\mathscr{E}$-number to matrix is a 'summed' transformation $\sum_j S_j(\ldots)R_j$. In II these summed transformations are examined as a new kind of tensor transformation for rotation of the frame. Physical applications of the theory are to follow these mathematical investigations.

A. D. Ritchie (1948). *Reflections on the Philosophy of Sir Arthur Eddington* (Cambridge).

H. S. Ruse (1954). On the geometry of $\mathscr{E}$-matrices. *Proc. Roy. Soc. Edinb.* A, **64**, 127.

The geometrical background of the matrices used by Milner.

E. Schrödinger (1953). The general theory of relativity and wave mechanics. In *Scientific Papers Presented to Max Born* (Oliver and Boyd, Edinburgh).

This essay was written in 1940. There is a discussion on Eddingtonian lines whether h/mc should be of the order $RN^{-\frac{1}{3}}$ or $RN^{-\frac{1}{2}}$.

N. B. Slater (1947). *Phil. Mag.* (7), **38**, 299.

N. B. Slater (1954). Recession of the galaxies in Eddington's theory. *Nature, Lond.*, **174**, 321.

An error in **F** §3 (see **3·1** above), noted in the 1947 review of **F**, leads to a change in the predicted galactic recession velocity in accordance with recent observational analyses.

A. H. Taub (1950). *Math. Rev.* **11**, 144.

A critical review of **F**.

Sir Edmund Whittaker (1945). Eddington's theory of the constants of Nature. *Math. Gaz.* **29**, 137.

Sir Edmund Whittaker (1949). *From Euclid to Eddington* (Cambridge).

Sir Edmund Whittaker (1951). *Eddington's Principle in the Philosophy of Science* (Cambridge). (Reprinted in *Amer. Scientist*, **40**, 45 (1952).)

The 1945 article is a very useful summary of the ideas and results of **F**; there is a chapter in the 1949 book of similar scope. 'Eddington's Principle' is put: 'All the quantitative propositions of physics, that is, the exact values of the pure numbers that are constants of science, may be deduced by logical reasoning from qualitative assertions, without making any use of quantitative data derived from observation.'

LIST OF SYMBOLS

The list is confined to symbols which are not defined always by the context, or have more than one meaning. Page numbers in the list indicate special meanings of the symbols.

A, B, classical, quantum designation, 162, 240, 253
A, B, observational, theoretical system, 32–5, 137–40, 175, 253–4
A, U, aether, uranoid system, 168, 224
A, energy tensor coefficient, 28, 40, 60, 61, 67, 83, 211
A, non-Coulombian energy coefficient, 18, 48, 145–6, 179, 249

B', classical electrodynamic system, 34, 137–8
B_{00}, $B_0{}^0$, $B^{\mu\epsilon}{}_{\nu\sigma}$, Riemann-Christoffel tensor

c, velocity of light

e, e', electronic charge in systems B, B'
E^0, E_r, $E_{\mu\nu}$, particle energy tensor (to p. 136)
E_μ, $E_{\mu\nu}$, E-symbol, 147 onwards
$\bar{E}_\mu$, *transpose* of E_μ as matrix
EF_μ, double-frame symbol
$\mathfrak{E}$, top exclusion energy

F_μ, $F_{\mu\nu}$, F-(duplicate E-) symbol
$F_{\mu\nu}$, electromagnetic field 6-vector, 177, 192, 194, 244, 254–8

$g_{\mu\nu}$, fundamental metric tensor
$G_{\mu\nu}$, Ricci tensor; $G = \Sigma G^\mu_\mu$
$G_{\mu\nu}$, $H_{\mu\nu}$, E-symbols, 189, 190, 226–7

$h \equiv 2\pi\hbar$ = Planck's constant
$\hbar_1$, $\bar{\hbar}$, γ, theoretical variants of $\hbar$
h, h', H, hamiltonians, 31, 121, 123, 126, 140–1, 189–90
H^0, total (field plus particle) energy

I, interchange operator

j, j_r and J, occupation factors and symbol

k, multiplicity
k, range constant of nuclear force, 18, 48, 146
k, $\mathfrak{k}$, surface-harmonic number, 43–4, 179–80

m, mass (usually proper mass) of particle
  m_0, standard carrier (V_{136})

m, mass of particle (*cont.*)
  M, external particle (V_{10}) or hydrogen atom
  m_e, electron
  m_p, proton
  m_s or m_k, particle of multiplicity k_s or k
  $\mathfrak{m}$, top particle
  $\mathfrak{m}$, $\mathfrak{m}_0$, mesotron, 182, 251
$\mathbf{M}$, M_0, mass of general and Einstein universe, 146
M, mass dimension, 46, 138, 144
M, unspecified factor, 184–5

N, total number of particles
N_1, planoid population

$\mathbf{p}$, p_μ, momentum (or operator $-i\hbar\partial/\partial x_\mu$)
p, component (e.g. p_1), 37, 52, 142
$p=\surd(p_1^2+p_2^2+p_3^2)$ generally
$p_4=\surd(p^2+m^2)$, one form of hamiltonian
$\mathbf{p}$ (quantum) $=i\mathbf{p}'$ (classical), 25, 92
P_α, external particle momentum
$P=\Sigma E_{\mu\nu}p_{\mu\nu}=\Sigma E_\mu p_\mu$ symbolic momentum vector, 148 onwards
$\mathbf{p}_{\mu\nu}$, total operator, 183, 206–7, 220
P, pressure, 39, 44, 60–1, 66–7

r, r', distance; *see* 18
R_0, radius of Einstein space
R_1, planoid radius (flat space)

$S=\Sigma E_\mu s_\mu$, strain vector

$T_{\mu\nu}$ (in **3**–8) energy tensor, usually *total* (field plus particle)
$T=\Sigma E_\mu F_\nu t_{\mu\nu}$ (p. 163 on) an EF-number, often identified with an energy tensor (see p. 163 for notations)

$\mathbf{v}$, v^μ, velocity vector
V_0, recession velocity, 18, 146
V, V_0, V_n (normalization) volume
V_k, particle of multiplicity k ($k=1, 3, 4, \ldots$)

W^0, $W_{\mu\nu}$, complementary field energy
$\mathbf{W}$, operator 186 *et seq.*

X, $X_{\mu\nu}$, generic energy
X, X_α, classifying characteristic, 22, 88, 90, 106
x_α, external particle coordinates

Z, strain tensor, Z_0^0 cross dual of energy tensor
Z_d, Z_m, Z_p, density, momentum, pressure terms of Z_0^0

LIST OF SYMBOLS

$\alpha = \frac{1}{137} = e^2/\hbar c$ fine structure constant (Eddington gives this name usually to $137 = 1/\alpha$)

$\beta = 137/136$; but
β denotes $\hbar$ in 62–7, 80–3, 121–8, 201–8
$\boldsymbol{\beta}$, observable, 273–4

$\delta_\mu{}^\nu$, Kronecker symbol
$\delta(r')$, delta function

η, η_0, hamiltonians

θ, ϕ, polar angles, 16, 74
θ, θ', phase

κ, constant of gravitation
κ_μ, electromagnetic potential

λ, cosmical constant
λ, scale-factor, 22

μ, mass of internal particle or intracule

$\varpi = \hbar/2\sigma$
ϖ, ϖ_α, internal particle momenta, 31, 98, 123–4, 127, 140–1, 189–90, 219, 226
ϖ_α, pentadic part of E-number, 155–6, 185, 221
ϖ_{04}, ϖ_{05}, magnetic terms, 156

ρ, ρ_0, density, proper density
ρ, internal distance, 127–8

σ, uncertainty constant
σ_ϵ, s.d. of scale fluctuation
$\Sigma_\mu{}^\nu$, strain tensor

ϕ, ψ, wave functions
ϕ, ψ, χ, ω, wave vectors

INDEX

Numbers denote pages. *Italicized* entries refer to surveys quoted from manuscripts.

INDEX